P9-DME-274

Maps and Mapmakers of the Civil War

MAPS

AND

MAPMAKERS

OF THE

CIVIL WAR

Earl B. McElfresh

Foreword by Stephen W. Sears

Harry N. Abrams, Inc., Publishers

in association with

History Book Club

To my Irish princess Michiko

Designer: Dirk Luykx

Library of Congress Cataloging–in–Publication Data
McElfresh, Earl B.
Maps and mapmakers of the Civil War / Earl B. McElfresh ; foreword by Stephen W. Sears.
p. cm.
Includes bibiliographical references and index.
ISBN 0-8109-3430-2
1. United States—History—Civil War, 1861–1865—Cartography.
2. United States—History—Civil War, 1861–1865 Maps.
3. Cartography—United States—History—19th century. I. Title.
E468.9.M135 1999
973..7'022'3—dc21   99-28098

Copyright © 1999 Earl B. McElfresh
Published in 1999 by Harry N. Abrams, Incorporated, New York
All rights reserved. No part of the contents of this book may be reproduced without the written
permission of the publisher

Printed and bound in Hong Kong

Harry N. Abrams, Inc.
100 Fifth Avenue
New York, N.Y. 10011
www.abramsbooks.com

History Book Club
1271 Avenue of the Americas
New York, N.Y. 10020

Contents

Foreword

by Stephen W. Sears

E XAMINING the maps reproduced in this unusual atlas, one is immediately struck by their qualities as works of art. Although the maps were intended for everyday, practical military use, to ease the way for generals making war, the mapmakers represented here, or at least a good number of them, managed to create a genuine form of art—the art of military cartography, let us call it. In the Civil War, anything beautiful is rare enough, and therefore most welcome wherever found. And there are few things more surprising, it is safe to say, than finding beauty in top-secret military documents.

Students of the war familiar with the United States War Department's *Atlas to Accompany the Official Records of the Union and Confederate Armies* (the *Official Records Atlas*), published in the 1890s, will doubtless recognize some of these maps. Included here, in such cases, are the original maps on which the lithographed versions in the *Official Records Atlas* were based. Because of the technical limitations of a century ago, true reproductions of these originals were not possible. The lithography process of the time could only show the surveying and engineering skills of the Civil War's mapmakers. *Maps and Mapmakers of the Civil War* reproduces these maps as they were originally drawn, revealing the delicate tinting, the meticulous lettering, the color penciling, and the watercolor effects that make them minor—perhaps not so minor—cartographic masterpieces.

The great majority of these maps have never been published before. Many of them are reproduced close to their original size. Details of oversized maps, such as Jed Hotchkiss's famous theater map of the Gettysburg campaign, enable the modern reader to see for the first time the exact information that Robert E. Lee relied upon as his army marched into Pennsylvania.

The 150 full-color reproductions in *Maps and Mapmakers of the Civil War* have been gathered from collections, obscure and famous, throughout the country. This is by far the largest, most comprehensive, most beautifully reproduced selection of Civil War maps ever published. The accompanying text is the only thorough examination of Civil War maps, mapmaking, and mapmakers ever presented. It documents the activities of the mapmaker in the field and identifies the seemingly arcane topographical details he searched out. It also illustrates the practical effect of the cartographer's art on actual military operations.

Maps were—or at least should have been—an integral part of every military operation in that war. The emphasis on maps varied with the individual commands and with their individual circumstances. A prime example of the impact of their presence, and of their absence, is the case of Stonewall Jackson. In Jackson's celebrated Shenandoah Valley campaign of 1862, he had the benefit of Jed Hotchkiss's expertly drawn maps and made scarcely a misstep in getting to where he wanted to go in that militarily complex terrain. Soon thereafter, when he joined Lee on the Virginia Peninsula, Jackson's missteps became frequent. Hotchkiss had been left behind in the valley to continue his mapmaking, and Jackson had to rely on maps furnished by others, which did not meet his military needs. Indeed, during the Peninsula campaign all too many Confederate generals all too often went astray within twenty miles of their own capital for want of adequate maps.

By contrast, in the western theater Generals William Tecumseh Sherman and George H. Thomas were map connoisseurs. They delegated their most talented engineers to perform map duties and nothing else. Then there was William W. Richardson, Company I, 104th Ohio volunteers, who performed more eclectic service. Richardson described his Civil War duties as "burying the dead, building fortifications, barbering, and drawing maps." However it was done, by 1865, where the armies had campaigned there was hardly an acre left unmapped.

This finely crafted atlas and its insightful textual analysis of maps, their contents, their uses, and their effects provides a unique and much-needed examination of the crucial role maps played in the waging, and in the outcome, of the Civil War.

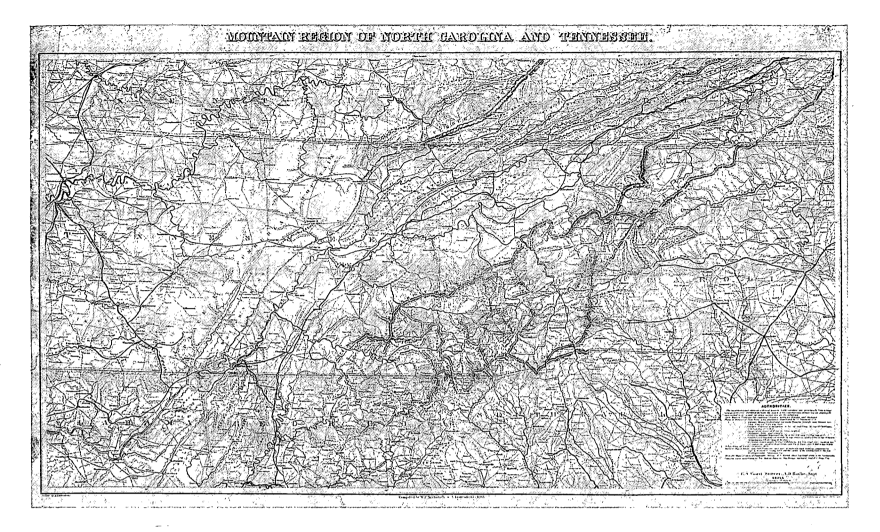

Abraham Lincoln's Map, Presented to Major General Oliver Otis Howard, September 1863

This United States Coast Survey map was mounted and used at the White House by President Lincoln. A trained surveyor, Lincoln made good use of maps and was quick to resort to them, as numerous anecdotes testify. An extensive number and variety of authorities are listed on the map, including the future president Andrew Johnson.

In September 1863, following the Union disaster at Chickamauga, Georgia, the Eleventh and Twelfth Corps of the Army of the Potomac were sent by railroad to reinforce the Union troops that had retreated to Chattanooga, Tennessee. Major General O. O. Howard commanded the Eleventh Corps and stopped in Washington to meet with Lincoln. In his autobiography, Howard describes the history of this map: "After an interview with my commanders, I paid a visit to the President. It was during that visit that Mr. Lincoln pulled down his map from the wall and, putting his finger on Cumberland Gap, asked: 'General, can't you go through here and seize Knoxville?' Speaking of the mountaineers of that region he declared: 'They are loyal there, they are loyal!' Then he gave me his mounted map, better for campaigning, and took my unmounted one, saying: 'Yours will do for me.' "

Introduction

THIS ATLAS contains a wide selection of the original maps either used or prepared by Federal and Confederate topographical engineers during the course of the American Civil War. There are examples of existing published maps that mapmakers relied on as a basis or as a frame of reference for their own works in progress. There are the field notes and sketchbooks used by the engineers to record their moment-by-moment observations in the field. There are crudely drawn, wildly inaccurate maps, and there are maps so precisely rendered that they remained in use well into the twentieth century. There are samples of maps prepared and then reproduced and distributed as sun prints, hand copies, engravings, woodblocks, lithographs, and photographs. There are maps drawn by topographical engineers and soldiers on both sides of the conflict and maps of the same battle from the perspective of both the Federal and Confederate participants. There are maps that are cut-and-paste reproductions consisting of printed maps with hand-drawn addenda. There are representative maps from the different map-producing organizations: the United States Coast Survey, the Bureau of Topographical Engineers, the Confederate Engineer Bureau, and others. There are theater maps showing thousands of square miles. There is also a map or plat showing every fence post, tree, and building on a single farm. There are maps that single-handedly deranged campaigns. There is a map stained with the lifeblood of the highest-ranking officer of either side killed in the war. There is a map President Abraham Lincoln used in the White House. There are the maps with which Confederate general Robert E. Lee planned the Gettysburg campaign. Some are as plain as pencils. Others are exquisite—true cartographic masterpieces. All of them are interesting, more or less, for one reason or another.

Many of the original manuscript maps selected for this atlas have never been reproduced or published before, certainly not in color. Some of them have probably not been viewed by anyone in the twentieth century until now. Very few of them have ever received serious attention or study.

Most of the topographical engineers who prepared these maps also have been overlooked. It was said that George Gordon Meade, commander of the Union Army of the Potomac, conferred more often and at greater length with a mere captain of topographical engineers named

William Henry Paine than he did with any of his corps commanders. On occasion, the movements of his entire gigantic army were stayed until Paine reported his findings and recommendations. Wartime photographs, brief biographies, and in most cases maps by a number of these mapmaking engineers are included in this atlas. Among them are such diverse characters—notorious, well known, or completely forgotten—as George Armstrong Custer and Ambrose Bierce, Washington Roebling and William Henry Paine, G. K. Warren and Jed Hotchkiss.

The text that accompanies the atlas is the story of Civil War mapping. It attempts first to establish the critical importance of detailed maps in the conduct of military operations. Armies of the time were extremely vulnerable to the lay of the land. The simple fact that Civil War armies were often too large to advance on one single road meant that maps were essential to maintain command and control of military operations in the field. An eminent British military historian of the late nineteenth century, G. F. R. Henderson, wrote that leading a large army on a long march required a greater talent than leading it in battle. Indeed, following in this vein, it is interesting to note that the Civil War armies that received the greatest acclaim were noted not so much for their fighting prowess as for their marching capabilities. Confederate general Thomas J. (Stonewall) Jackson's men proudly referred to themselves as "Jackson's Foot Cavalry." Confederate A. P. Hill's men were known as "the Light Division." The march of Union general William Tecumseh Sherman's army through the Salkehatchie swamps prompted his adversary to compare Sherman's soldiers to Julius Caesar's. It is almost a truism to say that good marching equated with good mapping, particularly under the conditions that prevailed for Civil War armies maneuvering in mid-nineteenth-century rural America.

The balance of the text is organized to follow in a general way the procedures used to prepare a map in the field. Each step in the process is described and explained. Military maps have a special provenance. Data are recorded that might seem superfluous, eccentric, decorative, or whimsical. But all of it is, in fact, grist for the military map mill. The text makes it clear that each pine tree had its place, the watercolors their purpose, the residents' names their particular function. The uses of these and other odd-seeming features of the maps are documented with the observations of soldiers, the opinions of historians, the written reports of officers, the memoirs or annotations that accompanied many of the maps, and, finally, the comments of topographical engineers themselves.

An eclectic set of resources was used in preparing this atlas. First and most important were, of course, the maps. They often contain a wealth of information about the "authorities," that is, the other maps and information used in their preparation as well as a brief history of the topographical engineers who had a hand in the mapmaking. One quite delightful aspect of these maps is the immensely creative way abbreviations are used on them. They become almost a decorative art form in the way letters are raised, lowered, underlined, dotted, omitted, added, or elongated. Sometimes the abbreviations are so elaborate they seem longer than the original word. But invariably their meaning is clear.

Information about the process of delineating or executing a Civil War military map came mostly in bits and pieces, culled from published sources—some well known, others deeply obscure. Quite a number of the manuscript map collections that were examined also contained the papers of engineers. Some of these papers provided what amounted to an itemized list of the materials used in preparing the maps: the paper, the mountings, the ink, the erasers, the watercolors, and the pencils. These workaday records also confirmed or clarified unintelligible references in published sources that would have been impossible to clear up any other way.

The memoirs, reports, letters, diaries, and other firsthand testimony of the American Civil War that were employed in researching this atlas are also a reminder of the extent to which the writings of the participants of the war constitute a substantial contribution to America's Victorian literature. The humor, the steady eye, the lack of pretension, the alert matter-of-factness, and the graceful writing styles of so many of these chance authors point up just how good and how extensive that body of work really is. These wartime writings are a joy to read and inspire a profound respect for the verve, the intelligence, the essential fairness, and the cracker-barrel humanity of mid-nineteenth-century Americans.

Some particulars of the atlas: it is not intended to be a synopsis of the American Civil War. Nor, except when absolutely necessary, are specific battles recapitulated simply because they are represented in maps. The emphasis is strongly on military maps, mapmaking, and mapmakers.

The maps appear for the most part in simple chronological order. The temptation to impose some sort of organizing principle has been resisted after careful consideration. It would have required a system of classifications and a formal designation of battles, campaigns, and theaters of war that would be fraught with difficulties while offering no commensurate advantages. When a set of maps has been reproduced, the maps in the set are usually presented together and in their order within the set. In a few instances, other criteria have been used to determine the position of an image in the atlas, but chronological considerations generally prevail. The titles assigned to the images in the atlas contain more information than the titles on the actual maps. All spellings in the atlas's titles have been made uniform, and spelling errors have been corrected.

The contemporaneous quotations used are intended to establish or corroborate statements in the text in a clear and helpful manner. To this end, editorial intrusions have been kept to a minimum. Each quote conveys what it was meant to mean in as straightforward a manner as is compatible with historical accuracy.

The rank of military figures reflects rank at the time of the episode in which they figure.

Some oversized maps could not be reproduced in total, yet they were simply too important or too impressive to be omitted. A good example is the theater map that Confederate mapmaker Jed Hotchkiss prepared of "the Valley of Virginia extended to Harrisburg, Pennsylvania" as a prelude to the Gettysburg campaign. The actual map measures thirty-two by fifty-two inches. It is most noteworthy for the massive amount of information it contains, all of which would be lost if the map were reduced and then reproduced in its entirety. The reproduction of the most important sections of the map offers a glimpse of the map as it was seen by Robert E. Lee and his lieutenants. However, the majority of the maps in the atlas are reproduced at, or very close to, their actual size.

The western theater of the Civil War is, in a break with tradition, very generously represented. By chance, several of the Federal generals in the West were particularly map conscious and saw personally to the establishment and maintenance of effective map organizations. The most prominent of these were George H. Thomas, William S. Rosecrans, and William Tecumseh Sherman. Thomas went to the greatest lengths and had the most dedicated (in the sense of specialized) mapping organization on either side during the war. Thomas had learned an early lesson on the value of military maps when he maneuvered his army in the vicinity of Mill Springs, Kentucky, in January 1862. Thomas was attempting to coordinate the movement of his force with that of General Albin F. Schoepf in the face of two enemies: the January weather and a Confederate army under General Felix Zollicoffer. Thomas and Schoepf were using two different maps—the roads, fording sites, place-names, and terrain features didn't match up—so coordinated movements were next to impossible. Thomas worked hard thereafter to avoid a repetition of that unnerving experience, so that some of the best-organized and most thorough mapping operations of the war were in the field recording Chickamauga, Chattanooga, the Atlanta campaign, Nashville, and the March to the Sea.

It was said that no army general studied his maps with half the industry or with half the intelligence that Abraham Lincoln gave to his. The maps included in this atlas can be studied with great profit, with industry and intelligence, but also with much aesthetic enjoyment. In many ways this is an art book as well as an atlas. Most of the maps and sketches reproduced are purely military documents, prepared for the practical purpose of enabling generals to plan marches for their armies to make. But it is likewise obvious that the topographical engineers who drew these maps were also artists who just happened to be working in wool uniforms of gray or blue. Each scrap of paper with a map or a map detail drawn on it is plainly, to one degree or another, a small work of art. The tiny embellishments—a carefully rendered tree or a miniature headquarters tent with its flagstaff—merely confirm the artistic bent of the engineers who drafted them.

When the Confederates surrendered at Appomattox (Virginia), Bennitt's Farm (North Carolina), Citronelle (Alabama), and Galveston (Texas), the guns were silenced, the war was ended, and the soldiers were mustered out of the armies. Northerners and Southerners returned to their homes and resumed the lives they had forsaken when their country or their way of life was threatened. In many instances, topographical engineers brought their maps, sketchbooks, and ephemera of Civil War mapping home with them. The original maps and sketches were apparently deemed to be the property of the engineers. Possibly they were considered redundant. Over the years, depending on circumstances, maps were lost, misplaced, destroyed, thrown out, confiscated, carefully preserved, or purchased. Even before the war ended, the Adjutant General's Office of the United States Army began to collect the data that would eventually become *The War of the Rebellion: A Compilation of the Official Records of the Union and Confederate Armies* (the *Official Records*). Maps were also located, borrowed, and otherwise drawn together to be reproduced in the *Atlas to Accompany the Official Records of the Union and Confederate Armies* (the *Official Records Atlas*). This six-and-one-half-year project, published between 1891 and 1895, encompassed some 821 lithographed representations of all sizes and sorts of Civil War maps. The original maps were then either filed away or returned to whoever had happened to possess them at the end of the war. Hotchkiss's own maps were nearly seized by the United States, but Hotchkiss instead agreed to make copies of his maps, many of which appear in the *Official Records Atlas*. The bulk of Hotchkiss's papers and maps eventually ended up in the U.S. Library of Congress. Other maps by other mapmakers ended up in small historical societies, museums, college collections, and libraries that had some connection to one of the topographical engineers or his heirs. Recently, a presentation about a Civil War mapmaker was given to a small audience at the historical museum in the county where he had resided. When asked if any of the original maps were still around, the presenter replied that their location was unknown. Whereupon a man in the audience stood up and said, "They're in the trunk of my car out in the parking lot." The maps were those of Ambrose Bierce. They are now faithfully reproduced in this atlas where they can be seen, studied, and appreciated in their original cartographic form, along with many other similarly compelling mapping artifacts of America's defining historical event.

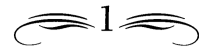

The Necessity of Military Maps

Walking over Strange Ground with Eyes Shut

I T IS SAID that the original American roads were the trails of animals—deer, elk, woodland bison—moving from dense cover to more open land where they could graze and water. They always took the shortest, easiest route between the hills and the river valleys. The animals' trails became the paths of the Indian. The same paths were later used by the white traders with their packhorses. Eventually, gradually, they turned into the roads on which the settlers came with their wagons and their families. In the mid-nineteenth century, in the eastern part of the country, when the turnpikes were built, they followed the identical routes except at the steepest ascents where they diverged, but only by a few yards, from the prehistoric trails.

In the East and in the West, routes that had seen the bison, the Indian, the packhorse, and the settler saw from 1861 to 1865 an altogether different set of travelers—the Civil War armies of Lee and Grant, Sherman and Johnston, Beauregard and McDowell, Jackson and Banks, and dozens of other greater and lesser luminaries of the central war of America's history.

The most sophisticated of Americans had never witnessed anything to compare with the sight of these armies. For most mid-nineteenth-century Americans they were genuine spectacles, thousands of soldiers stretched out for miles on roads that on a normal day would carry a dozen travelers: some boys, some buggies, some wagons, a couple of passersby, and a stray dog or two.

D. H. Strother was an American celebrity in 1861, a very well known travel writer as well as sketch artist for the most popular magazines of the day. Strother had toured extensively abroad. He had been a sightseer in Paris in February 1841, where he had witnessed one of the largest military pageants ever staged, the entombment of Napoleon I.[1] But nothing, as Strother was to write, that he had ever seen surpassed the "scenic splendor" of the advance by General Robert Patterson's army on Winchester, Virginia, in July of 1861. Almost a thousand men were deployed for half a mile on either side of the road, fanning out as skirmishers to protect the army from surprises on its front and flanks. The picturesque lines moved "irregularly, now over open fields of grass and clover, now wading waist-deep through yellow, waving lakes of uncut grain." The overall movement consisted of approximately eighteen thousand men, which included artillery batteries, some regular cavalry, and supply trains, all "stretching out in interminable perspective."[2] By later Civil War standards, Patterson's advance was a minor operation. But for Strother in 1861, it was a breathtaking experience and an epic day.

It was not unusual for a Civil War army to overflow sixty miles of country road and to be

Army Wagon, Six Mules, and a Teamster

This is what was known in a marching Civil War army as impedimenta, until the men needed rations, the animals needed forage, or the guns needed ammunition. The orders to bring up a wagon like this were not qualified with phrases such as "if possible" or "if practicable." The orders were peremptory. Military maps had to show a commander where his soldiers as well as his wagons could go. Confederate general Richard S. Ewell made the remark that "the road to glory cannot be followed with too much baggage." This has been repeated so often and so approvingly that one might be led to believe that it was always true. In the American Civil War it was not. General William Tecumseh Sherman led his Union army into military legend with nearly twice the wagons per man as his ineffectual fellow general George B. McClellan had in his. Built at a cost of about $150, this was a six-mule jerk-line freighter wagon. It could carry approximately 4,000 pounds on a good road, 1,800 pounds on a marginal road. This included 270 pounds of forage for the six mules. The civilian teamster managed the wagon from the left rear mule, the wagon itself having no seat. Tying a rag to a spoke of a wheel on one of these wagons created a primitive yet effective odometer for a topographical engineer. He had merely to count the number of rotations and multiply them by the circumference of the wheel to get quite accurate mileage readings.

so enormous that the havoc it wreaked in simply moving through an area would take years to repair. One Washington County, Maryland, resident reminisced: "It is hard to describe the change which is made by the encampment of an army. In an incredibly short time a splendid field of luxuriant verdure had been beaten down as hard as a turnpike road and every blade of grass had disappeared. It was years before the most careful cultivation could restore the land to anything like its former productive condition. When it was finally plowed, the land broke up in great clods and lumps which had to be pulverized with axes and mallets."[3]

The most famous Confederate army of the war was Robert E. Lee's Army of Northern Virginia. Some sense of the magnitude of this army can be gained by realizing that a division commander such as Lafayette McLaws was later to say he didn't know, had never spoken to, and probably wouldn't recognize by sight a fellow division commander in the same army, Jubal A. Early.[4]

Brigadier General Langdon C. Easton, who held the challenging position of quartermaster under William Tecumseh Sherman, wrote in his report that at the beginning of May 1864, "the effective strength of our army in the field was about 100,000 men with 28,300 horses, 32,600 mules, 5,180 wagons and 860 ambulances."[5]

An army of this size would fill a typical rural road and, moving day and night, take several days to pass. Even the supplying of the draft animals that accompanied the march would constitute an immense logistical feat involving hundreds of tons of forage. At Westminster, Maryland, as the Gettysburg battle raged, a local doctor took the time to count the mules, wagons, and guards that made up a reserve supply train parked in the village. There were five thousand wagons, thirty thousand mules, and ten thousand soldiers for an escort.[6]

One consequence of the size of Civil War armies was that maps—highly detailed, up-to-date military maps—could have a significant impact on their well-being while on the march. The motive power of these huge armies was draft animals, horses, and soldier's legs. The wagons were interspersed with the columns of troops and artillery because the supplies the wagons carried had to be accessible to the men as they marched. The need for wagons in Civil War military operations of any size was compelling. No matter who the general was or what reputation he had for traveling light, his wagon train would be miles long. The successful staging of a march was therefore very vulnerable to the lay of the land because even a slight rise in the grade of a roadway could result in delays amounting to hours that were exhausting and irritating to the men in the ranks. George T. Stevens, historian of the Union Sixth Corps, wrote: "No one who has not seen the train of an army in motion, can form any conception of its magnitude, and of the difficulties attending its movements."[7] Union artillerymen described the minor topographical features that presented significant delays in their hurried march toward Gettysburg: "After dinner we started again, and with constant halts pushed on till dark. Then for some inexplicable reason, the column halted and remained till nine, moving during that time about a quarter of a mile by fits and starts, then another halt of two hours. All the delay was caused by a hill ahead, up which our battery went with halting."[8] Edward Porter Alexander, Confederate general James Longstreet's reserve artillerist, explained the effect of these fits and starts on men and animals: "Of all tiresome things to soldiers there are few more disagreeable than to be ordered to march and gotten out to the road, and then to be held waiting indefinitely. And, besides tiring the animals, by keeping them in harness, it loses time when they might be grazing."[9]

Alexander's offhand mention of grazing fails to convey a proper sense of the thoroughness with which an army's animals ate. In middle Georgia a farmer marveled: "I had a noble field of corn, not yet harvested. Old Sherman came along, and turned his droves of cattle right into it, and in the morning there was no more corn there than there is on the back of my hand."[10] The Confederates, not surprisingly, had to eat too. Their "depredations" occurred on the local farms of their friends and countrymen but were no less thorough for that. An army surgeon wrote: "I

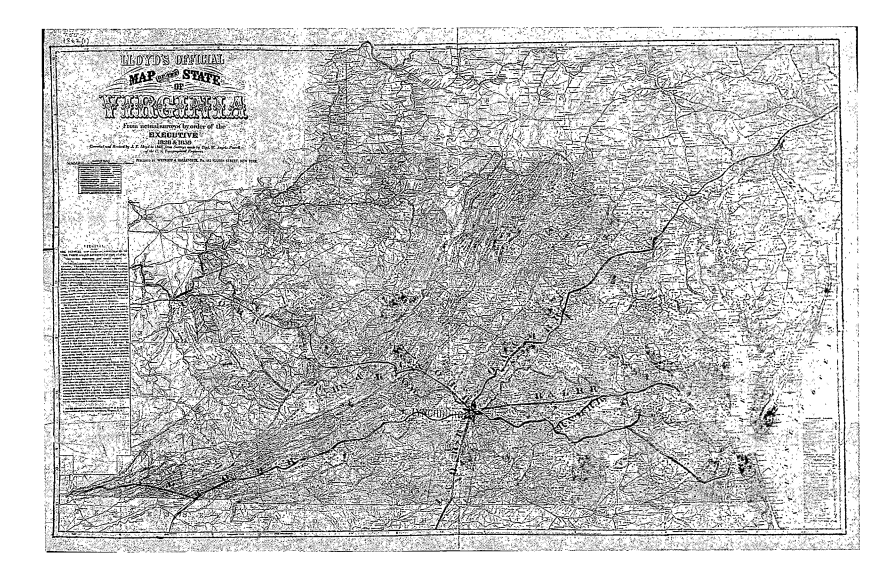

Lloyd's Official Map of the State of Virginia, Corrected in 1862 from Captain Powell's Surveys

There was nothing "official" about this commercially produced map of Virginia, published in New York. Based on what was known as the nine sheet map of Virginia by Danish born mapmaker Herman Böÿe, whose map was actually *published* in 1827 (Lloyd's 1828 is in error) and itself derived from county maps prepared by John Wood c. 1820–22, the Lloyd's map was sold to a public anxious to follow the course of the war in Virginia and to keep track of its soldiers. The lithographed map came out in several versions and was used by the military because, as Union brigadier Alpheus S. Williams wrote to his daughters, "for general purposes and for names of small places it is the best you can get." Two of the errors in Rappahannock County, just to name a few: Flint Hill is due south of Sandy Hook, not forty-five degrees southeast, and Newby's Crossroads is half a mile west, not four miles southeast, of Battle Mountain.

have often read of how armies are disposed to pillage and plunder, but could never conceive of it before. Whenever we stop for twenty-four hours every corn field and orchard within two or three miles is completely stripped."[11]

These Civil War armies were voracious visitors and in short order, wherever they halted, would eat and drink themselves out of house and home. The demands that such huge armies made upon the resources of the rural countryside through which they passed were enormous, relentless, and implacable. Thirsty men and animals had to drink. A horse or mule could drink ten gallons water a day. So armies needed to know where potable streams, springs, wells, and pumps were or were likely to be—information that would be carefully noted on military maps. Hungry men and animals had to eat. Military maps would locate orchards, cultivated fields, pastures and meadows, barns, and other likely food sources. Civil War armies couldn't very well afford to pause, retrace their steps, or follow directly in the path of a retreating or advancing enemy because the areas already visited by an army would likely be swept bare of food and forage and the wells and pumps would be wrung dry. It was estimated that George B. McClellan's army of approximately one hundred thousand men and tens of thousands of animals required six hundred tons of food, forage, and other supplies per day.[12] It could also be assumed that heavily laden soldiers and animals on a strenuous march used up twice as much food and forage, per capita, as civilians.[13] Even fencing was sometimes noted on military maps. It wasn't that tired soldiers had to brew coffee and used the rails to make their bright coffee fires. The actual purpose of this arcane information was to keep the army moving. Assuming a road was lined on both sides with fences, there were enough rails at hand to be used to corduroy the road; that is, there were sufficient rails to lay down a washboardlike pavement that would allow an army to make slow but steady progress over otherwise impassable, miry highways.

Ordinary published "political" maps—those showing townships, political boundaries, county lines—were quite common by 1861 and could generally be located and studied by army headquarters. Unfortunately, they were of little value and failed to meet the needs of a campaigning army actually in the field. The scale of the maps was much too small for practical military purposes. (A small-scale map is one that shows a large area but with minimal detail.) Most of the obstacles that could hinder an army's advance, and most of the resources that could help an army's advance, failed to appear on the inscrutable one-inch black line that represented fifteen miles of actual road on a typical county map. Simply labeling a town on a small-scale map could use up several square miles of ground. These published maps, when they displayed them at all, showed terrain and other physical features in a very rudimentary way. Even a slight rise could cause hours of delay, but these maps showed no terrain features short of a mountain range. A gap that altered the course of the war would not be represented at all. Routes that were prominently mapped would turn out to be overgrown and impracticable or fallen into disuse because a new canal or a railroad line had bypassed them and rendered them obsolete. Confederate general Braxton Bragg reported that he couldn't even locate on a map a place fellow general P. G. T. Beauregard had noted as being vital and strategic. "May it not have changed name or lost its place on the maps in these railroad days?" Bragg plaintively inquired.[14]

Union general Jacob D. Cox in his postwar *Atlanta* wrote that the maps in common use were erroneous and misleading to a degree that was exasperating. They gave the outlines of counties, the names of towns and villages, and some remote approximation to the courses of the principal streams. The smaller creeks and watercourses were drawn at random, as if to fill up the sheet, and were uniformly wrong. Union brigadier Alpheus S. Williams was gentler in his comments about "Lloyd's Official Map of the State of Virginia" when he wrote to his daughters: "If you have Lloyd's official map you can follow our lines of march pretty accurately, though the roads and relative positions of villages are by no means correctly laid down. For general purposes and for

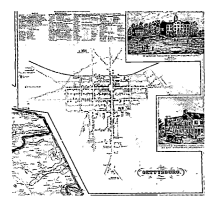

The county seat of Adams County, Pennsylvania, the busy market town of Gettysburg, is featured in great detail on the 1858 Converse/Hopkins map. Map is reproduced in full on page 139.

names of small places it is the best you can get."[15] Colonel Orlando M. Poe, Sherman's chief engineer, made the comment that the maps of North Carolina that he was able to procure "vie with each other in inaccuracy."[16]

Other available commercial maps were the county maps popular in the 1850s. They were prepared for a few counties in Virginia, Maryland, and southern Pennsylvania and were, for the most part, expert, accurate, and up-to-date productions. The most famous of these in Civil War terms is the Adams County, Pennsylvania, map of 1858, prepared by Griffith M. Hopkins and published by M. S. and E. Converse of Philadelphia. In the lower center of the map, as well as in a large-scale detail on the upper right, is the seat of Adams County, the bustling market town of Gettysburg. This map, and others like it—the 1862 Simon J. Martenet map of Carroll County, Maryland (which included Taneytown and Pipe Creek), and the 1858 map of Franklin County, Pennsylvania, surveyed by D. H. Davison and published by Riley and Hoffman of Greencastle (which included Chambersburg)—were in fact quite valuable in military terms. They showed details and presented some of the cultural information that was valuable to a marching army: roads, railroads, mills, forges—even such features as shoe shops. It was the lack of physical data shown on these county maps that limited their military usefulness. As has been noted, an army teamster approaching a hill with six tired mules was less interested in the name of the hill (which a county map might tell him) than in the steepness of the grade (which a county map would not).

In a paper presented to the American Philosophical Society in March 1864 in Philadelphia, Pennsylvania, R. Pearsall Smith stated: "At the last invasion of the State, preceding the battle of Gettysburg, the advance guard of the rebels swept the Great Valley clean of all its county maps; those of Franklin and those of Cumberland. The same fate befell those of Adams County."[17] Confederate general Jubal A. Early, leading one of Richard S. Ewell's divisions in the Gettysburg campaign, went so far as to say of these maps that they "were so thorough and accurate that I had no necessity for a guide in any direction."[18] Early's testimony is a bit suspect, however, because he was not adept with maps and had a difficult time orienting himself around them. The differences between the county maps that Early found satisfactory and true military maps were discussed by Captain G. H. Mendell in his *A Treatise on Military Surveying*, an 1864 work that was, in effect, a textbook on maps and mapping for Union officers: "The ordinary maps, upon which are found the positions of the principal towns, the location of the great routes, the courses of the rivers and canals, and the general form of the country, are sufficient to enable one to follow the operations of a campaign; but they are far from being in sufficient detail for the purposes of him who plans or studies the smaller operations of war. The most detailed maps are essential in these cases, because the merest trail, the most insignificant rivulet, the slightest undulation of the ground, may for a time become of the greatest importance, either for the offensive or for the defensive." Captain T. B. Brooks corroborated this from the field. He wrote from Morris Island, South Carolina, in August 1863 to the United States Coast Survey in Washington making the very same point: "Your charts of this coast are invaluable to us. Their great amount of detail and extreme accuracy leave nothing to be desired in a military map. Little high-tide marsh streams and hummucks [sic] no larger than a tent floor, which apparently could never possess the least interest or value on a map, have proven to be important landmarks and lines of communication for scouts and pickets."[19]

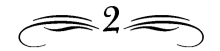

Familiar Territory

The South's Advantage

*M**AP* AND *MILITARY* are usually companion words. The first map mentioned in literature is on a bronze plaque being studied by two Grecian rulers as they confer and consider whether to join forces in a war against the Persians. Herodotus described both the conference and the maps circa 500 B.C. The conference broke up when the Greeks realized that, according to the bronze map, the Persian capital was a three-month march from the coast.[1] The very first maps were probably scratched in the dust by primitive man to guide his hunting companions to a kill or to a herd. It is intriguing that Stonewall Jackson was given to drawing maps in the ground with the toe of his boot. He did this at Savage Station during the Seven Days battles and again on the eve of his march through Thoroughfare Gap prior to the Second Battle of Manassas. The fact that Jackson was an abysmal draftsman and had absolutely no talent for drawing ("My hardest tasks at West Point were the drawing lessons,"[2] he cheerfully acknowledged to his aide and friend Colonel A. R. Boteler) may have had something to do with his dirt maps.

In any event, Jackson's map needs were thoroughly met by his topographical engineer Jed Hotchkiss, a New Yorker from the southern tier of the state who had spent his adult life residing in the Shenandoah Valley. Hotchkiss, nearly six feet tall, thoughtful looking to the point of sleepiness, was an indefatigable sketcher of maps and a brilliant amateur cartographer. His presence at Jackson's side ensured that the general had the map resources he needed to plan and execute his marches. Hotchkiss wrote to Jackson's biographer G. F. R. Henderson that his boss, like Napoleon, was very particular about having and studying maps. But he also gave the impression, though he didn't say as much, that Jackson's supposed "quick comprehension of topographical features" was in reality Hotchkiss's comprehension. The great Rebel captain's habit was to study his maps, then retire to pray.[3] Hotchkiss filled in from there, making graphic representations on the march or during engagements in answer to Jackson's requests for more information, more elaboration.

In having the services of Hotchkiss, with his talent for producing readily understandable and highly detailed maps, Jackson did not have any particular advantage over the Federals he was opposing. Their ranks were likewise graced with a number of equally competent, equally dedicated topographical engineers whose presence on Jackson's staff would have been as efficacious

The Nine Sheet Map of Virginia

This is the title section of the 1859 "Nine Sheet Map of Virginia," prepared by Herman Böÿe in 1826 and revised by Ludwig Von Bucholtz in 1859. The published maps available to desperate Civil War commanders in 1861 looked better than they were. Engineers gradually made corrections and developed reliable maps from these often inaccurate originals. Jed Hotchkiss was very pleased to obtain a copy of this map from W. B. Taliaferro. Hotchkiss's copy soon had numerous changes on it but this was a good map to start with and refer to.

as Hotchkiss's. Hotchkiss, however, possessed an invaluable attribute that the Federal mapmakers lacked. He knew the country, and they didn't. And there was no practical way for these Federals to learn the country without making intensive personal reconnaissance of the ground, which was clearly out of the question. The Federal topographical engineers had two alternatives: they could capture a map or capture the ground. Depending on the perceived importance of the desired information, the latter option was sometimes attempted, as we shall see.

Matthew Forney Steele in his 1951 *American Campaigns* points out that the American Civil War was unusual for a civil war in having a purely sectional basis. Allegiance in this civil war was decided by one's geographic location rather than class, religion, political allegiance, ethnicity, or other factors that usually set the battling factions in a civil war apart from one another. This meant, in practical terms, that in the American Civil War the sides fought not among themselves but arrayed against one another.

The Southern Confederacy's objective was simply to be left alone. The Union's determination was to deny them that forbearance. Thus, an "invasion" of the southern portion of the country, in Abraham Lincoln's blandly legal phraseology, to "subdue combinations too powerful to be supressed by the ordinary course of judicial proceedings," became the war's inevitable strategy.

Whatever the economic and cultural consequences of the war, in purely military terms this strategy gave the Confederacy an advantage: the enormous overall military advantage of familiarity with the terrain. It was not always clear-cut and there were exceptions to it, but this geopolitical fact was to have a great impact on the course of the first three years of the war, particularly in the East.

The first major campaign and battle of the Civil War, Bull Run (or First Manassas), provides a good example of the respective strategic and tactical positions of the combatants. Union general Irvin McDowell, under protest, moved into Virginia, reassured by his commander in chief that both armies were "green alike" and were thus operating under the same disadvantages. McDowell, however, had to make what was, by the standards of 1861, a long march to confront Confederate general P. G. T. Beauregard. Beauregard's men were settled in defensive positions of their own choosing on ground they knew well. Once in the immediate proximity of Rebel forces, McDowell had to spend two full days reconnoitering to find a way to get at his opponent. The delay gave the armies of Beauregard and Joseph E. Johnston time to unite and defeat McDowell—by a narrow margin. The subsequent rout of the inexperienced Union troops, deep in unfamiliar enemy territory, their cohesion broken, was only to be expected. Yet McDowell had managed to meet the criterion of Nathan Bedford Forrest's familiar military maxim. McDowell got almost where he wanted to be first and with the most men. The problem was that he didn't know where he was when he got there. He was lost geographically, and so he lost militarily.

The Union troops made strenuous and very imaginative efforts to remedy their lack of topographical knowledge. They had access to the various maps of the entire state of Virginia—

Lloyd's map, Ludwig Von Bucholtz's revised 1859 version of Herman Böÿe's 1827 nine sheet map of Virginia, which was itself based largely on county maps by surveyor John Wood, and Claudius Crozet's map of the 1830s. These maps were by no means primitive productions. In fact, one difficulty with their use was that they looked so impressive it was hard to believe they were sometimes quite inaccurate and/or misleading. Distances were wrong or topographical features were omitted, which hid the fact that some nearby places were inaccessible to each other. The common symbol for a road made no distinction between a practicable road, a footpath, or a trail through a swamp that was totally unfit for military use. The symbol for a bridge could depict a massive stone structure or a rickety footbridge. Nothing on the maps indicated impassable woodlands or other ground-cover features that could gravely affect military planning and operations.

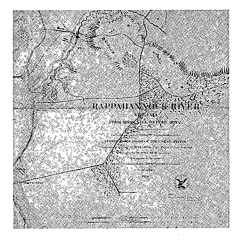

United States Coast Survey maps were the best maps available of any region they covered. Wartime topographical engineers adapted them to their own military purposes. A large-sized reproduction of this map or chart is on page 104.

The civilian United States Coast Survey prepared very detailed maps of coastal areas, as its name suggests. In some cases these maps corresponded with areas of interest to military map-makers and were used to augment or corroborate the other map authorities that were being relied upon. One Coast Survey map that was put to good use was of the Rappahannock River just downstream from Fredericksburg, Virginia. Army of the Potomac engineers used the printed Coast Survey map as a base map and added to it, even to the point of pasting a sheet of paper on the printed map to add their own chart of the Rappahannock. It is said that the monuments or official triangulation points set by the Coast Survey in the vicinity of Fredericksburg were in some instances pulled out of the ground by Union soldiers. They thought the markers were a facetious attempt by the Confederates to demarcate a new border between North and South because the monuments were marked "U.S." above "C.S." They mistook the initials as representing "United States" and "Confederate States," not realizing that they stood for "United States Coast Survey."

A more serious misunderstanding resulted from the use of a Coast Survey chart of the James River at the mouth of the Warwick River. Topographical engineer Thomas Jefferson Cram prepared a map of the Virginia Peninsula for Army of the Potomac commander George B. McClellan as the general prepared to shift the seat of the war from the outskirts of Washington City to the vicinity of Richmond. Cram's map, which was based on the best existing maps that he could find, depicted the Warwick River as running parallel to the James River. This made it a natural barrier that would protect the left flank of McClellan's army as he prepared to advance on Yorktown and sever its communications with Richmond. The Coast Survey map, showing only a narrow band of the shoreline, seemed to confirm, with its limited coverage, that the Warwick River did, indeed, flow in a course generally parallel to the James. The Coast Survey chart was thoroughly accurate. The problem was that just where the chart left off, the Warwick River veered to the right to flow in a course across the Virginia Peninsula. The river effectively blocked McClellan's advance instead of protecting it. This was the first snag in a campaign that eventually came completely unraveled.

Cram, to his credit, had used some imaginative resources in his efforts to gain information about the peninsula. He apparently dug through War Department files, surmising that fighting had raged on this same peninsula more than eighty years earlier—during the siege and surrender at Yorktown in the Revolutionary War. He found and studied maps of the siege of Yorktown, both British and French, that had rested quietly for decades in the files. Cram is also believed to have investigated various Revolutionary War memoirs and histories in hopes of finding additional topographical details to enhance his map's accuracy and usefulness. The unfortunate error in regard to the course of the Warwick River ended Cram's field services, but his initiative in seeking information was nonetheless laudable.[4]

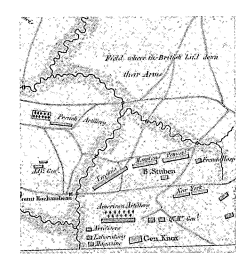

Topog T. J. Cram found Revolutionary War maps in the files of the United States War Department and gleaned relevant topographical details from them. This map is reproduced in full on page 85.

The situation in the Confederate camps during the same time period was, if anything, more dire in regard to maps. Confederate brigadier Richard Taylor, son of Mexican War hero and later president Zachary Taylor, commented with bemused wonder: "The Confederate commanders knew no more about the topography of the country than they did about Central Africa. Here was a limited district, the whole of it within a day's march of the city of Richmond, capital of Virginia, and the Confederacy, almost the first spot on the continent occupied by the British race, the Chickahominy itself classic by legends of Captain John Smith and Pocahontas; and yet we were profoundly ignorant of the country, were without maps, sketches, or proper guides, and nearly as helpless as if we had been suddenly transferred to the banks of the Lualaba."[5]

There were also some idiosyncrasies that made Virginia especially—and the South generally—difficult to map accurately and hard to get around in. To begin with, there was sometimes considerable confusion about the correct or accepted name of a given locality. Cold Harbor, Virginia, was sometimes called Coal Harbor, and there was also a New Cold Harbor and a "burned" Cold Harbor. Burned Cold Harbor was known by the locals as Old Cold Harbor.[6] Many of the roads were known by one of two names: the Market or River Road; the Williamsburg or Seven Mile Road; the Quaker or Willis Church Road. To add to the confusion, there were sometimes other nearby roads with the same or similar names that ran in completely different directions. A stranger looking for what he had heard called Darby Road would discover that this was the Virginia pronunciation of the name Enroughty.[7] Many of the little crossroads hamlets and small towns were likewise blessed with two names. When Union general John Pope asked his topographical engineer D. H. Strother the locality of Madison Mills, Strother responded that he didn't know, but "the cook Joe says it is the same as Liberty Mills." On another occasion, a frustrated Pope asked Strother about the location of Mechanicsburg—not the Mechanicsburg near Richmond, he sharply cautioned. "No," Strother returned mildly, "but the village in Louisa County, fifteen miles from the Courthouse, toward Gordonsville."[8] There was the confusion resulting from identical place-names and additional confusion from similar place-names. There was a Uniontown, a Union Mills, and a Union Bridge, all in Carroll County, Maryland. There were Middletowns, Middleburgs, Middleboros, and Middlevilles. Out-of-date maps caused confusion, too. Union general Philip Sheridan, map in hand, asked a cavalry officer, who had just reported to him:

> "Captain, how far is it to Green's Corners from this point?" The captain looked at him a moment and then answered: "What Green's Corners do you mean, sir?" "Why in the valley between Harpers Ferry and Martinsburg." The captain looked at the questioner a moment and said: "I have been in the valley since the battle of Antietam, but I never heard of Green's Corners before, and I don't believe there is any such place." The little man [Sheridan] jumped up with the map in his hand, nervously tapping it with his finger and said sharply: "Well, sir, I will show it to you on the map; here it is; Torbert, send this officer back to the regiment." He then turned to me and said: "Scout, do you know where Green's Corners are?" "No, sir, I never heard of it." His eyes snapped and he looked as though he was about to kick me out of the room, when General Torbert, who had been looking at the map, said: "Why, Sheridan, you are all wrong; you have got a department map thirty years old. The new map has it down as Smithfield. The name has been changed."[9]

Strother, who was in fact a Virginian and a mapmaker in the Federal army and in consequence quite a study in contrast with Hotchkiss, speculated that his native state had deliberately rebuffed Federal efforts in antebellum days to map the "Ancient Dominion." Virginia had "jealously maintained her constitutional impenetrability. No National Scow was ever permitted to rake mud out of her rivers, and no Federal engineer to set up his tripod on her sacred soil. The consequence was that reliable maps of the country could not be procured."[10]

The Topographical Engineer

Looking up the Country

THE MEN who set to work trying to procure the necessary elements for reliable maps—who went "looking up the country," as Union general Isaac Stevens termed the science of military map reconnaissance—were an interesting and disparate fraternity. The best known of them were self-taught amateurs, men such as Jed Hotchkiss, William Henry Paine, W. W. Blackford, and D. H. Strother. Each of them was a sedulous note keeper; three of them published diaries and/or memoirs after the war.

Hotchkiss is far and away the most famous of all of the war's topographical engineers. It is sometimes mistakenly asserted, as a result of his prominence, that he was the great mapmaker of the war. He was certainly among them. One of the oddities of Hotchkiss's Civil War career is that he seems to have spent the entire war as a civilian mapmaker, essentially hired by the Army of Northern Virginia to make maps. The members of the West Point–trained Corps of Engineers were considered the elite officers of their respective armies. But despite the advocacy of such prominent patrons as Robert E. Lee and Stonewall Jackson, Hotchkiss remained plain Mister Hotchkiss amid the majors, lieutenants, generals, and colonels on the various staffs with which he was associated. (The personal intercession of Lee and Jackson did shield Hotchkiss from being conscripted into the Confederate army as an ordinary private.) His pay was in the $93-per-month range, and he was reimbursed for his supplies.

Hotchkiss knew of and commented on but apparently never met his Union opposite, Strother. Strother's fame was such that nearly anyone who came into contact with him commented on the fact with satisfaction. Strother fails to comment on Hotchkiss and likely never heard his counterpart's name. The Virginian received an officer's commission in June 1862 as a lieutenant colonel in a "loyal, U.S. or bogus"—however one viewed the thing—Virginia volunteer regiment. Strother was paid at the rate of $125 per month and was relieved to get his relationship with the army regularized. He had served without pay for nearly a year, but a more serious concern was the fact that, as a civilian with the military, he might well have been taken as a spy were he to be captured, which had swift and unpleasant consequences in 1862. As a professional sketch artist, Strother was comfortable with the mediums he would be using as a topographical engineer. As a travel writer, he had featured the Shenandoah Valley in his articles and knew the lay of the land very well. Though he had no formal map training, the exigencies of the Civil War and the association in camp and on reconnaissance with trained topographical engineers such as Captain

James W. Abert quickly enabled Strother, naturally skillful with a pen and pencil, to become a proficient mapmaker as well.

Blackford had some training as a civil engineer. When war clouds appeared to be gathering, he helped raise a cavalry company in Washington County, Virginia, thereby becoming its lieutenant. The company eventually fell under the command of Jeb Stuart. Blackford lost his Confederate field command in the spring of 1862 when the army reorganized. As Blackford explained, "To solicit votes among the private soldiers was a thing I could not bring myself to do, and I made no effort in that direction, and Connaly Litchfield, the orderly-sergeant, was elected."[1] Despite the fact that he was not a West Pointer, Blackford, seemingly without difficulty, received a commission as captain of engineers. He spent a brief, unhappy interlude looking out of a fourth-floor office window, caught in the tame routine of a military bureaucracy, but before long he received orders to report again to Stuart, who had heard of his appointment and wanted the topographical engineer on his staff.

Paine came to the attention of Abraham Lincoln through a bit of derring-do. Union plans for campaigning in Virginia were necessarily based on imprecise information. Members of the War Department in Washington believed that the complete lack of knowledge about bridges— were they built? were they burned? were they standing?—made detailed planning impracticable. Paine, as an assistant to General Amiel W. Whipple, chief topographical engineer to the Federal Army of Northeast Virginia, volunteered to go scouting. In civilian clothes, he managed to find out the necessary information, quickly and thoroughly making his way through northern Virginia, reporting on the status of the bridges, and returning safely. The president was impressed. He had Paine appointed a captain of engineers, an honor, as we have seen, Hotchkiss was never accorded by his government.

Honors and promotion did not pile up for these soldiers, which is one of the reasons many topographical engineers made mapping work only a footnote in their careers. Most of them avidly sought out field commands, which were much more likely to bring them advancement and fame. Men such as cavalrymen George Armstrong Custer and James H. Wilson, artillerist Alonzo Cushing, and even George Gordon Meade had all spent time—from weeks to years—as topographical engineers. Though it was a dangerous business (one famous engineer was to write that he sometimes estimated distances by the fact that he was still alive[2]), it was not a particularly glamorous duty nor was it a natural vehicle for promotion. General Grenville Dodge explained the intricacies of army workings during the Civil War that kept immensely valuable soldiers, on whom the fate of a campaign sometimes rested, mere captains and majors while less deserving field and staff officers were promoted rapidly to colonel and brigadier. The mapmakers, wrote Dodge, were usually detailed from their regiments to serve on a staff. The state authorities were therefore reluctant to promote them in their regiments. Thus, men who, in deference to the important and specialized mapping duties they performed, were referred to as major were in point of fact still merely enlisted soldiers when their service ended.[3] When one of these men was sought after and enticed with a commission in a volunteer regiment, engineer captain James C. Duane expressed his exasperation: "Why President Lincoln can make a brigadier-general in five minutes, but it has taken five years to make you an Engineer soldier."[4]

The topographical engineers were not merely well trained. Most of them were very artistic, exhibiting not only fine technical skills but a genuine flair and an intuitive sense of color and composition. Their sketchbooks and field notes, along with their businesslike pages of measurements, compass bearings, and ground-cover assessments, contain whimsically but beautifully drawn pencil sketches of a single tree or a headquarters campsite. Céline Frémaux, whose father, Léon J. Frémaux, was a captain of engineers in the Confederate army, recalled their antebellum evenings at home in Louisiana. The children would play silent games around the dining room table while their father, at the head of the table, practiced his watercolor painting in the somber atmosphere of the candlelit room.

Strother was also an artist; indeed, he was a trained professional. Some of his map innovations exhibited his illustrative creativity. Rapid field sketches often were made to provide a straightforward, ground-level representation of an interesting locality drawn in isometric perspective. The usual practice was to identify points of interest by writing numerals above them; for example, below the number 1 was a gap, below the number 2 was a tavern, and so on. Strother thought that these numbers were a jarring intrusion and substituted birds in flight for the numerals. Below one flying crow was a gap, below two flying crows was a tavern. Apparently substituting crows for numbers brought some relief to his offended artistic sensibilities.

Strother also assessed the "literary and artistic atmosphere" of the topographical headquarters tents. He enjoyed his discussions about the relative merits of the moderns and ancients with Lieutenant John Meigs (the moderns, they concluded, were equals in most respects and superior in some).[5] And with his assistant, William Luce, Strother assessed and compared the qualities of Hawthorne and Longfellow. These mapmakers were, it should be remembered, Victorian-age craftsmen, steeped in poetry, novels, lectures, and art, and they would no more dash off a perfunctory map than a Shaker would slap together a chair.

The attractive maps that these men produced—certainly they are among the loveliest top-secret military documents ever produced—were not, however, embellished with colored pencil and watercolors for mere decorative purposes. Many of the generals lacked topographical expertise, and while they needed lots of information to make the maps valuable to them, they could grasp the information better if it was presented to them in clear graphic terms.

Hotchkiss made it a point "to be always ready to give [Jackson] a graphic representation of any particular point of the region where operations were going on, making a rapid sketch of the topography in his presence, and using different colored pencils for greater clearness in the definition of surface features."[6] The colors used by the engineers followed a formula: streams in delicate blue tints, trees and wooded areas in green, terrain features in brown, roads and railroads in red, and cultural details—the works of man—in black. The watercolor washes defined the ground-cover details, both simplifying their presentation and making them more representational. The meticulous lettering ensured legibility and clarity.

Mountain ranges, hills, ridgelines, and other elevations were most often recorded in a similarly representational or pictorial way. The use of contour lines was not alien to Civil War cartographers. This method of depicting terrain elevation, which had been inspired by the tide ripples on beach sand, was too cluttered and too esoteric for many Civil War generals to visualize and comprehend easily. Civil War cartographers tended to draw contour lines on their field sketches and in the initial rough drafts of their maps. The more traditional but far less exact depiction of elevation, by hachure marks, was employed on the finished map that was actually to be used by the generals, field commanders, and staff. Hachure marks were short, vertical, eyelashlike parallel lines that simplistically represented the slope and fall line of an elevation. The pictorial elevations that they created on a map were readily understandable and were perfectly satisfactory for the needs of a mid-nineteenth-century army commander. On some maps, pencil shading created a billowy contour effect that, like hachure marks, gave an unmistakably mountainous appearance to a range of hills that, again, would be comprehensible in even the most cursory inspection of the map.

And this was, ultimately, what these maps were designed for. Civil War generals seemed invariably to be studying their maps in the most awkward places, under intense time constraints, or with the dimmest of illumination. Confederate general Richard S. Ewell had been trying to figure out what Jackson was up to in mid-May 1862 as the Shenandoah Valley campaign began to

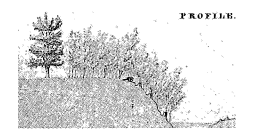

PROFILE.

Léon Frémaux's watercoloring skills are exhibited in this map. Map is reproduced in full on page 76.

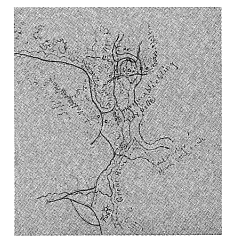

Jed Hotchkiss was always prepared to work up a quick sketch to help Stonewall Jackson grasp the topographical intricacies of the ground across which he maneuvered and deployed his troops. These sketches are reproduced in a larger format on page 160.

blossom. Hearing late at night that Union general James Shields was leaving the valley to move toward Richmond—the very maneuver that Ewell was supposed to prevent—"Ewell bounded from bed wearing only a nightshirt. Taking a map and crude lantern [other accounts specify a lard candle], he dropped to his knees and spread the chart on the bare ground."[7] In June of 1863, Ewell was seen at Chambersburg, Pennsylvania, studying maps in a buggy as his corps marched through. Similarly, Union general Winfield Scott Hancock rode part of the way from Taneytown, Maryland, toward Gettysburg in the back of an ambulance quickly reviewing his maps as he proceeded toward the battlefield to take interim overall command until Meade came up.

Jackson sat on a biscuit box at earliest dawn and went over with his chaplain, the Reverend Beverly Tucker Lacy, an outline map of Chancellorsville. Jackson hoped Lacy, who was familiar with the area, would be aware of a road that might lead to the Union flank. The two puzzled over the map by the light of "a little fire kindled under the trees."[8]

Late on the night of July 3, 1863, in the aftermath of Pickett's charge at Gettysburg, Confederate cavalry general John D. Imboden received a message that Robert E. Lee wished to see him. Lee was not at his headquarters. Imboden was directed a half mile farther on. "On reaching the place indicated, a flickering, solitary candle, visible through the open front of a common tent, showed where Generals Lee and [A. P.] Hill were seated on camp-stools, with a county map spread upon their knees, and engaged in a low and earnest conversation."[9]

During the Wilderness battle Ulysses S. Grant sat "on the ground, his legs tucked under him, tailor fashion," and looked over a map.[10]

Clearly, when campaigning in the field, the work of headquarters was carried on under conditions that were, at best, unpropitious. As it began its march from Savannah and before it touched dry ground at Robertsville, South Carolina, the Union Twentieth Corps marched through the Salkehatchie swamps, where its headquarters and its commander, General Alpheus S. Williams, spent nights and conducted business "safely ensconced in the forks of the trees."[11] Given the circumstances under which military maps were frequently used, clarity was valued highly. A cartographer such as Hotchkiss, whose printing was neat, minute, and legible, was prized because he could compress a large amount of information on his maps without having them appear cluttered and incomprehensible. One thing not required of the maps—and this may seem surprising—was great accuracy in terms of distances and elevations. Fighting in the Civil War was carried on, almost invariably, within sight of the enemy. Artillery ranges, for example, would be adjusted after observing the effect of the firing. If the enemy gave a startled look skyward, the firing was too high. Perhaps the most famous recorded incidence of this effect is recounted by Union general G. K. Warren. He found that Little Round Top at Gettysburg was held by only a couple of signalmen in spite of the fact that the possession of the hill was crucial to the Union line. Warren wasn't certain of the location of enemy units, but to determine it he says he tried the expedient of firing an artillery round where he thought they might be: "As the shot went whistling through the air the sound of it reached the enemy's troops and caused every one to look in the direction of it. This motion revealed to me the glistening of gun-barrels and bayonets of the enemy's line of battle."[12] The Confederates at Gettysburg used four English Whitworth rifled cannon. These guns had a maximum range of five miles,[13] but no sighting mechanism existed at the time to enable the Whitworths' crews to take aim at this range even if their maps had been accurate enough to provide a target.

Most of the active topographical engineers were capable of producing accurate surveys, but their experience with conditions in the field exposed the futility of spending time they didn't have to achieve accuracy they didn't need. Hotchkiss, in his first assignment mapping for the Confederacy, prepared an elaborate and fairly exact survey of Camp Garnett and Rich Mountain in western Virginia. He began work on the map on July 3, 1861. A week later he was leading a small column in a hasty nighttime retreat trying to reach the main Confederate line at Beverly. An attack by Union forces led by William S. Rosecrans had overrun the position. It had taken Hotchkiss longer to accurately map the vicinity of Camp Garnett than the army was able to hold it. Hotchkiss was forced to leave behind his tent, his map library, his engineering equipment, his personal effects—and his new, highly accurate, now virtually useless, map of Camp Garnett and Rich Mountain.[14]

Two highly trained mapmakers of the United States Coast Survey, "with their aides, accompanied the march of General [William Tecumseh] Sherman. The enemy made so little opposition, that one of the officers, in reporting his arrival at Goldsboro wrote to the Coast Survey Office in Washington: 'Our march had been so rapid, that our services have only been required at rare intervals.' "[15]

One exception to this acceptable inaccuracy was the valuable use Union admiral David Dixon Porter made of a Coast Survey mapping party as his mortar prepared to bombard Forts Jackson and St. Philip, some ninety miles downriver from New Orleans. Though the bombardment failed to subdue the forts, "the results of our mortar practice here," according to Porter, "have exceeded anything I ever dreamed of; and for my success I am mainly indebted to the accuracy of positions marked down, under Mr. Gerdes's direction, by Mr. Harris and Mr. Oltmanns. They made a minute and complete survey from the 'jump' to the forts . . . the position that every vessel was to occupy was marked by a white flag, and we knew to a yard the exact distance of the hole in the mortar from the forts, and you will hear in the end how straight the shells went to their mark."[16]

The experience of the Civil War armies in the field was to show that the timeliness of maps was more crucial than their accuracy. A fairly well drawn up map, accurate to within a couple of hundred yards in a mile and prepared within the course of a day or two, was obviously more valuable than an elaborate production that ended up in enemy hands.

There were some exceptions, however. A number of small-scale maps—theater maps—were inevitably time-consuming productions simply because of their physical size and the scope of their coverage. Two of the Civil War's most famous theater maps were prepared by Hotchkiss. Most famous of all was a map that Jackson offhandedly asked Hotchkiss to prepare. Hotchkiss began his career as topographical engineer of the Valley District of the Department of Northern Virginia and segued from that to topographical engineer of the Second Corps of the Army of Northern Virginia under the respective commands of Generals Jackson, Hill, Ewell, and Jubal A. Early.[17] Jackson couldn't order his hired civilian engineer to do anything, but he told Hotchkiss, "I want you to make me a map of the Valley, from Harpers Ferry to Lexington, showing all the points of offense and defense in those places."[18] It was, as Hotchkiss phrased it, a "big job." The map remained a work in progress for Hotchkiss throughout the rest of the war. In one of his last Civil War journal entries—Wednesday, April 5, of Appomattox week—Hotchkiss wrote that he "corrected Valley map some."[19] The map ultimately measured approximately four and a half feet by nine feet. It was a monumental production, but it was probably used constantly, as the gridlines that run through it clearly indicate, to prepare smaller tactical, or route, maps. The smaller-sized but larger-scale maps would start with the same basic information contained on the valley map—be it a section of road or a segment of a river—and would build on it. A new map, based on a detail of the valley map, would then be enlarged and elaborated, resulting in a much more detailed map containing the numerous cultural and physical data a commander had to have to plan and conduct a march through a particular area.

It is quite interesting to note that, on Hotchkiss's original, oversized valley map, there are no hachure marks. He used the more esoteric, for 1862, contour lines, almost as if the map were intended for his expert use only and was never meant to be used by the generals. Jackson was, as noted, never a terribly astute map user, and it seems that Early could hardly read one. Ten years after the war, Early wrote to John Bachelder referencing a map Bachelder had forwarded to him for the purpose of determining the Rebel general's positions at Gettysburg. Early admitted being baffled by the map, writing, "You know it is difficult to imitate accurately the configuration of the ground in topography. . . . I could much more readily recognize the positions of my troops from an examination of the ground itself than from a examination of the map."[20] Until the very end, as Hotchkiss continued to tinker with the valley map, his assistant, Sampson Robinson, worked on reducing the map—an activity last reported by Hotchkiss on April 4, 1865. Robinson was probably just trying to ensure the survival of the mapped information contained on the unwieldy master map—information he, Hotchkiss, and dozens of other topographical engineers, Federal and Confederate, had spent years gathering and recording.

Making the Maps

Pencils Were a Dollar Each in Richmond

THE BUSINESS of preparing military maps for a Civil War army usually started with a man and a horse. A Confederate mapmaker such as Jed Hotchkiss was more of a freelance operator than was a Union *topog* (as they were invariably called by the soldiers) such as D. H. Strother. When the South seceded, the existing, formally organized mapping establishments remained intact and in the Union. Only a few members of the Corps of Topographical Engineers went with their states—among them Joseph E. Johnston, who quickly became one of the top-ranking generals in the Confederate service. Thus, the Union engineers had the expertise and resources behind them of an established and functioning Corps of Topographical Engineers, as well as those of the civilian United States Coast Survey, the Smithsonian Institution, the Naval Hydrographic Office, and the Pacific Wagon Road Office, an agency within the Interior Department.[1] The Corps of Topographical Engineers adjusted to wartime conditions by replacing wooden filing cabinets with iron cabinets and by putting solid shutters on the first-floor windows of its offices in the Winder Building in Washington to thwart the "designs of the evil-disposed."[2] The Confederates, by contrast, had no files to protect, no office windows to secure. In short, they had no engineering department and had to start one from scratch.

The list of supplies required to begin work on a map in the field was not very extensive, but the Confederate mapmaker had no one to present a request to until the late spring of 1862. Necessities such as telescopes, field glasses, prismatic compasses, odometers, tape measures, measuring chains, barometers, boxes of colors, tracing paper, field notebooks, cotton backing cloth, pens, pencils, and thumbtacks were hard to come by for Confederate engineers. Worth an entry in Hotchkiss's diary for April 15, 1863, was the news that one pencil cost $1 in Richmond. Considering that it was possible to hire a draftsman for $3 a day, that price for a pencil was quite high. When engineers could not requisition these items, they had to resort to one of three expedient measures: They could advertise for them in the local paper and perhaps be satisfied with opera glasses in lieu of field glasses; they could wait and capture what they needed from that profligate supplier to the Rebel army, the United States government; or they could use their own supplies and put in a claim for reimbursement. An actual Confederate "topographical organization," as Albert H. Campbell styled it, came into being on June 6, 1862, hastened along by Robert E. Lee, with the commissioning of Campbell as a captain of engineers[3]: "Campbell's office in Richmond came to be recognized as the map bureau of the engineering department, and under his supervision the army's map problem was ultimately solved."[4] Whether Campbell in fact solved anything would be disputed by Hotchkiss. The soul of courtesy and the embodiment of the Christian ideal, Hotchkiss bitterly remembered Campbell for the "large amount of bad work he had done and for the long time he took to do it in."[5] In any event, a Confederate topographical office began to function. Drafting, coloring materials, and other engineering items were ordered from England. The purchases included "drawing instruments, horn protractors, triangles, steel rules, T-squares, India ink, watercolors, drawing paper, tracing cloth, pantographs, pens, and pencils." An examination of Hotchkiss's effects indicates that his supplies were obtained closer to home than England. He had secured a fine Federal wall tent at General Robert Milroy's abandoned camps at McDowell, Virginia, in May 1862. By the autumn of 1864 he wore a captured Federal overcoat. He used a compass stamped *U.S. Coast Survey,* some of his maps are drawn on paper embossed *Mt Holly Paper Co., Pa,* and the cover of his field sketchbook was a blank Federal commission, again courtesy of General Milroy, this time from his headquarters in Winchester, Virginia. Hotchkiss carefully varnished the cover to protect it and the contents of his sketchbook.

Though far better situated in this regard than their Rebel counterparts, Union mapmakers had supply difficulties of their own. Strother spent six weeks in the late fall of 1861 trying to recover one of his watercolors, "a cake of very fine ultramarine, used in coloring . . . maps." He resignedly searched an old, picked-over campsite: "The first object that met my eyes was a little square package of mouldy paper lying among the straw and leaves. I dismounted and took it up. It was my lost ultramarine."[6] Civil War mapmakers had a variety of watercolors available to them: ultramarine, vermilion, burnt siennas, ochers, lead-tin yellows, Chinese white, and more. The watercolors came in paper-wrapped two-by-one-inch rubbing cakes that were somewhat weather resistant. These cakes had to be ground into a powder using a pumice stone, and then a wash was created by adding water and stirring the mixture.

Winsor & Newton's watercolor catalogue of 1857 explains Strother's concern to recover his lost ultramarine blue. The other colors—the violet carmine, the emerald green—cost from one English shilling six pence to five shillings. Ultramarine cost a whopping twenty-one shillings. (According to John McCusker of Trinity University, twenty-one shillings or a guinea would have approximated half a Union private's monthly pay in 1862.) The reason for this anomaly in price is that ultramarine was made from lapis lazuli, a semiprecious stone mined in Afghanistan. It was ground up and mixed with mastic resin, stand (a thick linseed) oil, and beeswax to form a soft, claylike lump. This lump was wrapped in several layers of cheesecloth, submerged in clear water, and kneaded like dough to provide a brilliant blue color—admired as the blue of the Renaissance. Kneading the same lump in a fresh clear water would give a dimmer color, ultramarine ash-blue. These colors were perfect for maps because of their luminous translucent quality.[7]

Though the Union engineers started out with more and better equipment than their Rebel counterparts, they didn't have to proceed very far into Virginia or Tennessee or Georgia before the handicaps they worked under became obvious. The Union topographer was in unfamiliar and hostile countryside. The Confederate was among friends. A common problem plagued both sides as they conducted their reconnaissance, made their notes, investigated farm lanes, and sketched their preliminary maps: they were subjected to continual arrests. In the South, "one of the most annoying features was the frequent arrest of these [topographical] officers while diligently at work by our own outposts and scouting parties, involving a ride under guard to headquarters and a serious interruption to the work. Even when at work in the rear of the army these officers were frequently regarded with general suspicion. One officer was held up by a party of six zealous citizens, who had been watching his motions for some time and who insisted upon escorting him about twenty miles to army headquarters, two in front, two in rear, and one on each side, with guns ready for action. The officer, however, calmly continued to take his notes as he went along, and was allowed that privilege."[8]

Strother, a Federal, had a less cordial encounter. A Union soldier approached him in a barroom and demanded to know his name, business, destination, and why he was traveling with a map of Washington County in his haversack. Strother then turned on the soldier: "I asked him sharply who he was, what he was doing here, and upon what authority he undertook to question travelers. He answered, that he was a soldier of the Potomac Home Brigade, and he considered it his duty to find out whether a man had a right to travel about with a map of Maryland in his pocket. In return, I informed him that I was in the United States service, and attached to the Topographical Corps of Patterson's army. 'To-py——top-py——to-pee——to hell!' he exclaimed, staggering with the effort to accomplish the knotty polysyllable. 'I believe yor're a dam'd rebel spy.'"[9]

As they set out to prepare their maps, the topographical engineers were usually accompanied by an aide, such as Hotchkiss's orderly, S. Howell Brown. Hotchkiss described him as "a big stout fellow, accommodating and pleasant, but a mass of facts; painfully matter-of-fact man, at times fearfully exact."[10] Brown provided Hotchkiss with companionship, assistance, and protection. Hotchkiss, as a rule, carried no weapon. As a civilian with the Army of Northern Virginia, he would have had to do some quick thinking and provide a convincing explanation as to his status if he were ever captured. Spies and bushwhackers got short shrift from regular uniformed army units, and a civilian Hotchkiss with a weapon could certainly have been taken for one or the other. It is also true that with the constant mounting and dismounting, and drawing with his leg crossed on his saddle, a weapon might have been more dangerous to Hotchkiss than to any conceivable foe. Most topogs engaged in preparing the rapid field sketches that provided the initial research—the basis for most Civil War maps—were deliberately lightly equipped. They had to move along briskly. The information was usually needed in a hurry, and the preoccupied engineer was a vulnerable figure as he made his way through the wartime countryside. The aide or escort would perform topographical chores: following bypaths and farm lanes to their end, inquiring and ascertaining the identities of each and every inhabitant along the route of march, sidling his horse into streams and watercourses in an attempt to determine the depth of the water and the feasibility of fording there. Finally, he would maintain an alert watch while his chief concentrated on the tools of his trade and their application to the job at hand.

The typical Civil War mapmaker would ride with a drawing board resting on the pommel of his saddle. His first sketching would be done with a soft lead pencil. Author and naturalist Henry David Thoreau had discovered that a softer pencil could be made by decreasing the amount of Bavarian clay in the pencil lead (a baked mixture of graphite and clay). Soft pencils hold the paper best and will not be jogged off-line by any movements of the horse. Soft lead also erases easily and does not engrave itself on the paper. The pencil might well be tied with a piece of string to a buttonhole so it could not be dropped. For an eraser, the mapmaker used a piece of hard India rubber, also attached to him by a string. In a pinch, a piece of stale bread would suffice as an eraser. A wooden ruler, handy for making straight lines, for measuring, and for balancing on an extended finger to estimate a distant grade, would be safely stowed in a top boot when not in use.[11] A field sketchbook would be used to record the information that the topographical engineer noted at the moment of passing. Engineers with the Union's western armies had the best-designed, most specialized topographical notebook. It had a black, untitled cover and was pocket-size (five by eight inches). It was ruled with a vertical line down the center of each page. On the inside of the front cover was a lengthy illustrated list of the various standardized symbols to be used by the engineer. This included appropriate renditions of different types of fences (stone, rail, and post and rail), roads of varied surfaces, orchards, pine trees, deciduous trees, buildings, cultivated fields, hills, woods, cuts, and fills. Bridges were to be distinguished by their construction, for that would have provided some indication of their wartime longevity as well as their practicability for heavy wagons

Schmalcalder Compass

The Schmalcalder or prismatic compass was invented and patented in 1812 in London, England, by Charles Schmalcalder. This version was made for the U.S. Engineers by J. Green, New York, circa 1860. The compass is made of brass. Because of its portability, this instrument was ideal for military use, in which, as an 1856 catalogue of instruments says, "little more than a sketch map of the country is required." Quite a number of Civil War maps specifically cite the Schmalcalder compass under "Instruments Used." See *Official Records Atlas* (plate 55). The compass card is divided into 360 degrees, inscribed on the South point because the eye and prism look due North. The figures on the card are printed in reverse because they are seen in reflection. The use of this instrument enabled a quick-moving topographical engineer to read degrees in the prism and magnifying lens at the same time he took his vertical sightings. The compass card has a jeweled pivot. A button under the vertical sight operates a spring-and-metal strip that immobilizes the card. The actual diameter of the compass card is 2½ inches. The Schmalcalder compass could be "carried in the pocket without inconvenience."

and artillery. Examples were provided of stone, wooden, suspension, and trestle bridges. The most poignant symbol—a small chimney drawn with a tiny cavity to represent a fireplace—stood for a burned house. As a rule, the left page of the notebook contained compass headings combined with the distance between, for example, points A and B. On the right was drawn a map with a road marked A and B to identify the data recorded on the opposite page. The top and bottom of a sketched road were marked to indicate where it resumed on the next page.

The compass favored by the topog in the field was the prismatic, or Schmalcalder, compass, which was simple, sturdy, light, and portable. It consisted of a brass case, a glass lid with a hinge, and a hinged sight in the rear containing a mirrored prism. This prism enabled the user to obtain a reading in degrees on the free-turning magnetic dial while simultaneously sighting the compass on a specific point—a tree, a telegraph pole, a well—resulting in quicker, more accurate measurements.[12] As a precaution, the engineer would also carry a small pocket compass, the simplest, smallest, and most basic direction-indicating piece of equipment of all.

Another staple was the aneroid barometer, the most delicate and expensive instrument generally carried on a day-to-day reconnoiter. This nonliquid barometer was used to measure the changes in air pressure in ascending or descending a hill. It provided a rudimentary but reasonably satisfactory estimate of altitude. This slightly hit-or-miss approach to altitude underscores how mid-nineteenth-century military tactics gave mapmakers some leeway when it came to providing exact measurements.

That timeliness was more important than accuracy also is reflected in the techniques used to measure distances. When engineers were mapping out a possible route of march, time was invariably of the essence, and a minimum of exposure was desirable to avoid both detection and capture. The engineer's activities gave to prying eyes a clue to an army's intentions and possible objectives.

Several mechanisms and methods could be used to measure distance. The odometer was most common and consisted of a wheel of known circumference attached to a counter. This instrument provided accurate measurements but was cumbersome and slow. A faster but still surprisingly good way to measure distance was by keeping track of horse paces. The walk of a horse is very consistent and an engineer would be sure to know, for example, that his horse covered 400 yards with every 436 paces. It was not unusual for even the most highly trained and experienced topographical engineers—Major General G. K. Warren is one—to use horse paces as the basis for the scale of their maps: "1050 horse paces equal to one inch" is a typical scale on a Warren map. Lieutenant John R. Meigs's habits while making a routine military survey in the Shenandoah Valley on October 3, 1864, were described as follows: "He was riding along in a walk and counting the footsteps of his horse, as was his custom. He got his measurement of distance in this way. When he changed his course, he made a note of the number of his horse's steps, took his new bearings and started counting again." The habit of counting became reflexive. Meigs counted paces whenever he rode, "whether he noted them down or not."[13] An engineer would also train himself to keep a steady pace. Ambrose Bierce's sketchbook made note of the fact that with eighteen foot paces he covered fifty feet. This scale would be adjusted when measuring on an incline because the length of a man's pace shortened when he was moving up or down a hill.

Other conveniently handy means of measurement also were used. The length of an engineer's horse would be used to figure out the width of a lane or a bridge. As noted earlier, the horse could be ridden into a stream so its depth could be estimated, and the rider would not get wet or have to dismount. An engineer also would know the exact span of his own arms and fingers.

A topographer would use the known size of familiar objects—a cow, a fence post, a row of corn—in his line of sight to help him gauge distance. He would know what a church steeple looked like from ten miles away, what a house looked like from five miles away, what a chimney looked like from two and a half miles away, what a marching column looked like from one mile away, and what a single pane of glass in a mullioned window looked like from five hundred yards away. These frames of reference were very valuable since atmospheric conditions, terrain features, the angle of the sun, and the level of an object relative to the observer could sometimes drastically affect the judgment of distance. It was noted that in the extremely dry atmospheric conditions prevailing in the West, Easterners would misjudge badly, thinking objects ten miles away to be only two or three miles away.

Any existing map of an unfamiliar region would be taken along by the mapmaker as he made his initial inspection of the terrain. Almost any map would be of some assistance, and the first order of business would be to ascertain just how reliable any extant map or survey happened to be. If a decent map was on hand, it was easier to concentrate on the numerous cultural and physical details in a maneuvering army's environment.

The deep concentration given to the work at hand could be dangerous. Under circumstances never completely settled, Union lieutenant John R. Meigs was killed by a party of Rebels while he was mapping behind Union lines in the Shenandoah Valley in October 1864. The survivors of Meigs's little party claimed they were bushwhacked and that Meigs was killed without provocation or warning. Soldiers on the other side contended that Meigs was killed in a stand-up fight that they tried to ride away from. Meigs was the son of the Union army's quartermaster general, Montgomery C. Meigs, and he was General Philip Sheridan's chief of engineers. His death,

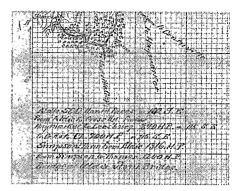

The footsteps of a horse are very consistent. Topographical engineers counted them almost instinctively and used them to establish approximate distances, as indicated on this sketch map. Sketch map is reproduced in full on page 132.

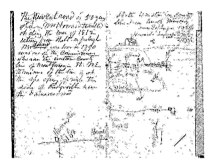

The circumstances of Union topographical engineer William Luce's capture were detailed by D. H. Strother in his diary and postwar writings. Luce's field book is reproduced in a larger format on page 80.

or assassination as Sheridan insisted, was therefore not an isolated and overlooked incident. A number of houses, in a radius of five miles, were burned in retaliation, though Sheridan's orders to burn them all were very leniently interpreted by General George Armstrong Custer, the cavalryman detailed by Sheridan to carry out the planned reprisal.[14]

Strother had great respect for his assistant, a tall, thin draftsman named William Luce, but he considered Luce's surveying style "recklessly venturesome." When Luce was late returning to camp, Strother always supposed he had been captured. On March 23, 1862, near Berryville, Virginia, Strother got the word that Luce had been captured, along with his wagon, horses, and instruments. Ten days later, on April 2, Strother was interrogating two Confederate prisoners. One, a Lieutenant Duff of Colonel Turner Ashby's cavalry command, acknowledged that he was the man who had captured Luce. Duff had been told there was a Federal engineer approaching, so Duff "laid in wait for him with his squad, and presently Luce came riding along, absorbed as usual in his note-book and compass." When the lieutenant advanced, Luce assumed that he was being challenged, as usual, by zealous or officious Union sentinels and tried to brush away the disturbance until a cocked pistol was brought to bear on him and he was made aware of his mistake by being taken prisoner. The lieutenant, apparently, was sufficiently alert to the value of Luce's effects to pass them along to Turner Ashby. He in turn passed them on to Hotchkiss, in whose possession Luce's notebook and instruments remained at war's end.[15]

Meigs had been surveying with two escorts. Luce had been surveying with a teamster. They were, at the time they were, respectively, killed and captured, performing quite routine mapping assignments. Robert E. Lee, according to topographical engineer Samuel R. Johnston, had found while in Mexico "that he could get nearer the enemy and do more with a few men than with many."[16] Strother's August 21, 1861, diary entry carefully enumerated a topographical foray: "To-day Captain Abert started with his party to reconnoiter and survey the roads toward the Potomac River. The company consisted of Captain Abert, chief, with Mr. Luce and myself as assistants, then came our followers: Benjamin the Swiss valet, Swizert the groom, Henshaw the teamster, and Frisby the big cook, with two of Thomas's dragoons as escort—in all, nine men and twelve horses."[17]

W. W. Blackford, Jeb Stuart's chief topographical engineer, agreed with Lee that less was more. Described in a report by Stuart as "always in advance, obtaining valuable information of the enemy's strength, movements and positions, locating routes and making hurried but accurate topographical sketches,"[18] Blackford said he "had already found that for obtaining information a small party is the best for many reasons, and ultimately I never took but three unless it was necessary to send back reports before I returned, and then I only took as many more as would be required for this purpose. With three, besides myself, I would ride as near as possible to the enemy without attracting attention, and if the view could not then be commanded I would dismount and take two dismounted men with me, leaving the other man to hold all the horses. We would then advance, availing ourselves of all cover and inequalities of the ground, until a good view of the enemy could be had from some commanding point. While I examined them with my glasses, the men watched to see that we were not flanked and to call my attention to anything they saw in front."[19]

At Culpeper, Virginia, in May of 1863, Blackford was ordered to reconnoiter the Rappahannock River from Chancellorsville in the direction of Warrenton. General Lee wanted a topographical map prepared of the militarily significant points along the river—which was occupied on the north bank by Union forces—so when the time came, Lee could choose where to force a crossing. For this urgent and risky scouting in close proximity to the enemy, Blackford took along his assistant engineer officer, Lieutenant Frank Robertson, and twenty-five handpicked cavalrymen. With an escort of this size, Blackford could fight just long enough to get safely away from almost any enemy force he encountered.

Ambrose Bierce made vivid use in his postwar writings of his Civil War experiences as a topographical engineer in the Union's western armies. This passage from his story "George Thurston" would probably seem overly dramatic were the facts it expressed not further documented in Civil War memoirs and regimental histories:

> Whether in camp or on the march, in barracks, in tents, or in bivouac, my duties as topographical engineer kept me working like a beaver all day in the saddle and half the night at my drawing-table, platting my surveys. It was hazardous work; the nearer to the enemy's lines I could penetrate, the more valuable were my field notes and the resulting maps. It was a business in which the lives of men counted as nothing against the chance of defining a road or sketching a bridge. Whole squadrons of cavalry escort had sometimes to be sent thundering against a powerful infantry outpost in order that the brief time between the charge and the inevitable retreat might be utilized in sounding a ford or determining the point of intersection of two roads.[20]

On September 11, 1861, Union colonel Isaac Stevens, a former topographical engineer who was to be killed almost a year later as a major general at Chantilly, Virginia, led a classic reconnaissance in force at Lewinsville, Virginia, with two thousand troops. Apparently unable to capture a map, Stevens chose the alternative of capturing the ground:

> With skirmishers in advance, and exploring the ground on both flanks to the distance of a mile, the command [many of whom were to gain individual or unit fame as the war progressed] advanced steadily to Lewinsville, the enemy's cavalry pickets falling back without resistance, and occupied the village at ten A.M. Cavalry pickets were thrown out on all the roads; three guns and some five hundred skirmishers were posted well out to command the approaches on all sides; and the position was held for five hours, during which Lieutenant Orlando M. Poe, of the engineers (afterwards General Poe), and Mr. West, of the Coast Survey, made a topographical map and sketch of the place and vicinity. Colonel Stevens, with Captain Griffin and Lieutenant Poe, thoroughly examined the whole position of Lewinsville, of which he reported, "It has great natural advantages, is easily defensible, and should be occupied without delay."[21]

Upon withdrawing from this immense fact-finding operation, the units were attacked by Jeb Stuart's cavalry and suffered some of their first casualties of the war.

Strother, serving with Union general Nathaniel P. Banks at Port Hudson in March 1863, was volunteered for and reluctantly agreed to accompany a reconnaissance organized to unmask or determine the location of some artillery batteries of Port Hudson's Confederate defenders. Someone was needed to prepare sketches of the road as the expedition proceeded. The escort was a squadron of cavalry and a regiment of infantry—well over a thousand men and hundreds of animals. Strother's massive expedition reflected the relative importance of the information he sought. They passed a Rebel camp where the fires still burned. Carved on a tree was a message: "Beware Yankees: this is a hard road to travel." It was indeed, and Strother's report on the poor condition of the road led Banks to postpone an attack. "I do not mean to lead my men to a fool's fight and a dog's death," Banks wrote.[22]

5

Roads

A Calamity at Their Worst

ACCORDING to Captain William Willoughby Verner in his instructive 1889 *Rapid Field Sketching and Reconnaissance,* "it should be the aim and object of the reconnaissance officer to discipline his powers of observation and that he instinctively notice everything of military importance." This was what Ambrose Bierce meant when he described "observation entered in red": fixing in the memory a picture of a locality and its militarily significant features.

"Everything of military importance" was a lot. As we have seen, in the case of D. H. Strother's reconnaissance, it was the road that was evaluated most attentively. The report of its poor condition precluded an advance and attack by General Nathaniel P. Banks. Roads were the milieu of a Civil War army. Other elements on the ground—forests, rivers, hills, and the like—had an effect on a maneuvering army, but none of these were the least bit relevant when the roads that led through them, up them, over them, or around them were not passable. For this reason, map-makers made very exact inquiries about roads and studied them with care. At issue was not so much where a road led but what it could carry; specifically, whether a road was *practicable* (a very popular wartime adjective) for wagons and artillery. Stonewall Jackson did not merely ask Jed Hotchkiss to find a road to the Union flank at Chancellorsville. He asked him to find a road that was practicable for artillery.[1]

Strother's analysis of the road would have taken a great amount of information into consideration. Banks was contemplating a deployment of artillery, so the criteria for an adequate road would have to include factors such as the steepness of the road's inclines and the nature of the road's surface. Artillery and caissons could manage, at the most, a ten-degree incline—and that meant ascending or descending. The width of a road would be carefully noted. (A member of Confederate General John Bell Hood's division noted humorously that during the Gettysburg invasion the issue of gills of whiskey after fording the Potomac had the weaving, pixilated men more concerned about the width of the road than about the length of it.) If a road narrowed in passing over a bridge, through a defile, sturdy fences, or woods, or beside a watercourse, the progress of a military operation could be slowed considerably or obstructed completely. After his May 8, 1862, victory at McDowell, Virginia, Jackson started in pursuit of retreating Federal units under Robert Milroy and Robert Schenck. He wanted to make certain that they and Banks could not connect up. He sent Hotchkiss to block the roads through the mountain passes by which the two beaten generals might attempt a union with Banks. Pushing boulders down into the passes, felling trees, and destroying bridges, in short order and with relatively few men, the novice engineer Hotchkiss was able to close the possible connections between the separated Federal units.

Grenville M. Dodge provided Major General James Birdseye McPherson with a topographical engineer's assessment of a number of different roads "leading from the Mississippi line to the Coosa Valley." His comments make it very clear what the important factors were when considering the passability of a road: "The Byler road leaves the valley at Leighton, runs up Town Creek, crosses mountain in Low Gap, and forks at New London, one branch going toward Columbus, Mississippi and one direct to Tuscaloosa. It is an old road, well settled, well watered, fair for forage, crosses the streams high enough to avoid much difficulty, and is one of the best roads over the mountain. . . . Between the roads mentioned there are by-roads and mountain paths, over which cavalry can travel and probably light trains, but army transportation would stick."[2]

Dodge's characterization of a road as "old" and "well settled" meant that the road would probably remain passable under most conditions. The roads on the Virginia Peninsula, as the Union army discovered too late, were, as General O. O. Howard relates, "never reliable. Fair to the eye at first, with the rain and travel of heavy trains, the crust, like rotten ice, gave way, and then horses, mules, and wagons dropped through into sticky mud or quick sands."[3]

John S. Clark studied and mapped roads in the vicinity of Culpeper, Virginia, for General Banks in the summer of 1862. The map he prepared had separate symbols for "stone turnpikes," "roads impracticable," and "roads very bad." Clark accompanied the map with further commentary: "No roads between Sperryville and Amissville practicable for artillery or trains—except turnpike via Washington, all are gullied rocky and mountainous."[4]

Hotchkiss's rough sketches were dotted with written descriptions of the condition and character of the roads he was mapping: south of Winchester, Virginia, "at Mrs. Craig's—the plank used up—merely a dirt road—badly cut up—clayey and sandy." On another sketch of the Luray Turnpike and vicinity, he noted that the road was "winding—well-graded McAdamized [macadamized] road—rocky in some places." Though a macadamized road was as close to paved as a Civil War road got, it was not without some drawbacks, especially to the poorly shod men and horses of the Rebel armies. It was built of uniformly sized broken stones, humped in the center to facilitate drainage, and with ditches to catch the runoff, but the road's broken stones would lame a horse if it cast a shoe and could quickly hobble men who had marched along cheerily on dirt and sandy roads.[5]

The detailed information about the condition of roads was important to the men of the marching army, but it was critical for the artillery and the wagon trains. George Gordon Meade commented that his trains and artillery cut up even "the best roads in Pennsylvania."[6] As John D. Billings pointed out in his 1887 "unwritten story of army life" *Hard Tack and Coffee,* the wagon trains were called "impedimenta" until the troops got hungry or the ammunition ran low. When the wagon train was needed, "it must be there. There was no 'if convenient' or 'if possible' attached to the order. The troops must have their rations, or more important still, the ammunition must be at hand in case of need." In many cases, the infantry didn't even march on the roads. For their wagons to be accessible to them while on the march, instead of being strung out far to the rear with perhaps a quarter of the troops detached to guard them there, it was not unusual for the infantry to cede the road to the wagon trains and march beside them. Confederate forces did this as soon as they entered Pennsylvania. They "opened the fences and enlarged the roads about twenty yards on each side, which enabled the wagons and themselves to proceed together."[7] In these cases, the artillery traveled on the harder ground of the uncultivated pastures and meadows while the infantry made its way through the soft, plowed ground of the corn, wheat, and rye fields.

The surface and general condition of a road also made a great impact on the logistics of a march. On a macadamized road, an army wagon could carry more than two tons of forage, rations, equipment, or ammunition. On a dirt road or a country lane, the maximum load would be less than a single ton. If the road surface was not made known to the quartermaster via the information conveyed on the maps, the army could mistakenly march with too few half-filled wagons or the army would try to move with wagons that would promptly stick in the road.

The condition and practicability of the roads varied dramatically in the mid-nineteenth-century environment of the American Civil War. The Valley Turnpike in the Shenandoah was an excellent road, almost unique for its time in being laid out to run in straight lines for miles at a stretch—a shocking sight to many soldiers who had never seen anything but whimsically meandering roads in their lives. More familiar perhaps were the roads in the neighborhood of Somerset Mills described by Union general Alpheus S. Williams in a letter to his daughter: "Marched at about 3:30, as soon as light showed me the road down a muddy ravine. . . . These Virginia roads wind about in all directions; seem to be the chance paths of farms connected by lanes; never repaired; and all the little brooks flow into and along them, cutting out gullies and forming immense mud holes. It is incredible that heavy loaded wagons get over them."[8]

As Williams's letter brings out, the distinction in parts of the South between roads and watercourses could be slightly blurred. It was not unusual to find even principal roads utilizing the natural right-of-way of a stream. At Fredericksburg, Virginia, the Richmond Stage Road ran a considerable distance in the streambed of Hazel Run.

Similarly, what was referred to as "the New Madrid Canal, so called" during the Island No. 10 operations on the Mississippi River, was in fact an old Tennessee wagon road. The Union engineer looking at the road running through an overflowed cornfield and into "an opening in the woods back" thought to himself, "The Engineer Regiment has talent enough to take a fleet of boats through those woods." That engineer was Colonel Josiah W. Bissell of the Missouri Engineer Regiment, and he accomplished the feat of taking "steam transports across the fields and through the woods, around the Rebel fortifications at Island No. 10 to New Madrid."[9]

Bissell quibbled with the word *canal* for his project because nothing was dug except a cut in the Mississippi River levee, and the steamboats and barges that were used simply sailed down the route of the submerged wagon road. The feat in the whole project was the cutting of trees four or five feet below the level of the water.

Obviously, an army topographical engineer would normally regard water or watercourses as an impediment to the army's progress rather than as an avenue. The availability of water was, however, important. With hardworking mules and horses drinking ten gallons a day, plentiful sources of water were important and often would be noted along a route of march. Union major general George H. Thomas's map journal carefully noted, among many other cultural and physical details, the availability of potable water. But nearly all of the engineer's attention to water concerned itself with determining where an army could cross a river, stream, creek, or run with the least disruption to the command.

6

Military Fords

We've Struck This River Lengthways

PROBABLY no other aspect of marching in the Civil War gets so much detailed attention in soldiers' diaries, memoirs, and letters as the difficulties and discomforts of arriving at the edge of a river and having to ford across it.

A military ford was much more than a shallow area in a streambed. Engineers had a number of tactical and geologic assessments they had to make when they evaluated the suitability of a fording site. The approaches to and from a ford were important. A steep riverbank meant that wagons and artillery particularly would have difficulty negotiating a ford. The fact that men alone could cross a waterway was not satisfactory in military terms. If the wagons could not cross and deliver ammunition and rations safely and dry on the opposite bank, an army could not cross. While a cavalryman could cross in five feet of water and an infantryman in four feet of water, a loaded wagon was restricted to two and a half feet of water. Thus, unless a ford shallow enough for the rations and gunpowder to cross dry was found, the entire army was brought to a complete halt. An army of men separated from the impedimenta—the wagons and caissons—is essentially disarmed and stranded.

During the Confederate operations along the Rappahannock River in August 1862, as Stonewall Jackson trudged along the south bank of the river in search of an undefended crossing, he sent Jubal A. Early's brigade across on a tumbled-down dam at Warrenton Sulphur Springs. A heavy downpour stranded Early on the Yankees' bank of the river from late afternoon of August 22 until late afternoon of the next day. Jackson spent one gloomy hour after another, sitting on his horse, staring, glaring at the offending river and across at Early's stranded, vulnerable command. At one point, Jackson sat his horse in "the swirling water."[1] The river began to recede and Early safely backtracked, but the dire peril of his command had been painfully evident.

That potentially disastrous situation, which had the normally imperturbable Jackson gazing fixedly across the swollen waters at Early's isolated brigade, was what Union major general Ambrose E. Burnside sought to avoid while he commanded George B. McClellan's left wing at Antietam. The narrow, sturdy stone bridge known as Rohrbach Bridge until September 17, 1862, spanned Antietam Creek south of Sharpsburg, Maryland. A Confederate brigade commanded by Georgian Robert Toombs looked squarely down on the bridge from a steep and wooded riverbank. Burnside spent many hours and many men trying to cross at the bridge, bringing up the

A good example of the close attention an engineer gave to a fording site. See enlarged reproduction on page 133.

James F. Gibson's Stereoscopic View of Rohrbach or Burnside's Bridge across Antietam Creek, Sharpsburg, Maryland, September 21, 1862

Built in 1836–37 by John Howard for Washington County at a cost of $2,300, this was known as Rohrbach Bridge until September 17, 1862. Union general Ambrose E. Burnside has been ridiculed for not simply fording this obviously shallow creek. However, the steep, high, and wooded bluff on the Confederate-held bank meant that only soldiers could have negotiated the climb. Their artillery, caissons, and wagons would have had to be left behind on the opposite side of the Antietam, which would have rendered the infantry helpless.

question: Why couldn't Burnside's soldiers simply splash across the obviously shallow creek (Alexander Gardner characterized it as a "miniature river"), thereby avoiding the tactical bottleneck at the bridge?

The reason was that where Burnside wished to cross his wing of the army was not, in military terms, a practicable ford. The precipitous bluffs on the south or west side of the creek were heavily wooded and easily defended. Men could scramble up them without terrible difficulty, but the wagons, artillery, and caissons, elemental to the support and success of large-scale operations, could not. As averred to by eminent Civil War historian James I. Robertson, Jr., in his 1997 *Stonewall Jackson,* infantrymen who have advanced beyond the covering fire of their own artillery are helpless. Because of this fact, because he had to have his wagons and his guns, Burnside had to have his bridge.

A topographical engineer's assessment of the fording possibilities in the area of Rohrbach Bridge would have taken a number of physical details into account. The creek bottom would be observed. If it was firm and regular, sandy or gravelly, and without boulder-sized rocks, that was helpful. Boulders would be a serious obstacle to the passage of wagons and could trip up and disorder the infantry. The speed of the current and the depth of the water were the obvious elements in evaluating a ford, but as we have seen, the depth that was manageable when a significant military force was crossing a ford was much shallower than that when a squad of soldiers or a cavalry reconnaissance was in progress. In a pinch, the flamboyant cavalry leaders—Jeb Stuart and

George N. Barnard's Photograph of Sudley Springs Ford, Manassas Battlefield, Virginia

Some Civil War fording sites were as well known as battlefields. Fords such as Kelly's, Welford, Ely's, United States Mine, and Germanna were the scenes of constant skirmishing, pitched battles, and numerous crossings by both Federal and Confederate forces. Knowledge of practicable fords was a primary responsibility of topographical engineers. There was nothing, as is obvious from this wartime photograph, outwardly notable about a good ford. Sudley Springs Ford, however, has the prerequisite characteristics of a military ford: gentle riverbanks to allow the passage of wagons and artillery; a shallow, slow-moving current so that their loads, especially the ammunition, stayed dry; a firm, smooth streambed so the wagon would not snag and the soldiers would not stumble. There also were no readily available positions from which an enemy could resist a crossing.

Nathan Bedford Forrest—could manage fairly difficult fording exploits: "Forrest developed an unusual ability to get his entire command across, including artillery, by swimming the horses with their riders and then pulling the guns across by means of a rope attached to a double team on the other side. The gun itself would be completely under water but was not permanently harmed. The ammunition was held above water by the men or taken across in small boats."[2] Similarly, during his infamous detour in the midst of the Gettysburg campaign, Stuart came to Rowser's Ford "of the Potomac," as W. W. Blackford phrased it: "The ford was wide and deep and might well have daunted a less determined man than our indomitable General, for the water swept over the pommels of our saddles. To pass the artillery without wetting the ammunition in the chests was impossible, provided it was left in them, but Stuart had the cartridges distributed among the horsemen and it was thus taken over in safety. The guns and caissons went clean out of sight beneath the surface of the rapid torrent, but all came out without the loss of a piece or a man."[3] What Stuart did lose, of course, was time and lots of it, and part of the reason for finding and using a good ford was to speed the progress of an army.

There were several expedients that could be tried to facilitate the passage of a ford. Cavalrymen on their mounts might be stationed upstream as infantry crossed to help blunt the force of the currents, and additional cavalry would be downstream to scoop up any soldiers swept off their feet by the water. Stephen W. Sears described the advance of General Charles Griffin's division toward Chancellorsville and its negotiation of Ely's Ford on the Rappahannock: "Ely's was a rough and rocky ford and the current was swift, and men were tumbled off their feet, only to fetch up against a lifeguard line of cavalrymen stationed across the river downstream."[4] Sometimes the tallest and strongest foot soldiers performed the same service for their smaller, lighter fellows. It was said that when Richard S. Ewell's men were able, just barely, to get across the flooded Potomac River at Williamsport after the Gettysburg campaign, "the tallest men formed two lines from shore to shore with their guns interlocked to mark a strong and stable lane across the river. Through this passage, the rest of the men waded in water up to their necks to the safety of the far shore."[5]

A ford that was not suitable could, with a little work, be made serviceable. During the Antietam battle of September 17, 1862, Brigadier General Winfield Scott Hancock, at "about 9:30 o'clock A.M. crossed the Antietam at the ford constructed by our engineers."[6] Another example was White's Ford, of the Potomac, several miles above Leesburg, Virginia. Early discussed and described the ford in his memoir of the war: "This ford was an obscure one on the road through the farm of Captain Elijah White, and the banks of the river had to be dug down so that our wagons and artillery might cross." With such improvements, White's could not remain an obscure ford for long. On June 23, 1863, as the Federal forces were preparing to move from northern Virginia to counter the Rebel invasion of Pennsylvania, inquiries were sent out requesting "information . . . as to White's and any other practicable fords."[7]

The difficulty of finding a suitable ford was a problem that mostly affected the Federal forces. They were less likely to have men in the ranks who were familiar enough with the area of operations to know the various river and stream crossings. There were often telltale signs. A path that wound its way to a river might simply dead-end at a watering place, or it might not. John H. Worsham was with the Twenty-first Regimental Virginia Infantry under Early as he advanced on Washington City early on July 9, 1864. At the Monocacy River, near Frederick, Maryland, Worsham and the men of his division were enjoying the unusual role of spectators when a courier "leaving a streak of dust behind him" rode up and the lolling men were hurriedly ordered into place. According to Worsham, "[We] filed to the right, going through fields and over fences until we came to the river. . . . We found a small path on the river bank leading down to water, and on the opposite bank a similar one, denoting a ford used by neighbors for crossing the river."[8]

At First Manassas, believing that the Confederates had mined the stone bridge over Bull Run, Colonel William Tecumseh Sherman searched the left bank of the stream for a crossing place: "Early in the day, while reconnoitering, Colonel Sherman had noticed a horseman descend the opposite bank, ride to our side of the run and shake his sword as if defying us to cross; this officer proved to be Major Wheat of the Louisiana Tiger, who served our brigade a good turn by showing us the road [i.e., the ford]."[9] According to Sherman's subsequent report of the action, his whole brigade crossed at Farm Ford without difficulty, but the steep bluff on the opposite bank did make it impossible to cross the artillery, and the guns stayed behind. Two long years later, facing Confederate Joseph E. Johnston on the Chattahoochee River near Marietta, Georgia, Sherman's topographical engineers mingled with fraternizing Rebel and Yankee pickets along the river. They picked up topographical information along with the other usual soldier gossip. Johnston attributed Sherman's unmasking of fords to these informal truces.[10]

Another recorded example seems to confirm Johnston's suspicions. Sam R. Watkins, memoirist of the classic *Co. Aytch,* recounted the following anecdote of a cagey evaluation of a fording site by a Union officer. Watkins may have exaggerated a bit, but the incident is probably true. As was usual when bored, freewheeling, basically friendly, and intensely practical men stood opposite one another long enough, Watkins and his partner Sergeant John T. Tucker began calling out to their Federal counterparts on the right bank of the Tennessee River near Missionary Ridge. Wrote Watkins:

> We had to halloo at the top of our voices, the river being about three hundred yards wide at this point. But there was a little island about the middle of the river. A Yankee hallooed out "O Johnny, Johnny, meet me halfway in the river on the island." "Allright," said Sergeant Tucker, who immediately undressed all but his hat, in which he carried the *Chattanooga Rebel* and some other Southern newspapers and swam across to the island. When he got there, the Yankee was there, but the Yankee had waded. I do not know what he and John talked about but they got very friendly, and John invited him to come clear across to our side, which invitation he accepted. I noted at the time that while John swam, the Yankee waded, remarking that he couldn't swim. The river was but little over waist deep. Well, they came across and we swapped a few lies, canteens and tobacco, and then the Yankee went back, wading all the way across the stream. That man was General Wilder, commanding the Federal cavalry, and in the battle of Missionary Ridge he threw his whole division of cavalry across the Tennessee River at that point, thus flanking Bragg's army, and opening the battle. He was examining the ford, and the swapping business was but a mere by-play.[11]

When the depth of water at a ford couldn't be determined beforehand or circumstances and the press of business made a ford practicable by sheer necessity, the results could be unfortunate—and perhaps far-reaching. En route to the Gettysburg battle, Confederate general Joseph Kershaw crossed the Shenandoah with his brigade: "the river was swollen and the passing most difficult." The Third South Carolina infantry lost 2,370 rounds of small-arms ammunition when it waded across. More ominously, Major Mathis W. Henry's artillery battery lost four hundred rounds of ammunition as it made the cold crossing. Ten days later, July 3, 1863, on Gettysburg's Seminary Ridge, General James Longstreet wanted to halt Pickett "right where he is" and bring up additional ammunition to replenish the artillery supply. He was told, "General, we can't do that. We nearly emptied the trains last night."[12] Some of the rounds that might have been used to cover the disastrous Rebel charge were resting on the river bottom at Berry's Ford.

Given the size of the Civil War armies, however, even the most ideal ford under the most perfect of conditions could turn into an impassable morass. Union general Franz Sigel reported by telegram on July 6, 1864, that "the enemy [in fact, it was Early on his way to the outskirts of Washington] have been crossing at Antietam Ford for forty hours."[13] What this simple statement of fact meant for the soldiers actually making the crossing can be gleaned from the following anecdote about one army at a ford:

> At one place [in the vicinity of Mine Run, Virginia] we forded a branch and the road ascended a steep clay hill, the wet shoe of the soldier after coming out of the branch and treading on the clay had made it perfectly slick, and many a fall was the consequence. We had a boy recruit just from home and this was his first march. He wore wooden bottom shoes, and, poor fellow, he slipped back into the branch, getting out a step or two, so often that some of his comrades finally undertook to help him. Frequently they went with him two or three yards from the branch, when he would commence to slip, pulling them all back together into the water. He was finally told to sit down on the road side until daybreak, when he would be able to see his way, and could join us.[14]

This single infantry anecdote, replicated by tens of thousands of others, gives some idea of the effect of a ford on the progress of a march. Still to come were the artillery and wagons, even slower than the soldiers, nearly as numerous, and in need of more specialized treatment.

Jackson's irreverent and badly overworked quartermaster, Major John Harmon, clearly enjoyed the opportunities a ford gave him to offend the pious sensibilities of his general. It seems genuinely and inexplicably to have been the case that rugged cursing motivated a tired-out mule as nothing else could. When Jackson's wing of the Army of Northern Virginia approached Cedar Mountain, just south of Culpeper Court House, Virginia, on August 8, 1862, it crossed Crooked Run at a ford. The steep north bank of Crooked Run, soaked by the wet infantry, was "hanging up" the heavy, vital wheeled vehicles: cannon, caissons, and ambulances. Jackson came upon Harmon "in full feather and the air was blue with his oaths." Harmon desisted when his commander diffidently inquired if the swearing was essential, but the next heavily laden wagon got stuck fast in the deep, churned-up ruts on the high bank. Jackson turned away, and before he had moved fifty yards off, Harmon's "fluent damnation so startled the mules and their Negro driver that the wagon was jerked out of the stream and was alongside the retreating general in half a minute."[15]

The fame of some of the fording sites used by the maneuvering armies is the most compelling evidence of their importance to military operations. Many of them are as well known as the battle sites. Sudley Springs Ford, Kelly's Ford, Ely's Ford, Germanna Ford, Boteler's Ford, Raccoon Ford, and United States Mine Ford all have the same obscure and homely sounding American place-names that, like Antietam and Shiloh, have enriched the nomenclature of military history.

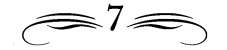

Woods and Forests

Will Anybody Come through There?

WOODS were another topographical feature that the engineers paid particular attention to and mapped carefully. Because of his topographical engineering expertise, Major General G. K. Warren of the Union Army of the Potomac probably exerted more influence in the councils of the army and in the field than any other man except the commander himself. Warren wrote unequivocally in his papers, and then underlined it, that "the forests are the most important military feature in the battles of Virginia." He amplified his opinion when he came to write his report of the May 1863 Battle of Chancellorsville, which was accompanied by one map of the field and another, smaller-scale map. The latter map

> contain[ed] all our known topography in the entire region from the Potomac to the James River, and from the Blue Ridge to the Chesapeake, a region whose characteristic is a dense forest of oak or pine with occasional clearings, rarely extensive enough to prevent the riflemen concealed in one border from shooting across to the other side; a forest which, with but few exceptions, required the axeman to precede the artillerist from the slashings in front of the fortifications of Washington to those of Richmond. No pains have been spared to make the forest topography on this map as complete as possible. It will be of great assistance in future operations, and it will aid those seeking to understand why the numerous bloody battles fought between the armies of the Union and the secessionists should have been so indecisive. A proper understanding of the country, too, will help to relieve the Americans from the charge, so frequently made at home and abroad, of want of generalship in handling troops in battle—battles that had to be fought out hand to hand in forests where artillery and cavalry could play no part, where the troops could not be seen by those controlling their movements, where the echoes and reverberations of sound from tree to tree were enough to appall the strongest hearts engaged; and yet the noise would often scarcely be heard beyond the immediate scene of strife. Thus the generals on either side, shut out from sight or from hearing, had to trust to the unyielding bravery of their men till couriers from different parts of the field brought word which way the conflict was resulting before sending the needed support.[1]

Armies do not like surprises, and the woods, as they say, are full of them. Control of a Civil War army depended upon sight, a commodity decidedly lacking in dense thickets and heavy foliage. The death of the highest-ranking officer to fall in the Civil War, Union army commander James Birdseye McPherson, occurred in the woods on the outskirts of Atlanta, Georgia. Stonewall Jackson's mortal wounding, possibly the most strategically significant single incident of the war, happened at night in a thick tangle of woods. One of the litter bearers carrying Jackson back to safety tripped in the dark woods, and Jackson fell to the ground, a trauma that probably exacerbated the general's already serious injuries and made them fatal.

The topographical engineer would make a close examination of any woods and forests, noting their location, their extent, and indicating their nature—scrub trees, clear forests, tangled with brush and vines—and the general visibility within them. The course of streams through a woods would be mapped, as well as the direction of any paths in them—be they bridle trails, deer, cattle, or hog paths, wood roads, or bypasses. With this knowledge, a battle line, advancing in the woods, "would be formed with a flank resting on a trail or woods-road, a ravine or watercourse, the flank regiment in such cases acting as the guide."[2]

Cavalry never showed to advantage in wooded areas. William Tecumseh Sherman explained, "Infantry can always whip cavalry and in a wooded or mountainous country can actually thwart it and even at times capture it. . . . I have not seen in this war a cavalry command of 1,000 that was not afraid of the sight of a dozen infantry bayonets, for the reason that the cavalry to be effective have to have a road or smooth fields, whereas the infantryman steps into the bushes and is safe."[3]

A Confederate soldier echoed Sherman when he described the reaction of some of Jackson's men to the appearance of a half-mile-long column of charging cavalrymen of the Eighth Pennsylvania on the Plank Road at Chancellorsville: "An encounter with cavalry in that dense country seemed to be as unlikely as an attack from a gunboat."[4] The cavalry's temerity was not rewarded. The Rebel infantry quickly recovered from its surprise. They leveled their muskets and wiped out a good percentage of the leading battalion with a single volley. The cavalry squadron broke in complete confusion, the survivors running, hiding, or surrendering.

The other branch that was entirely out of its element in wooded areas was the artillery. It was difficult to get guns through the trees, it was hard to deploy them, and it could well turn out to be impossible to develop a field of fire. And often enough, enemy infantry could get closer to an artillery piece without presenting a target than a gun crew felt comfortable with. Also, normal artillery tactics were to move the cannons after several rounds had been fired, preventing the enemy from pinpointing the location of the piece and bringing accurate fire to bear on it. When the artillery was deployed in a small woods clearing, this type of movement was not usually possible. Robert E. Lee in May 1864 hurried his Army of Northern Virginia forward in order to intercept the advance of Ulysses S. Grant. Grant, who had been brought East, promoted to lieutenant general, and placed in overall command of the Union armies, made his headquarters with the Army of the Potomac and, in effect, shared command of that army with its nominal commander, General George Gordon Meade. This Union army had been reinforced and reequipped and heavily outnumbered the Confederates, particularly in the artillery branch. Lee intended to neutralize the Federal preponderance in numbers, especially the artillery, by bringing on a battle in an area of western Spotsylvania County, Virginia, known in local parlance as "the Wilderness." "No one," wrote Warren, "can conceive a more unfavorable field for the movements of a grand army than it presents."[5] Lee's tactic worked, and the Battle of the Wilderness would have been a defeat for any Union commander in the East prior to the advent of Grant.

The Union forces were not alone, however, in failing to bring superior numbers to bear when fighting in woods. Jackson's attempt to cut off the retreat of Union general John Pope after

Chancellorsville, Virginia

In the foreground of this photograph are the planks that formed the surface of Orange Turnpike. In the center of the photograph is the intersection of Orange Turnpike with Bullock Road, which led to United States Ford Road. In the background are some of the Virginia forests that Union general G. K. Warren, chief topographical engineer of the Army of the Potomac, called "the most important military feature in the battles of Virginia." They exhausted, baffled, and frightened the soldiers and made effective command and control of their troops impossible for officers.

Second Manassas on September 1, 1862, brought on the Battle of Chantilly, Virginia. Attempting to deploy in the thick woods lining the Little River Turnpike, Jackson's entire corps of fifty-one infantry regiments was opposed by nineteen outnumbered Union regiments, that came into the line of battle in the open fields and clearings of the Millan farm. Though two top-notch Union generals, Major General Philip Kearny and Major General Isaac Stevens, were killed on the field, the Union could rightfully claim a victory over the Confederate corps, whose military performance was marked by confused, uncoordinated, ill-timed maneuvering and clumsy fighting "along a front over half a mile wide in heavily wooded, somewhat broken terrain where control was difficult."[6]

Control also was difficult for Union general William S. Rosecrans at the Battle of Chickamauga, Georgia, on September 18 to 20, 1863. Writer Ambrose Bierce described the battlefield: "the forest was so dense that the hostile lines came almost into contact before fighting was possible,"[7] and the battle: "a hidden battle roaring and stammering in the darksome forest behind a smoking gray front."[8] War Department observer Charles A. Dana was with the Union commander at his headquarters in the widow Glenn's home. They could see nothing of the "conflict being fought altogether in a thick forest, and being invisible to outsiders. The nature of the firing and the reports from the commanders alone enabled us to follow its progress."[9]

While some generals lost their lives in the woods, others lost their reputations in them. Rosecrans was one of the latter. A Confederate breakthrough at Chickamauga resulted in a panic on the Union right. This tumult communicated itself to Rosecrans. Unable to see what was happening on the left line, where Union resistance was solid, the commanding general left the field of battle to save the army's supply trains and to see to the defenses of Chattanooga, to which he and his army retreated. Though his actions were, in a strict sense, justifiable, Rosecrans's behavior nonetheless smacked of panic and flight and he was shortly after removed from command.

Union major general Fitz-John Porter's case was more complicated. He was relieved of his command by General Pope, commander of the short-lived, ill-fated Army of Virginia, after that army was defeated at the Battle of Second Manassas in August 1862. Pope blamed his defeat largely on what he termed Porter's "disobedience, disloyalty, and misconduct in the face of the enemy." What in fact Porter was "in the face of" was an impenetrable woods, which precluded Porter from obeying the orders he was cashiered for disobeying. At one point, Porter summed up the situation he found himself in: writing at his desk, he glanced up toward Pope's staff officer D. H. Strother and asked him how to spell the word *chaos*. One of the problems that heavy timber could cause was the inability of separated units to come to one another's support or to cooperate effectively in planned deployments. John Pope intended vaguely for Porter to link his right up with the balance of the army. As it was, Porter's Fifth Corps was isolated, the woods in front harbored an indeterminate number of Confederates, and to Porter's right "was a very wide gap, hidden by a wood through which Generals McDowell and Porter were unable to pass on horseback."[10] "The country on the right was found to be impassable for infantry in any order, and absolutely so for artillery. It was rough, thickly wooded and full of ravines."[11] Disgruntled, uncomfortable with Pope's command, unwilling to press forward, and unable to move sideways, Porter ended the day doing nothing but listening to the sounds of distant battle, a lethargy that would wreck his Civil War career.

Dense woodland could derange the tactics of even the finest generals while in the midst of their best-fought campaigns. Grant was an intuitive military genius. He seems to have studied neither tactics nor strategy particularly. One topographical engineer, Lieutenant Colonel James H. Wilson, said that Grant "appeared to take little interest in this work [i.e., topographical engineering]" and that Grant's great achievements "were won mainly by attention to broad general principles rather than to technical details."[12] Grant, of course, did well enough for himself, maps or no, but his attention to maps and mapping was probably a bit less offhanded than Wilson indicated. Grant's headquarters kept hold of maps that others wanted but also complained officiously when other commands retained maps that it wanted. Lieutenant Colonel Horace Porter, aide-de-camp to Grant, observed the commanding general's use of maps with quizzical interest:

> It was always noticeable in a campaign how seldom he consulted them, compared with the constant examination of them by most prominent commanders. The explanation of it is that he had an extraordinary memory as to anything that was presented to him graphically. After looking critically at a map of a locality, it seemed to become photographed indelibly upon his brain, and he could follow its features without referring to it again. Besides, he possessed an almost intuitive knowledge of topography, and never became confused as to the points of the compass. He was a natural "bushwacker," and was never so much at home as when finding his way by the course of streams, the contour of the hills, and the general features of the country.[13]

Grant himself said he lost his bearings only once in his life, at Cairo, Illinois, where he was bewildered by the looping configuration of the Mississippi River.

Toward the end of May 1864, Lee, after spending most of three weeks locked in combat with Grant, withdrew his Rebel army from Spotsylvania. Lee wrote to Confederate president Jefferson Davis that "in a wooded country like that in which we have been operating, where nothing is known beyond what can be ascertained by feeling, a day's march can always be gained."[14] In other words, Grant could always steal a march on him when woods blocked the view.

It was likewise true that Grant could have a march stolen on him. Champion's Hill was fought in May 1863 halfway between Vicksburg and Jackson, Mississippi. It was the biggest field battle of the Vicksburg campaign—probably the best-conducted and most brilliant campaign of the entire war. But the heavy woods in the Champion's Hill region severely hampered Grant's operation and his control of the battle. His opponent, General John C. Pemberton, took a strong position, facing east, with his left protected by ravines, Champion's Hill itself, and Baker's Creek; all of this area was "grown up thickly with large trees and undergrowth, making it difficult to penetrate with troops, even when not defended."[15] Pemberton was astride the three parallel roads upon which the Federal troops were advancing. On the two southernmost roads, Grant's divisions moved cautiously, unsure what they were confronting in the thick woods. This movement stalled out in uncertainty.

Meanwhile, on the third, northernmost road, another of Grant's divisions was fiercely engaged, and the fighting here raged back and forth indecisively. A fourth division was sent north of the northern road in an attempt to flank the Rebel defenders. This division too was operating blindly, a belt of woods hiding the situation from Grant, who had accompanied the flanking division. At the same time, off to the south, a Union brigade was coming up to reinforce the flank of the southernmost Union divisions. The fighting lasted until Grant allowed part of the flanking column that he was with to go back to the northern road to reinforce that beleaguered division. When this reinforced Yankee front advanced, the Rebels quickly broke and retreated.

Unbeknownst to Grant, the deployment of his troops had briefly surrounded the Rebels. Furthermore, the three roads upon which the Union troops were operating were two roads more than Pemberton could use. The Union forces were wrapped around him, and only the middle road of the three was available to him. Grant's northern flanking movement had effectively blocked that middle road, but because of the obscuring trees, Grant did not know this. Not only was Pemberton surrounded but his retreat route was cut off. When he thought he was reinforcing his division on the northern road, Grant in reality was loosening his grip on the "only road over which the enemy could retreat"[16]—which they immediately proceeded to do once Grant pulled his troops back. Grant had held the Confederate army in the palm of his hand but was unable to see the troops for the trees.

The decisive effect of woodlands on Grant's Champion's Hill battle was to be reprised in a more general fashion a year later during his operations in the East. From the Wilderness fighting to North Anna, from Cold Harbor to Petersburg, woods fashioned and framed the strategies and tactics of both the Federal army under Grant and Meade and the Confederate army under Lee.

In one instance, woods served as a singular expedient. Union general Robert C. Schenck had his great moment in the Civil War under unlikely circumstances. He was making a hasty retreat from the western Virginia battlefield of McDowell in early May 1862. Jackson's fierce marching was bringing his victorious troops closer and closer to the rear of Schenck's tiring columns as both armies struggled through the difficult, narrow valley between McDowell and Franklin: "Men in the ranks anticipated another fight but thought this bleak country would offer no hardships surpassing those already endured—then the forest ahead of them began to smolder. The Yankees had set the woods afire to cover their retreat."[17] The smoke choked Jackson's men and made the Confederate advance like a night attack—bewildering and dangerous. In a unique instance, he complimented his adversary for a "most adroit expedient." Jackson might have added that it was a most effective expedient as well, for by the time he got to Franklin, Schenck had had time to fortify and Jackson had to relinquish any idea of an attack.

G. F. R. Henderson, like Warren, considered the woods and forests of America's interior as among the most important strategical and tactical influences on the war. Henderson differed from Warren in regarding this mid-nineteenth-century American landscape with awe. While Warren saw the woods and forests as pragmatic difficulties to be overcome, Henderson thought there was a primeval quality to this vast wilderness that was the theater of the Civil War. Henderson regarded the movement of a substantial army over the "few and indifferent roads that penetrated the wilderness a military achievement almost without example."

So encompassing were these forests that their presentation on the military maps was often a fairly exact science. The cartographer carefully sketched a scalloped outline to render the forest's perimeter and filled in the forest depths with a tracery of treelike lines. A thin green watercolor wash completed the effect. Topographical engineers often added a further detail to their forests: "Pine and spruce are a dark green hue and stand out boldly over all other foliage.... When these trees are scattered in clumps in a forest, they should be rendered very exactly, both as to form and place, for they catch the eye at once and are very useful for orientation."[18] In other words, stands of pine trees form natural landmarks in the midst of deciduous forests and were included for that reason among the map's details.

Topographical engineers distinguished between deciduous trees and pine trees because the darker, distinctive pine trees constituted a natural landmark. For full reproduction of this map sketch, see page 127.

Local Knowledge

A Right Smart Distance
I Reckon

F AMILIARITY with local landmarks, both physical (streams, fords, woods, hills) and cultural (farms, byways, bridges, wells), came from the reconnaissance and personal observations of the topographical engineer. But it also came from questioning the locals, as Union mapmaker William E. Merrill described them: "all persons familiar with the country in front of us. It was remarkable how vastly our maps were improved by this process."[1] The process of getting the desired information and the interrogation of the people who might have it usually proved frustrating and amusing by turns. Interrogating effectively required shrewd resourcefulness and perceptive analysis. By reputation, Sergeant N. Finegan (sometimes misspelled as Finnegan) of the Fourth Ohio Cavalry, the Union Army of the Cumberland, was "exceedingly expert in extracting information."[2] So valuable was the intelligence he gleaned that the preparation of maps for General William Tecumseh Sherman's campaign from Chattanooga to Atlanta was delayed while Finegan finished questioning his spies, scouts, refugees, travelers, prisoners, preachers, and peddlers—his "motley crew," in Merrill's words. It is not known what Finegan's specific techniques were, but one can probably assume that he was simply better at doing what others like him did.

Rebel mapmakers always got local cooperation much more readily than their Yankee counterparts. If the details they got were not necessarily accurate or reliable, the informants were not being deliberately misleading. But when white Southerners were questioned by Northern engineers, they were not inclined to be helpful or truthful. Staff officer Colonel Horace N. Fisher, Union topographical engineer in the western theater, was ordered by General Alexander McDowell McCook to secure information as the Army of the Cumberland advanced toward Murfreesboro, Tennessee, and the Battle of Stones River. The first important intelligence required was the lay of the land and then the determination of where the Rebels, retreating slowly in the Union's front, were heading for and where they had been. Fisher began by securing local Tennessee guides. He rounded up a dozen farmers and put them on spare horses. He secured their cooperation by driving an implacable bargain:

> I told them that there was "just one thing that they all could agree upon—the truth." Also, I should ask certain definite questions of each one of them separately; and that, if they lied, it would be at their peril. After examining them satisfactorily,

> I gave them fair warning that, if they led us into any trap, the orderly alongside of
> each one had orders to shoot him on the spot; but that, if they did their duty, they
> would be taken back to their homes safe and sound. Otherwise they would be left
> dead by the road. You never saw a dozen men more anxious than these men to tell
> of every path, trail, or lane near our line of march.[3]

Fisher then sent part of his escort to pick up some Confederate stragglers. The escort "corralled a score," and Fisher questioned them, his notebook in hand. A few questions gave Fisher the names of the units he was following, and a quick look in Rebel haversacks indicated that they were only carrying one day's cooked rations. Fisher concluded rightly that the only logical objective of a one-day march would have to be Murfreesboro, a fifteen-mile march, and not Shelbyville, twenty-six miles away.

A preliminary sketch, or key map, as Fisher called it, of what became the Murfreesboro or Stones River battlefield was drawn the next day, December 29, 1862. It was partly based on Fisher's own recollections of the area from an August march when he went through Murfreesboro en route to Nashville. A local physician, captured by Union cavalry scouts, was closely questioned by Fisher. The key map was examined by the reluctant collaborator and pronounced accurate. Thus, Fisher had, in short order, mapped the advance of his own army, determined the intentions of the enemy, and mapped the site of the probable field of battle. Fisher incidentally had been tapped as a topographical engineer by the mere fact that he had read a book on the subject before the war. He therefore learned as he went along, but he obviously learned quickly and he learned a lot.

Fisher had discovered earlier in the year that the country doctor was a uniquely valuable topographical resource. Fisher was with the advanced units of General Don Carlos Buell's Army of the Ohio. These troops were the reinforcements desperately needed by Ulysses S. Grant to help turn the tide of battle at Shiloh, raging some eight miles upstream on the opposite bank of the Tennessee River. Time was of the essence, but it appeared that the only practicable route to the landing, where the river could be crossed, was to make a long countermarch to a road that followed a ridge to the landing. One of the staff was alert enough to find the local doctor, who knew of and led three Union brigades along a narrow path through densely wooded swampland to the landing. The timely arrival of these brigades shifted the balance in the decisive last hours of fighting and set the stage for a successful Union counterattack the next morning, which resulted in an expensive but decisive Union victory on April 7, 1862.

At the Battle of Chickamauga in late September of 1863, Confederate general James Longstreet, who had been transferred to the West to reinforce General Braxton Bragg, maneuvered with difficulty through the unfamiliar Georgia countryside until he secured a guide. Tom Brotherton had farmed the land across which Longstreet was advancing, and according to Chickamauga battle historian Glenn Tucker "knew every pig trail" through those woods. Longstreet's subsequent attack broke through a dishevelled Union line at an opportune moment in one of the climatic incidents of the war.

Union general John Pope came East after a successful early showing in the western theater. A former topographical engineer, Pope knew the value of local knowledge, but in one instance he got too much information from a Virginian who didn't know "every pig trail" but knew everything else. As D. H. Strother described the incident: "I reported at General Pope's room, and found him engaged in questioning an old Virginia loyalist in regard to the roads and topography of the country toward Gordonsville. The responses were correct, and in the main satisfactory; but at every crossroad and stream our lower country gentlemen would stop to expatiate on the character and genealogy of the residents, until the General, with some impatience, turned to me for a more brief and military description of the country."[4]

General George B. McClellan sent for a local man, George C. Rohrer, to help him interpret the action during the Battle of Antietam on September 17, 1862: "A large map was shown Mr. Rohrer and whenever they would see puffs of smoke Mr. Rohrer would locate them on the map for the General." McClellan's headquarters are described as a tent, so this assistance was probably rendered when McClellan was observing the fighting from General Fitz-John Porter's headquarters.[5]

Southern blacks were a godsend to the Union army. It was the considered opinion of one knowledgeable, perceptive, and generally objective Rebel officer, Charles M. Blackford (brother to mapmaker and memoirist W. W. Blackford), that "the enemy has a better chance of invading us because our land is much more sparsely inhabited and because of the information they derived from the Negroes."[6] Strother, with the insights of a Virginian, recognized a fascinating sociological conflict at work in the white Southerner's attitude toward blacks: "The Negroes I knew were both faithful and willing, and strange to say, were trusted on the other side with a persistence that amounted to fatuity. While every white man's motions and actions were watched with a most jealous scrutiny, the Negroes were permitted to run hither and thither as if they had been merely domestic animals."[7] In the West, Union general Grenville M. Dodge noted the same thing: Southern pickets seldom stopped and questioned blacks—so these freedmen made ideal, completely trustworthy messengers. Strother was surprised himself to find, when he was gathering topographical data, that

> while the whites are usually more comprehensive in their knowledge, the Negroes are far more reliable for local details. They know nothing of maps, but a limited district, which they have traversed night and day visiting, hunting raccoons, and robbing hen-houses, they will describe with great accuracy, naming every house, blind path, bridge and ford. When I got to the limit of one fellow's beat I engaged him to bring an acquaintance from the adjoining estate, or village, and in this manner I was enabled to get a very satisfactory description of a whole district into which our troops had not yet penetrated. Having had an opportunity of comparing this sketch with a map of the same region, afterwards captured from the enemy, I was myself astonished at its accuracy.[8]

Sometimes, however, the eagerness of blacks to be helpful would be counterproductive. Strother went on: "Negroes were continually running to us with information of all kinds and they are the only persons upon whose correct truth we can rely. Of course, we cannot always rely upon their reports for lack of judgment and means of obtaining military information. With that obsequiousness of spirit, born of slavery, they have too often a tendency to tell us what they think will be agreeable to us rather than what they know."[9] Union admiral David Dixon Porter mentioned this tendency in his memoir of the Civil War: "[Blacks would] come in crowds with wild inventions . . . if they could not find something real to tell. . . ."[10] The same overwrought tendency on the part of the naive local blacks was noted by an aide to Federal general Ormsby Macknight Mitchel when operating in Alabama in 1862. "It was not a rare sight to see a Negro dash up to headquarters on a horse covered with foam, and gasping for breath. 'They'se cumen, massr!' 'Where?' 'Crossen de riber.' 'How many?' 'Three hundred thousand!' It would probably turn out that some thirty guerrillas had made a crossing. The uneducated outpost did not know the difference between thirty and three hundred thousand."[11]

Perhaps the most astonishing example of a black man providing reliable and crucial intelligence occurred along the Mississippi River on April 29, 1863. Having gotten his men downstream of the Confederate stronghold of Vicksburg, Mississippi, Grant had to decide where he could safely land his troops on the east side of the river. Grant described in his memoirs what happened: "When the troops debarked, the evening of the 29th, it was expected that we would have to go to Rodney, about nine miles below, to find a landing; but that night a colored man came in who informed me that a good landing would be found at Bruinsburg, a few miles above Rodney, from which point there was a good road leading to Port Gibson some twelve miles in the interior. The information was found correct, and our landing was effected without opposition."[12] Thus, eighteen thousand men on transports, ironclad gunboats, river steamers, and barges changed course—and military history—on the advice of an anonymous black man along the riverbank. As Grant said, he had his army "on dry ground on the same side of the river with the enemy" and, having done that, "felt a degree of relief scarcely ever equalled since."[13]

On occasion, it was necessary to take matters in hand and not wait for providence to supply a guide. During the Overland campaign of May 1864, as they marched southward after fighting in the Wilderness, the Seventy-ninth New York Highlanders found it necessary to secure the most direct route to Spotsylvania. From the skirmish line—which marked the farthest advance of the Union army—a line officer pointed out to some of the men a dwelling a half mile off. "Now boys," he said, "I want you to take a good look at that house and the surroundings, and get the bearings well fixed in your minds, for I want you to go there after dark and capture a rebel for me."[14] This was promptly done; a guide was captured and closely guarded through the night. At daylight, the passive prisoner was informed, "If you prove false . . . you will be shot as soon as we discover your mistake."[15] The dutiful guide did as he was asked, directed the advance to the Rebel lines, and was then allowed to return safely to his own lines.

Stonewall Jackson, already noted as being obsessively secretive, carried his security concerns a bit too far in late June 1862. As he tried to fathom his way through the bewildering fighting and topography of the Virginia Peninsula, Jackson sought assistance from a trooper in the Fourth Virginia Cavalry who had grown up on a farm nearby and knew the area well. Jackson murmured to his new guide that he wanted to go to Cold Harbor. With no further enlightenment and no further ado, the cavalryman sensibly chose the largest and most direct of the roads leading to Cold Harbor. Unfortunately, there was a New Cold Harbor crossroad and an Old Cold Harbor crossroad. Jackson's corps was being led toward New Cold Harbor, away from the firing and out of the fight at Old Cold Harbor. In his concern for security, Jackson had been unwilling, in effect, to tell his guide where he wished to be led. An angry Jackson remonstrated with the trooper, who replied stiffly, "Had you let me know what you desired, I could have directed you in the right direction."[16]

The inclusion on military maps of each farm and residence with the owner or occupant carefully identified (and often with the wartime sympathies, Union or Rebel, helpfully noted) was not a matter of thoroughness on the one hand or whimsy on the other. Civil War–era roads, particularly in the South, were seldom marked with road signs or other route markers beyond the occasional milepost. In fact, the names of roads, like other major topographical features, were sometimes unknown to the local residents. A Marylander said, "I never heard it [the battle site of South Mountain] called South Mountain till after the battle." Nor had he heard of Turner's Gap, one of the principal features of the battle: "We always called it just the gap in the mountain."[17] Rural inhabitants of the 1860 countryside and small towns could be utterly oblivious to the world beyond their own fence lines and woodlots. Sometimes their opaqueness was feigned. Near the Southside railroad at Prince Edward Court House in Virginia, Union general Philip Sheridan inquired of a very dignified old gentleman, who had spent his whole life in the locality, how far it was to Buffalo River. The old gentleman replied disdainfully, "I'm sure I don't know, sah." Sheridan was a busy man and failed to appreciate this facetiousness. He turned to an aide. "Orderly. Take this man down to Buffalo River, and show him where it is."[18]

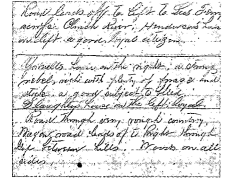

Topographical engineers minced no words in describing the wartime sympathies of residents along the route of march. This field journal is reproduced in full on page 155.

On many other occasions, however, the ignorance of these rural citizens was real and profound. Compounding it was the bewildering and fearsome presence of military operations being carried on in astounding numbers at their very doorsteps—more men were passing in an hour than many of these locals had seen in their lifetimes. The circumstances were calculated to overwhelm their perspective and common sense. Confederate general Jubal A. Early found when he tried to get topographical directions that "all men, except the old ones, had gotten out of the way, and the latter, as well as the women and children, were in such a state of distress and alarm, that no reliable information could be obtained from them."[19] Union general Alpheus S. Williams wrote home about his observations of the people along his route of march toward Gettysburg: "The people are rich but ignorant of everybody and thing beyond their small spheres. . . . Altogether they are a people of barns, not brains."[20] A Rebel sergeant in Jackson's brigade "stayed all night at a house way up in the mountains, and the people were so ignorant that they did not know that the war was going on. When he began to explain it to them and told them that he belonged to General Jackson's command, they said: 'Oh! Yes; we have read about General Jackson and his army!' He got them to show him the book. It was about old Andy Jackson, in the war of 1812."[21] A Union cavalryman, John P. Sheahan, recounted his impression of Marylanders: "They don't know anything atall, they don't know a mile from two miles, ask them how far it is to such a place they will at once say, 'well a right smart distance I reckon,' and that is all that you can get out of them for that is all they know, and you can't get more out of anyone than what they know." Another Yankee added, "I don't believe the inhabitants even know the day of the week."[22] One Union soldier was taken aback when Southern civilians had never heard of his home state of Wisconsin.[23] Robert E. Lee's adjutant general Walter H. Taylor noted in his memoir that "the country people who were relied on as guides seemed to have no knowledge . . . beyond the immediate vicinity of their homes."[24]

Further fascinating anecdotal evidence of the difficulties of navigating in the antebellum and postwar South is contained in two travel books: Frederick Law Olmsted's *The Cotton Kingdom,* and John Townsend Trowbridge's *The South: A Tour of Its Battlefields and Ruined Cities,* respectively. Olmsted's attempt to locate the home of a "Mr. W . . ." near Petersburg, Virginia, is recounted in a ten-page odyssey of impassable roads, impossible instructions, unfordable streams, fractured phrasings, and absolute nonsense. There were "long miles" in the directions he received, as well as "short" miles and "middling" long miles. A road that was to be followed "straight" would, in the course of a mile, contain twenty forks. "The best-traveled road" was another meaningless instruction, for the possible routes were indistinguishable and what appeared to be a good plain road could begin to "ramify" into a dark, wooded path. Directions might include such "helpful" information as "there's a big house—but I don't reckon you'll see it neither, for the woods." And finally, the phrase that has been the bane of their existence for travelers of all places and times, "You can't mistake it."

Trowbridge made a tour of the South fourteen years after Olmsted, in the year after the Civil War ended. The war, it is clear from his account, had done nothing to improve the roads of the South, and the directions Trowbridge received were as convoluted as any that Olmsted had tried to follow.

The most effective remedy for this militarily dangerous confusion was to get the name of each and every resident correctly on the map. The erasures and corrections evident on manuscript maps indicate the care with which this was done by the topographical engineers. The idea was—and it proved to be true—that if a local knew anything at all, he would know the way to the next farm or to his immediate neighbor's house. The names worked as de facto route markers. By consulting the names on his map, a commander could proceed with some degree of confidence in the correct direction of his march. The map's annotation about the inhabitants' loyalty served as a practical alert as to the relative dependability of any directions or information provided by them. In a "Memorandum as to Maps," General G. K. Warren explained the additional value and importance of the residents' names on the maps as a basic frame of reference: "The names referred to in orders were intended to indicate positions generally. Thus if one was directed 'to advance to Miss Hargraves and halt' it meant to move about to the position Miss Hargraves's house was supposed to occupy, and halt whether there was any such house or person there or not."[25] The idea was that even if the name was wrong or was located incorrectly, so long as everyone was using the same map and making the same mistakes, all would be well.

The parochial limitations of these rural residents were a factor for both sides to the very end. On the last leg of its career, Lee led his Army of Northern Virginia toward North Carolina in a march that ended in fact at Appomattox Court House. Lee was in a desperate race: away from the pursuing Federal army and toward rations for his battered, hungry troops. On April 5, 1865, Lee pressed toward expected supplies at Burkeville Junction, but Federal cavalry were found to be astride the road and they were believed to be supported by two corps of infantry—an unassailable host. Lee, unfamiliar with this southside region of Virginia and anxious to learn what his options were, "called in the farmers from the neighborhood and talked to them of the country ahead, but he found they knew little of it."[26] In a final, parting irony, two days later at Farmville, Lee failed to credit both his own map and "a very intelligent looking gentlemen . . . at a house nearby . . . who looked at the map and pronounced it entirely correct."[27] Lee unaccountably chose the wrong and the longer road. The pursuing Union army took the correct and the shorter road. Both roads led to Lee's surrender at Appomattox Court House.

Union general A. A. Humphreys described the countryside as *terra incognita* for the Federal army as it pursued the Rebel army to Appomattox: "From the people of the country, few in number and silent, no information could be gathered. The Corps followed Lee as one follows the chase of game in an unknown country."[28]

A second factor in noting and naming each resident was the sense this gave the map reader of the population density of the region: "It has been found that the matter of supplies is dependent upon the population. Where the country is thickly settled supplies are abundant . . . where sparsely, the country cannot be relied upon to support an army."[29] This according to Orlando M. Poe, Sherman's chief engineer, a man who knew a thing or two about supporting an army off the land.

One reflection of the population density would be the number of what might be called "road services" available to a mid-nineteenth-century army. These were meticulously recorded by the topographical engineer on his maps and constitute a directory of rural 1860 America's amenities for the traveling public. The shops and resources that a marching army might need to avail itself of were numerous and varied. Out of the tens of thousands of horses and mules, soldiers and teamsters, caissons and cannons, some of them would be needing things as obvious as a well or pump or spring, or as strange as an undertaking establishment—the latter for the proprietor's woodworking talents rather than his mortuary services. Routinely labeled on a map were taverns, hotels, stables, coopers, wheelwrights, general stores (their inventory generally included dry goods, coffee, salt, tobacco, lead and gunpowder, shoes, and medicines), saddlery and harness shops, blacksmiths (when an observer from the British army, traveling with the Rebels toward Gettysburg, lost two horseshoes "there was no opportunity to replace them; the army was using all the blacksmith shops along the way"[30]), gristmills, sawmills, woolen mills, and shoe shops. Also noted, depending on the scale of the map, might be pastures, orchards, cornfields (which were shown on military maps because they were a source of forage and sometimes because they were a possible place of concealment), and farms. A Dutch farm near Chambersburg, Pennsylvania, was inventoried by a duly impressed North Carolinian: outdoor bake oven; barrel of sauerkraut, pile of mighty cheeses; ducks, turkey, guinea fowls, peacocks; orchard, sass garden; smokehouse, milk house, wood house, toolhouse; beehives; calves, pigs, sheep, horses.[31] The transactions carried on with these establishments were as varied as the individuals involved. Jed Hotchkiss was in Chambersburg, where he said, "I procured maps, and engineering supplies and purchased some goods." His low-key approach contrasts strongly with the more dramatic behavior of a Rebel cavalryman in Hanover, Pennsylvania: "A member of [Lieutenant Colonel Elijah V.] White's staff rode down Baltimore Street to Peter Frank's blacksmith shop . . . and summoning the smith asked that his horse be shod. Frank said he wasn't working because it was a holiday on account of the Johnnies [i.e., the Rebels] being in town. The officer laid his hand on his pistol holster, whereupon Frank started fanning up a fire with his bellows and went to work. The Confederate paid $2 in greenbacks for two shoes."[32] In the end, both parties to this hard bargain were probably fairly content with it.

Hills, Mountains, Rolling Terrain, Gaps, and Impregnable Positions

Not Only Length and Breadth but Thickness

WHILE carefully incorporating the relevant cultural, or artificial, features of a region on his map, the topographical engineer would continue to inspect and record the physical, or natural, features his army would encounter.

Hills and mountains had a strategic impact on an overall campaign and a tactical effect on the outcome of a battle. Stonewall Jackson's knowledge of and finesse in using the mountains of the Shenandoah Valley—the Massanutten, the Blue Ridge, and the Allegheny—were the strategical basis of the brilliantly successful 1862 Valley campaign. The possession of Little Round Top and of Culp's Hill by Union forces at Gettysburg ensured that they could maintain possession of the dominant terrain on the battlefield.

The campaign and battle of Chickamauga, Georgia, fought by Federal general William S. Rosecrans and Confederate general Braxton Bragg in September 1863, involved a strategy keyed to mountains and came down to battle tactics centered on a hill. Rosecrans was intent on maneuvering Bragg out of Chattanooga, Tennessee, where a railroad passed through the Appalachian Mountains. Swinging below and west of Chattanooga, Rosecrans intended to move his army along a forty-mile line, marching through widely separated gaps in Lookout Mountain. This strategy flushed Bragg out of Chattanooga, but instead of retreating to Atlanta and allowing Rosecrans to enjoy the spoils of Chattanooga, Bragg chose to attack the vulnerable Union columns. He failed in this effort, barely, and failed again, barely, to completely rout the Union army on September 20, 1863, at Chickamauga. Union general George H. Thomas made a stand on the tactical oasis of Snodgrass Hill after a break in the Union line had sent most of the army, including Rosecrans, on a swirling retreat to Chattanooga.

Other, much more minor terrain, too indistinct to be dignified even as hills, also could have a bearing on the success or failure of military operations. The vocabulary of the Civil War contains numerous examples of otherwise mundane topographical features that for a brief period took on immense importance in the conduct and outcome of a battle or even a campaign. Names that now reverberate—the Miller Cornfield, the Wheatfield, Bloody Pond, the Round Forest, Orchard Knob, the Rail Road Cut, the Unfinished Rail Road, the Copse of Cedars, Hazel Grove, the Sunken Lane—failed to appear on any maps prior to the battles in which they briefly figured as objectives, rallying points, tactical positions, or heavily fought over ground. (An odd exception is Gettysburg's Sherfy Peach Orchard, which does make an inexplicable prebattle appearance on the G. M. Hopkins 1858 Adams County map, page 139.)

It was W. W. Blackford's habit, whenever possible, to examine the field immediately after a battle. Where the stark postbattle photographs depicted tragic human wreckage, Blackford's topographical eye saw "an interesting and instructive study for a soldier. There is to be seen, by the results, the relative strength of positions, and the effect of fire; and nothing cultivates the judgment of topography, in relation to the strategic strength of position, so well as to ride over the ground while the dead and wounded still remain as they fell. You see exactly where the best effects were produced, and what arm of the service produced them."[1]

South Carolinian J. F. J. Caldwell recognized the significance of the terrain in the relative effectiveness and execution of infantry fire: "We had somewhat the advantage, for the enemy, descending to attack us, naturally fired too high, while we had either a level or a rise to fire on. I need not explain this, for it is a universally known fact that men fire above their own levels." Later the Yankees had the advantage of the terrain, and it was Caldwell's turn to suffer, rather than deliver, the accurate fire: "The Federals fired with unusual accuracy. It was to be expected; for we stood in full relief upon the crest of the hill."[2]

Lieutenant Peter C. Hains, very early in the war, saw the firsthand effect of hilly countryside on the movement of artillery. Hains, just out of West Point, was given command of a detached piece of heavy artillery—a thirty-pound Parrot gun—that he led in the Union advance to First Manassas in July 1861. This was a huge and magnificent cannon. There was some hopeful speculation that this three-ton piece of ordnance would, in and of itself, end the Civil War. Moving from Washington City to Alexandria, the gun, its untrained crew, its ten untrained horses, and its two-hundred-man escort came to a hill. Hains tells the story:

> The ten horses threw themselves into their collars. The gun started up a bit, then the pace slowed, paused, and—then the giant gun began slowly to drift backward down the grade. We quickly blocked the wheels, as there were no brakes. I rode up and down the line, cheering on the men. The drivers yelled, and lashed their horses; the ten animals strained and tugged—but the gun remained motionless. "Get out the prolonges," I ordered, and these lines, of about three-inch rope and knotted together to about a hundred feet in length, were quickly hooked to the axle of the gun. Two hundred men instantly trailed onto them. With wild yells and cheers they started that gun forward, the ten horses and two hundred men soon dragging it upward to the crest. It was great. And most of us were very young indeed.
>
> But arriving upon the crest of the hill was not the climax at all. There was the other side of the question—to go down that hill. Without resting, away they went. At first the gun followed gently. Then it began to gain headway. After a hundred feet it began to push my good wheelers a bit, and they crowded upon the forward horses. Six thousand pounds on wheels had started down the grade. Away the line went, and before they arrived at the foot the wheel-horses were galloping frantically for their very lives, with that monstrous engine of destruction thundering after them with a rumbling roar and a cloud of dust. But no one even grinned. It was as it should be, and my gallant team kept up the desperate pace until the leveling roadway slowed down the wild chase. Covered with dust and foam, the horses recovered themselves a mile farther on, while the two hundred men, myself among them, came panting in the rear.[3]

In the rout from Manassas, the big gun, now without its escort of two hundred soldiers, came to a steep hill just east of Cub Run on the Warrenton Turnpike. Immobilized at last, Hains spiked the stranded gun and walked back disconsolately to Washington.

It was not the Rebel army that defeated the thirty-pounder; it was the Rebel countryside. It was with unfortunate losses such as Hains's in mind that topographical engineers carefully noted and estimated the steepness of road grades and other elevations that an army might need to traverse. The general rule was that a large body of infantry could manage a slope as steep as forty-five degrees. Cavalry was limited to a thirty degree slope. Artillery and wagons could negotiate a ten degree slope, and as we have seen, the angle of the descent was every bit as important as the angle of ascent. There were several procedures that could be followed to quickly and accurately estimate the angle of a slope. The level foundations of structures along the road provided a handy comparative reference, as did a common ruler balanced on the end of a finger. A glance at the fence posts or telegraph poles along the side of a steep road also gave a quick idea of the grade. A panoramic view of railroads and rivers provided a true level that hills, valleys, and atmospheric conditions could sometimes otherwise obscure.

Rolling terrain was an oddly dangerous land formation that a mapmaker would analyze and represent or annotate on a military map. It was deceptive ground because it was difficult to judge distances across, there were not usually any commanding heights from which to see very far ahead, and the folds of earth can effectively hide not only troops but large numbers of them. As intense a commander as Jackson was badly misled by the undulating ground at Cedar Mountain, south of Culpeper, Virginia. He went into battle almost casually. He had a poor opinion of his Union opposite, Nathaniel P. Banks, and he initiated the battle with the majority of his troops still strung out en route to the battlefield. That done, Jackson opted for a nap. Jubal A. Early, in the advance of Jackson's corps, discerned before him a tributary of Cedar Run, a thick woods through which the road traveled, Slaughters or Cedar Mountain to the right, and an "undulating valley, consisting of several adjoining fields." Enemy cavalry alone was visible on a ridgeline opposite the position the Confederates were occupying: "It was on and behind this ridge the enemy's batteries were posted, and it was in the low ground beyond that I supposed, and it subsequently turned out, his infantry was masked." Early's uncertainties about the Union deployment were shared by fellow general William B. Taliaferro, who reported that "the undulations of the country made reconnaissances very difficult."[4] Early thought he saw "covered ground" where the Rebel left could move to flank the Union artillery, "but, in a very short time afterwards, the glistening bayonets of infantry were discovered moving stealthily to our left."[5] Lulled into carelessness by the seeming emptiness of the rolling ground, Jackson found he was about to be outflanked. Up now from his nap, he made his most dramatic appearance of the war, waving his unbuckled sword, scabbard, and belt all at the same time, helping save the battle for the South. But it had been a very near and unnecessary thing, and Jackson had been engaged with an army only half the size of his own. Undulating terrain mislead as astute a general as the Union's William T. Sherman. His poor performance on the left flank at Chattanooga, Tennessee, on November 23, 1863, came about because the ground he thought he had to seize to gain Tunnel Hill was, in fact, one valley beyond where he thought it was, and his attacking columns faltered because of the misunderstanding.

Probably the most notorious example of rolling ground affecting the disposition of troops was on July 2, 1863, at Gettysburg. Union general Daniel E. Sickles pushed his Third Corps out in a convex bulge that deranged the Union defensive line and left a key topographical feature, Little Round Top, undefended. Sickles was apparently mesmerized by the gently rolling land that rose gradually to the west and the amateur general instinctively moved to deploy his soldiers on the highest rise, thereby forgetting his tactical responsibilities.

Hills and mountains also had a tactical significance. Confederate Bragg, whose experience with Rosecrans gave him firsthand knowledge, declared: "A mountain is like the wall of a house full of rat holes. The rat lies hidden at his hole, ready to pop out when no one is watching. Who can tell what lies hidden behind that wall?"[6] The "rat," or army in question, would "pop out" of practicable gaps in the mountains. A topographical engineer would find and examine these gaps from the perspective of the offensive and defensive. Union general Philip Sheridan was given the job, in the fall of 1864, of finally neutralizing the Shenandoah Valley—so often the scene of Federal humiliations and the avenue of Confederate invasions. In his memoir, Sheridan described the Valley Pike as the Valley's "Main Street," so to speak, along which were located the villages and towns: "[From the Valley Pike there are] lateral lines of communication extending to the mountain ranges on the east and west. The roads running toward the Blue Ridge are nearly all macadamized, and the principal ones lead to the railroad system of eastern Virginia through Snicker's, Ashby's Manassas, Chester, Thornton's, Swift Run, Brown's and Rock-fish gaps, tending to an ultimate center at Richmond. These gaps are low and easy, offering little obstructions to the march of an army coming from eastern Virginia and thus the Union troops operating west of the Blue Ridge [i.e., in the valley] were always subjected to the perils of a flank attack."[7]

A gap was relatively easy to defend and, conversely, was relatively difficult to force. But a gap that was somehow overlooked could have immense consequences. The first Union maneuver of the momentous Atlanta campaign almost brought the campaign to a decisive end. General George H. Thomas, probably the most conscientious map reader on either side, brought to William Tecumseh Sherman the intelligence that the Confederate defensive position at Dalton, Georgia, could be turned by taking an army through Snake Creek Gap. There was a defile on his flank and rear that Rebel Joseph E. Johnston was somehow unaware of or whose significance he somehow overlooked. General James Birdseye McPherson led his Army of the Tennessee through Snake Creek Gap—"a wild and picturesque defile, five or six miles long"—unopposed. "The road was only such a track as country wagons had worn in the bed of the stream or along the foot of the mountain."[8] Once through this defile, however, uncertainty manifested itself, and McPherson cautiously chose to simply hold the gap rather than attack. A terribly disappointed Sherman adapted to realities, but one of the war's great "what ifs" remains the Monday, May 9, 1864, march of McPherson's army through Snake Creek Gap and the failure to grab the great opportunity it met with there.

A good grasp of the defensive possibilities of a gap was displayed by Union general David Hunter. Hunter was the last of the Union generals to be roughed up, defeated, and sent packing in the Shenandoah Valley. Repulsed at Lynchburg, Virginia, in mid-1864, Hunter made excellent use of Buford's Gap. As Early described the situation in his memoir: "The enemy was pursued into the mountains at Buford's Gap, but he had taken possession of the crest of the Blue Ridge, and put batteries in position commanding a gorge, through which the road passes, where it was impossible for a regiment to move in line. . . . As the enemy had got into the mountains, where nothing useful could be accomplished by pursuit, I did not deem it proper to continue it farther."[9]

The successful defense of high ground also required an engineer's eye because the true line of defense was not the crest of a hill but the "military crest" of a hill. Put very simply, a military crest is the position on an elevation from which all the ground in front of it is in view. The most spectacular failure to occupy a proper military crest during the Civil War occurred at Missionary Ridge overlooking Chattanooga, Tennessee, on November 25, 1863. "The ridge here was some two hundred feet high, with steep slopes broken by many ravines and swales, and at this time, obstructed by the stumps of recently felled timber. The Confederate engineers had placed the upper line of works, which had been begun only two days before, on the natural instead of the military crest; this mistake 'left numerous approaches up ravines and swales entirely covered from the fire of the breastworks' "[10] The practical results of this error were described by Alabamian

Confederate sketch of Big Creek Gap in Tennessee's Cumberland Mountain. Map is reproduced in full on page 97.

Lieutenant W. M. Boroughs: "By some management, or engineering, Anderson's brigade had their works so far behind the crest of the ridge that they could not see an enemy approaching in their front, until within twenty or thirty feet of the line. Knowing the material of which that brigade was composed, I remarked to someone near me, 'Whenever that Yankee regiment reaches the crest of the ridge, they will be swept out of existence in the flash of a gun.' However, one of those incomprehensible things happened which frequently turned the tide of success in our Civil War. When this regiment reached the crest of the ridge, there did not appear to be more than fifty or sixty huddled around their flag. The gallant band hurled themselves with a yell upon the lines of Anderson's old brigade of Mississippi veterans. Not a shot was fired but with one impulse they swept them out of their works, and the little band of Federals took possession of the battery, waved their flag over it, and trained the guns so as to rake our lines."[11] One Confederate described the whole tactical question succinctly: "You Yanks had got too far into our inwards."[12]

Some of the Rebel difficulties at Missionary Ridge, apart from the faulty deployment along its crest, also may be blamed on the false sense of security felt by its defenders. Ulysses S. Grant is supposed to have been asked after the battle why Missionary Ridge was believed to be impregnable. Grant replied that it was impregnable. In instance after instance throughout the war, generals and troops were lulled into complacency because they held what was believed to be an impregnable position or because they grew careless with a portion of their line supposedly protected by impossible, impassable terrain. It was the topographical engineer's province to be alert to carelessness of this sort, which could be both exhibited and exploited by the same general at different times. During the Battle of Fredericksburg, Virginia, in December 1862, the one significant Yankee breakthrough came in General Jackson's sector. Jackson's confidence as he finished inspecting his line was such that he actually began to whistle, but "in reality, a glaring gap existed in the center of the line between the brigades of Archer and Lane."[13] This area was a triangular shaped woods, wet, cut by a deep ravine, full of scrub trees and dense underbrush, and it was undefended. Ironically, when the Union attack came, the withering Confederate firing actually pushed a Federal division under General George Gordon Meade into the haven of this undefended woods/gap in Jackson's line. The Yankees broke through but were unsupported. They reeled back, and the Battle of Fredericksburg resumed its briefly interrupted course as a Federal debacle.

Four and a half months later, and only a few miles to the west at Chancellorsville, Union general O. O. Howard commanded the Eleventh Corps on the extreme right of the Union army: "Throughout its length this line was on dominating high ground and faced to the south. Some 500 yards beyond Talley's [farmstead] on the Turnpike it simply ended, where Howard ran out of men."[14] Howard's flank, in military parlance, was "in the air." "Clearly," wrote Stephen W. Sears, "General Howard and his lieutenants were relying on the Wilderness as their first line of defense." An engineer tried to caution Howard about the vulnerability of his line, but the Eleventh Corps commander "pointed out that everywhere the forest was 'thick and tangled; will anybody come through there?' "[15] Coming through "there," the late afternoon of Saturday, May 2, 1863, was Jackson with his entire command: three divisions—twenty-eight thousand men and eighty guns. The thick and tangled woods were not impenetrable after all. The Rebels came through the matted vines, the pricker bushes, and the underbrush as quickly as a gale of wind.

There were occasions when the unexpectedness of an army's advance was so complete that it elicited the excited admiration of its foes. Such was the case in January and February 1865 when Sherman began to move his soldiers from Savannah by groping and wading through the Salkehatchie swamps. It was the wet season of the year, but Sherman was making for the rear of Lee's Army of Northern Virginia, held fast in its trenches at Petersburg, Virginia, by his friend Grant. "The good people in Savannah," wrote one of Sherman's staff officers, "thought the under-

George N. Barnard's Wartime Photograph of Missionary Ridge from Orchard Knob, Chattanooga, Tennessee

From this vantage point, Union generals Ulysses S. Grant and George H. Thomas could see the enemy's headquarters on Missionary Ridge. Confederate Braxton Bragg's staff officers could be seen coming and going all day, November 25, 1863. Grant described Bragg's position on Missionary Ridge as "impregnable." The Confederates, however, failed to occupy the military crest of the ridge, that is, a line from which every inch of ground to be defended could be seen. Furthermore, too many Confederate troops were deployed in a line of rifle pits at the foot of the ridge. Their retreat back up Missionary Ridge caused the defenders to fire high to avoid killing their own men. Thus, as Grant wrote, "the Union soldier nearest the enemy was in the safest position." Without orders, the Federals took advantage of the strange tactical situation to continue on to the crest of the ridge to carry the day and redeem their defeat at Chickamauga two months earlier.

taking hazardous in the extreme—in fact, impossible."[16]

Sherman had hoped that decent weather would accompany the march through inland South Carolina, but it was not to be. Instead, the rain poured down. Still Sherman's army moved relentlessly onward under the incredulous observation of the Confederate generals, including Johnston and Lafayette McLaws, whose commands were delegated to obstruct and defeat this march. It seemed blatantly clear to them that the Salkehatchie River, flooding and spreading out into fifteen unfordable streams, was doing their work for them. Sherman's own soldiers thought, "For once Sherman's stuck sure."[17] "The South Carolina low country turned into an immense swamp and the Confederate authorities considered it impossible for the Yankees to make a campaign now."[18] Johnston received a telegram from fellow general William J. Hardee as the two of them waited and the unpredictable Sherman disappeared into the great swamp. The telegram read, "The Salk is impassable."[19] Johnston agreed. "My engineers reported that it was absolutely impossible for an army to march across lower portions of the State in winter."[20]

But they were mistaken. Stationed nearby in antebellum days, Sherman knew the lay of the land to some extent. More important, he knew intimately the capabilities of his own men. Johnston also found out: "When I learned that Sherman's army was marching through the Salk swamps, making its own corduroy roads at the rate of a dozen miles a day and more, and bringing its artillery and wagons with it, I made up my mind that there had been no such army in existence since the days of Julius Caesar."[21]

⚬10⚬

Procuring the Maps

Requisitioned, Captured, Bought, Borrowed, Bartered, and Stolen

A N INTEGRAL and notorious aspect of General William Tecumseh Sherman's marches was his advance contingent, a special and unique group of soldiers known to their army and to posterity as bummers. These bummers were conceived in Sherman's orders in November 1864 as he began to lead his army on the March to the Sea, from Atlanta to Savannah. He knew that with no lines of communication, no base of supplies, and no railroad, his army of sixty thousand could not hope to supply itself. Hence, they would "forage liberally on the country during the march." Formal foraging arrangements were made, but there quickly developed self-employed foragers who roamed away from the advancing army at will, for days at a time, and constituted a sort of alternate army of freebooting quartermasters. These "raiders on their own account" kept Sherman well supplied. They also kept him apprised of the enemy's whereabouts. In addition to hams, meal, potatoes, vinegar, and beehives, they brought back letters, newspapers, and maps. Officers were detailed to receive and review these materials.[1]

The procurement of existing maps was of vital importance to the preparation of detailed, accurate military maps. Sherman's bummers might find a county or state map hanging on a parlor wall; they might tear open a letter and find a self-drawn map of a soldier's camp or fortification that could be set aside for future use.

In 1863, Jed Hotchkiss bought maps at Chambersburg, Pennsylvania, and he requisitioned maps at nearby Greencastle, Pennsylvania, during the course of the Gettysburg campaign. A year later Hotchkiss rifled through United States postmaster general Montgomery Blair's papers and appropriated his maps shortly before that cabinet member's house was burned to cinders by the Rebels on July 13, 1864. Blair's map collection most likely included some excellent and usually overlooked maps—route maps of the United States Postal Service that were prepared at the topographer's office in the Post Office Department in Washington. They were "designed to show all the permanent routes, distances and post offices in the United States,"[2] invaluable knowledge for a military mapper. Hotchkiss was involved in a more genteel map transaction with General William B. Taliaferro. They exchanged some maps, and Hotchkiss received a copy of what is known as the "Nine Sheet Map of Virginia,"[3] a very general map but a fine framework for preparing smaller and much more detailed maps.

D. H. Strother took leave of the scenes of hostilities and made rather suburban-seeming trips into Washington to get the maps he needed. On August 20, 1862, the near capture of Rebel cavalry chief Jeb Stuart in the environs of Verdiersville, Virginia, netted Strother what he described as a "very valuable map of the vicinity" that was found in a confiscated haversack belonging to the general.

Roles were reversed and fortunes changed two days later, when a chagrined Stuart attacked Union headquarters and, among other indignities that he inflicted, "thoroughly cleaned out" the Topographical Department, taking its maps, equipment, and its eight-thousand-dollar payroll.

Some maps were taken from soldiers who had been killed in battle. One of James Longstreet's staff officers, Moxley Sorrell, was given a "good general map of Virginia, of use afterwards" that was among the personal effects of Union general James Wadsworth, mortally wounded and left behind Confederate lines during the May 1864 Wilderness fighting. Union cavalrymen found a finely detailed map of Richmond's defenses on the body of Confederate brigadier John R. Chambliss, killed at Deep Bottom, Virginia, on August 16, 1864.

During their stay in Frederick, Maryland, in September 1862, during the Antietam campaign, the Southern commanders Robert E. Lee and Stonewall Jackson procured maps of Pennsylvania as well as of Frederick County, Maryland.

Railroad maps were highly valued for their accuracy. General George B. McClellan received "a remarkably clear map of the Baltimore and Ohio R.R. and the tributary roads" from the Coast Survey in Washington in May 1861. He also received a timely and very detailed map of Frederick County, Maryland, in the midst of the Antietam campaign in September 1862. The latter map was prepared by Isaac Bond, a post office topographer, and forwarded to McClellan by Postmaster General Blair.[4]

An engineering circular issued by the office of the topographical engineers of the Union Army of the Cumberland on September 9, 1863, describes a unique western theater map resource:

> The northern portion of the State of Georgia known as the Cherokee Purchase and a small part of S.E. Tennessee having been divided into Townships and Sections thereby giving additional facility for mapping the country, the following regulations will be observed: At every house endeavor to find out the exact number of the section and township lot upon which the house is located. If the occupant does not recollect ask for the tax receipts and get the exact location from them. Write this on the map by the side of the house. In this manner we can obtain fixed points and thus obviate one of the greatest difficulties in mapping a new country.
>
> Be very particular about the locations of all towns. They can be fixed as accurately as houses. The clerk's office in county towns will frequently contain section maps or lots of property with their exact locations. Such maps and lots are very valuable. They should be seized at once and forwarded to this office for assistance in making information maps and correcting surveys.[5]

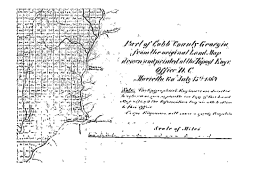

Detail of Land Map copied by Union forces campaigning in Georgia in 1864. Map is reproduced in full on page 185.

In addition to these orders to seize the maps found in county clerks' offices, William E. Merrill, chief topographical engineer of the Army of the Cumberland, also printed on a card the names and addresses of known local civil engineers so the cavalry would know where to look and whom to ask for area maps.[6]

The corrected Map of Northeastern Virginia (plate 7 in the *Official Records Atlas*) includes "Data Used in Compilation." The data consist of, among a multitude of other items: surveys by William Henry Paine; reconnaissance by topographical engineer Lieutenant Henry Larcom

Abbot about Blackburn's Ford; information provided by interviews with Generals George Archibald McCall, George Gordon Meade, and John F. Reynolds about the vicinity of Dranesville and Difficult Creek; maps of the Great Falls Aqueduct, the Mount Vernon estate, and the Orange and Alexandria Railway; manuscripts of the Accotink Turnpike by General George Washington; and additions and corrections from the office of the topographical engineers of the Army of the Potomac. Clearly, no resources that could help complete the picture of the lay of the land were overlooked in the preparation of military maps.

Preparing the Map
and the Map Memoir

Worked at Maps All Day

ONCE the process of gathering information was complete, once the physical features had been observed and the cultural details recorded, the topographical engineer would begin the actual cartographic process of producing a map. There were all styles and sorts of maps. Varying scales and different techniques were employed by mapmakers for commanders with more or less capability of grasping what the map was about. In an effort to standardize the maps in his department, Engineer Circular Number 3 was issued by "Wm. E. Merrill, Capt U.S. Engrs., Chief Top Eng. Dept. Cumb'd." Entitled *Information Maps,* it was, in effect, a complete instruction manual for the preparation of Civil War military field maps in the Department of the Cumberland:

> These are the most important maps issued from the T.E. [Topographical Engineer] office. Compiled from the best maps and the latest information obtained from residents of the country, they are the maps upon which all movements of the army are based. From the method of construction it is impossible that they should be minutely accurate but at the same time they are sufficiently so for military movements. The T.E.'s of this army should examine these maps as they pass over a route, and correct them as they go.
>
> *Method of Construction.* Rule a sheet of strong paper into inch squares. One set of lines will be north and south, and the other east and west. Information maps are generally made at the scale of one inch to the mile, unless only a small area is to be embraced and then the scale may be any number of whole inches to the mile.
>
> From the most reliable maps of the county to be mapped that can be procured, transfer to the information sheet a map of the county to be examined, carefully enlarging or reducing to the scale of the proposed information map. This transfer must be made in *lead pencil only,* so as to permit alteration. Having the information sheet thus prepared, send for all persons who know the country to be investigated, and question them carefully about it. Whenever the map is found to be incorrect, rub out the pencil marks and put in the corrections with colored pencil.

When the information received corroborates the original map, put down the parts confirmed in color. The conventional colors used are red for roads, and blue for streams. Be careful not to put in the colors too heavily, as better information subsequently obtained may compel a change in the first corrections and the marks of colored crayons are difficult to erase. Railroads are generally put in with the common lead pencil. Observe the same rules as if you were actually surveying the country, carefully locating all houses and giving the owner's names. Such minor features as woods, fences, etc. may be omitted, except on local maps. In general, find out all you can which can affect military movements in the county investigated, particularly the character of the road and the supplies of grass and water.

Be careful to letter distinctly. Where it is possible to indicate mountains approximately it should be done. Contour lines are generally more expressive than hatching to indicate mountains. They should not represent differences of elevation of less than 50 feet on maps of one inch to the mile; on maps of larger scale, lines may be drawn for proportionally less differences of elevation.

In locating roads the distances given must be written on them for future reference. Great care must be taken about roads in mountainous regions, as they are always much longer than the air line between the points which they connect.

The judgment of the Engineer will here come into play very largely. The turns should be layed down if possible and allowances should be made for the grade of windings; practice will soon enable this to be done with great accuracy.

It is desirable that all sketches sent to this office should be in color so as to prevent confusion between roads and streams.

As a general rule, all information maps will be made at Department Head Quarters where there are greater facilities for such work, and Corps, Division and Brigade Topographical Engineers will not make such maps unless specifically ordered, or unless they have surveyed and mapped all the country within their reach. Topographical work in this Department will be divided as follows: Surveys by the Engineer Officer in the field. Information maps, by the HdQ. Top. Eng. Office.[1]

It was also a good practice for the topographical engineer to submit, in addition to the completed map, a document known as a map memoir. Obviously, all of the data the engineer sought out could not be included amid the graphics of the military map. A map handed to a general during the hectic planning or execution of a military movement could not be an encyclopedia of the surrounding country. The presentation had to be plain enough to be looked over and quickly grasped and legible enough to be clearly understood no matter what circumstances of ambient light or confusion the map was being examined under. A written memoir therefore was prepared to annotate and supplement the map. As an example, while a fording site would be located on the map, the characteristics of the ford would be contained in the memoir. The memoir either would be submitted with the map or would be referred to by the topographical engineer when he reviewed the map with the appropriate staff members or field officers.

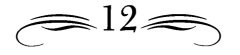

12

Map Reproduction

Each Had Received a Bound Copy of the Map

ONCE a map was drawn and completed, the next step was to get the map copied and distributed. There were a variety of methods for doing this, each having its advantages and disadvantages. The most primitive technique was pinpricking. Several sheets of blank paper were laid under the finished map. A pin was then pressed hard at very short intervals along the principal roads and watercourses of the original map, making a pinhole in the blank sheets. These holes or dots were connected, and rudimentary map copies were produced.

Tracing the finished map was perhaps the most common reproduction method used in the field. Skilled copyists laid translucent paper over the original map, making quick and detailed copies of the map available, but on very crinkly, fragile paper.

A draftsman could also simply copy the map, relying for accuracy on the setting up of a grid on the blank paper to correspond to a grid lightly drawn on the original map. Each square inch would therefore be self-correcting and ensure that the copy was true to the original. The advantage to this more time-consuming method was that the map was reproduced on more suitable, sturdier, and, for the Confederates particularly, more readily available paper.

Photographic reproductions of maps were common, principally in the Union army. The photographic advantages were speed and exactness. The drawbacks were that the equipment was bulky, sun conditions had to be just right to make reproductions, and photographic copies distorted at the edges. This last drawback made it difficult to connect maps of adjacent sections because the roads and streams were distorted where they were supposed to meet. Finally, the use of the photographic copies in sunlight caused them to fade. Photographically reproduced maps were, as William E. Merrill wrote, "objectionable on many accounts."

The Federal and Confederate armies both devised rather complicated methods of reproducing maps based on what could loosely be called sun printing. The Confederacy's engineers devised their technique out of desperation. They lacked the facilities and resources to meet the demand for maps and came up with a technique by which an original map, precisely drawn on tracing paper in black India ink, was laid on glass face-to-face with paper previously treated in an elaborate series of chemical baths. When it was exposed to the sun, a dark print or negative was produced on the treated paper. From these negatives, positive photo reproductions were possible.

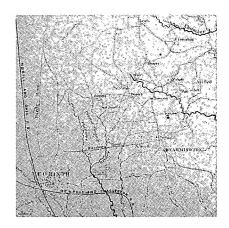

An example of a sunprint map. Map is reproduced in full on page 92.

Mathew Brady's Photographic Headquarters, Petersburg, Virginia, March 1865

A camera and set-up of this type was used to reproduce maps photographically in the field. The process was expensive and had a number of drawbacks. Sunlight was required to take the photograph, but exposure to sunlight made the maps fade. The borders of the maps were out of focus, just where they needed to be accurate to connect with maps of adjacent areas.

Nathaniel Michler reported that the "notes" of topographical engineers were "at once photographed in the field, and distributed for use. Revised editions of these photographs were published as fast as any new information was procured."

Union general William S. Rosecrans allowed a photographer (almost certainly George N. Barnard, some of whose photographs are reproduced herein) to remain with his army taking lucrative photographs of soldiers so long as his "photographic apparatus" was available to photograph maps.

A similar process was devised by Captain William C. Margedant of the Topographical Department of the Union Army of the Cumberland. Chemically treated paper was placed under tracing paper on which a map was drawn in heavy black ink. In the sun, the treated paper stayed white under the black ink, while the rest of the treated paper blackened from the sunlight. The resulting black maps, as they were called, had white rivers and roads. The roads were hand-colored red and the streams hand-colored blue to avoid confusing the two. Updates and corrections could quickly be made on the tracing paper, and new black maps reflecting the latest information could be rapidly prepared and distributed.

The premier method of map reproduction was the lithograph. The lithographic stones and presses were heavy and cumbersome—"too heavy for an active campaign, they were left at the depot nearest to the front," according to Merrill. But for producing a large number of detailed images, quickly and with quality results, in all types of weather and under all conditions, the lithograph was unsurpassed. As Captain Merrill described the process, draftsmen traced the completed maps with autographic or nonsoluble ink on thin paper, and when the traced map was finished it was transferred to a lithographic stone. This transfer process was different from the technique used with the normal lithograph. It saved time because the map was drawn on the thin paper exactly as it was meant to be printed. This was especially helpful because of the amount of lettering on a map and because the map was finished in its final form on the paper and got transferred to the stone in a single step. The lithographic process is based on the "coincident and disparate affinities of a certain Bavarian stone for water and grease."[1] Thus the map image traced in autographic ink is placed facedown on the dampened stone. The press squeezes the water from the stone, which fixes or transfers the nonsoluble image onto the stone. Since the image is reverse-fixed on the stone, the prints made from the inked stone are "right reading" and ready for distribution. Maps could be lithographed onto paper, linen, or even handkerchiefs so that cavalrymen and officers in the field had handy maps that would not fall apart when they got wet or worn; indeed, they could be washed and wrung out to dry when they got dirty.

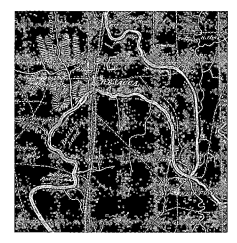

Known as a "black map," this easy and inexpensive method of reproducing maps in the field was used by the Union armies in Georgia in 1864. Map is reproduced in full on page 173.

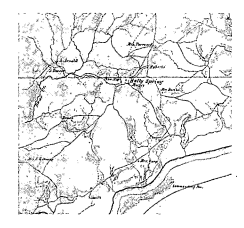

An example of a map lithographed in the field by Union topographical engineers in Georgia in 1864. Map is reproduced in full on page 176.

THE

MAPS

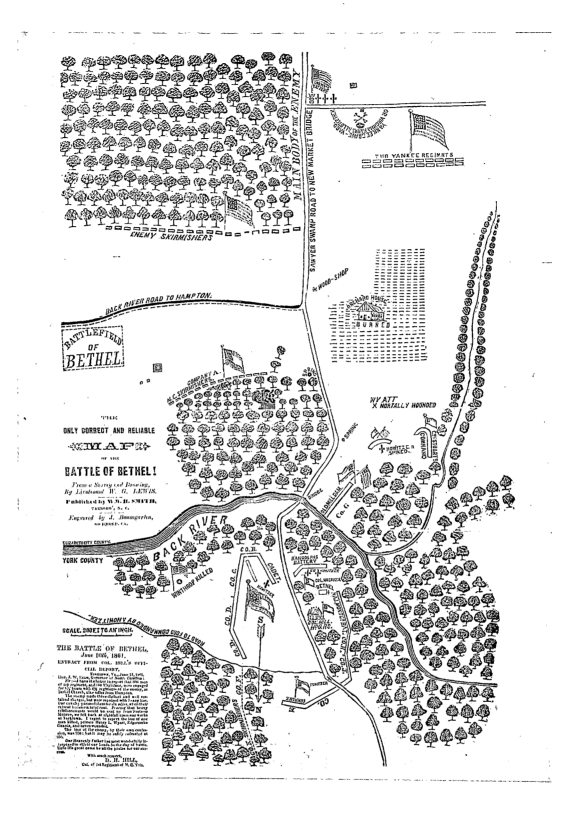

Battle of Bethel, Virginia, June 10, 1861

A wood engraved map, printed by stereotype. Also known as Big Bethel, this action was by later standards of the war a mere skirmish but nonetheless has become known as the first land battle of the American Civil War. The Union loss was seventy-eight, the Confederate loss eleven. The map is oriented with north off by 180 degrees. Note the editorialized description of the upside-down "Yankee Grave-Yard" at the top of the map. A free black named W. H. Ringgold had a copy of this map, which made its way to Union intelligence and Allan Pinkerton. Pinkerton headed up General George B. McClellan's Intelligence Department. Pinkerton described the map as being "of Southern manufacture and diction."

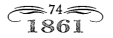

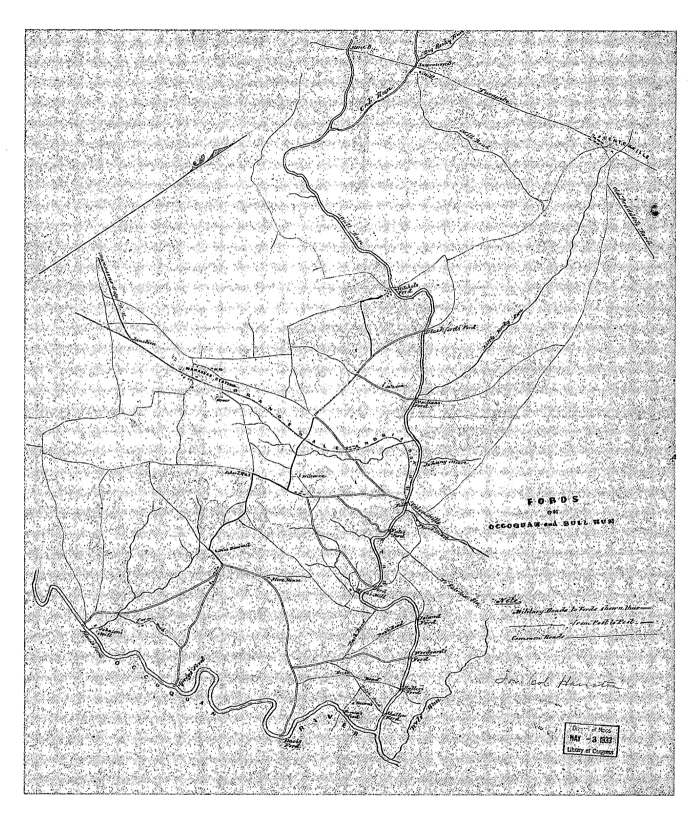

Fords on Occoquan and Bull Run, 1861, probably by Léon J. Frémaux

This very accomplished but minimal map is neither credited nor dated. Centreville is misspelled, but in every other respect the map is the work of an expert cartographer, very likely Captain Léon J. Frémaux of the Eighth Louisiana Volunteers, who was on duty at Manassas keeping watch over Occoquan River fords. A civil engineer by profession, Frémaux drew maps unofficially at Manassas, though he was eventually ordered to Richmond and received a commission as captain of engineers in the Confederate army, a singular honor for a non–West Point graduate. He was soon posted to the Mississippi Valley (see pages 76, 83, and 106).

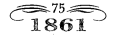

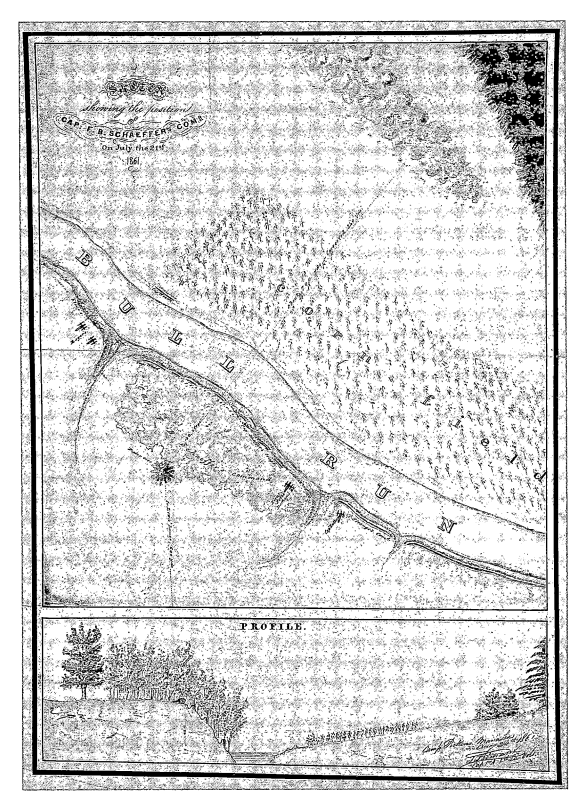

Léon J. Frémaux's Sketch, Captain F. B. Schaeffer's Command, Bull Run, July 21, 1861

Arguably the most attractive cartographic artifact of the Civil War, this manuscript map/sketch in ink and watercolor shows the position of the New Orleans Cresent Blues and Captain Schaeffer's command on the principal day of the First Manassas battle. The sketch depicts Bull Run just downstream from the Stone Bridge at Lewis's Ford. The profile on the bottom of the sketch presents a graphic diagram of the tactical advantage enjoyed by the Confederates on the right bank (left side of profile) of Bull Run. Frémaux was later to superintend the defenses of Port Hudson on the Mississippi River and of Mobile, Alabama. He returned to his watercolors and to New Orleans at war's end. A careful examination of the details of this map will be richly rewarding.

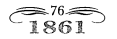

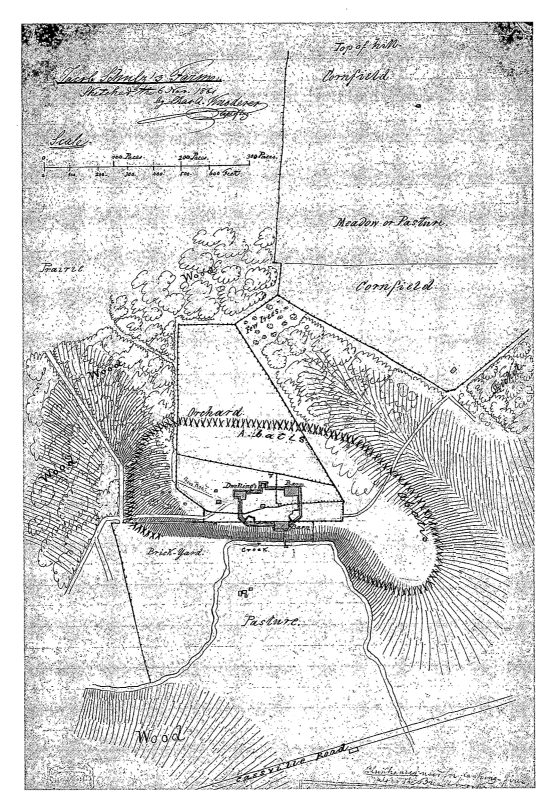

Charles A. Knoderer's Sketch of Jacob Shultz's Farm, November 6, 1861

Located near Wilson's Creek battlefield in the vicinity of Springfield, Missouri, this sketch is unique in its focus on a single farmstead in such complete detail. Not only are the fences shown post by post and the abatis of felled trees meticulously re-created but the ground cover differentiates among cornfield, orchards, woods, trees, thickets, pasture, and meadow. The sketch is done on plain notebook paper in India ink. There is a Jacob "Schultz" farm shown in the 1860 census for Campbell Township, Greene County, Missouri. A large farm, located two miles southwest of Springfield, it was very likely occupied and fortified by Union general Franz Sigel's division. Charles A. Knoderer was a graduate of a polytechnic school in Baden, Germany, and served on General Sigel's staff. Knoderer, a colonel, was mortally wounded on the Virginia Peninsula on January 30, 1863, at Deserted House. He died on February 15, 1863.

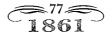

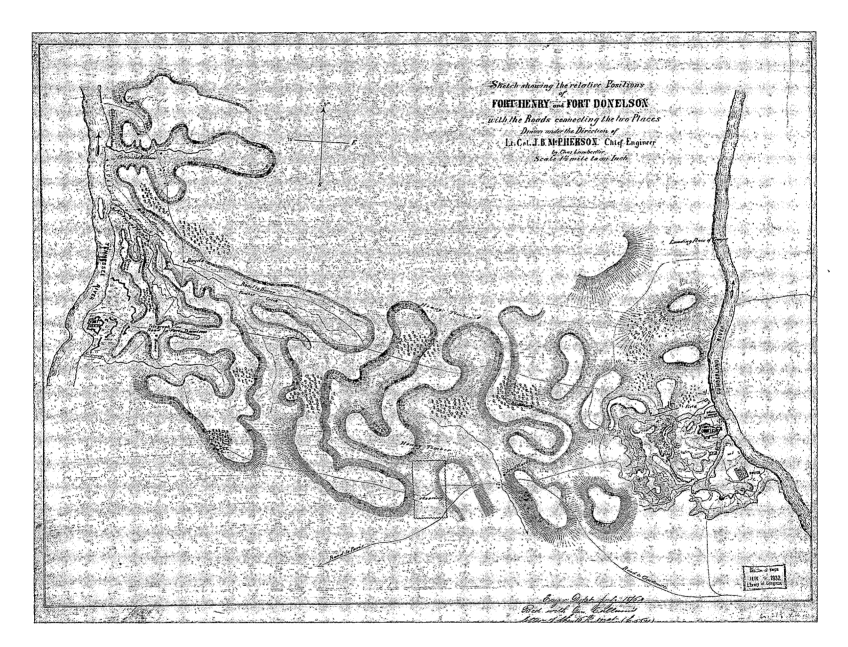

James Birdseye McPherson/Charles Lambecker: Sketch of Relative Positions of Fort Henry and Fort Donelson, February 1862

This rather spare manuscript map, with a thin blue wash on the rivers, includes a naval as well as an army operation, symbolic of the cooperation achieved by Union general Ulysses S. Grant and Admiral Andrew H. Foote. The northward flow of the rivers encouraged the use of naval gunboats to bombard the forts because if the vessels became disabled they would drift north and out of danger. The move from Fort Henry to Fort Donelson, Grant thought, would occupy two days, including the taking of the second fort. The job turned out to be longer and harder. The fall of Fort Donelson, which Grant had considered a mere formality, was actually a near defeat away, but when the smoke cleared, he was famous across the nation as Unconditional Surrender Grant and Confederate strategy in the West was in tatters. A lithographed version of this map is in the *Official Records Atlas* (plate 11, no. 2).

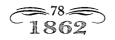

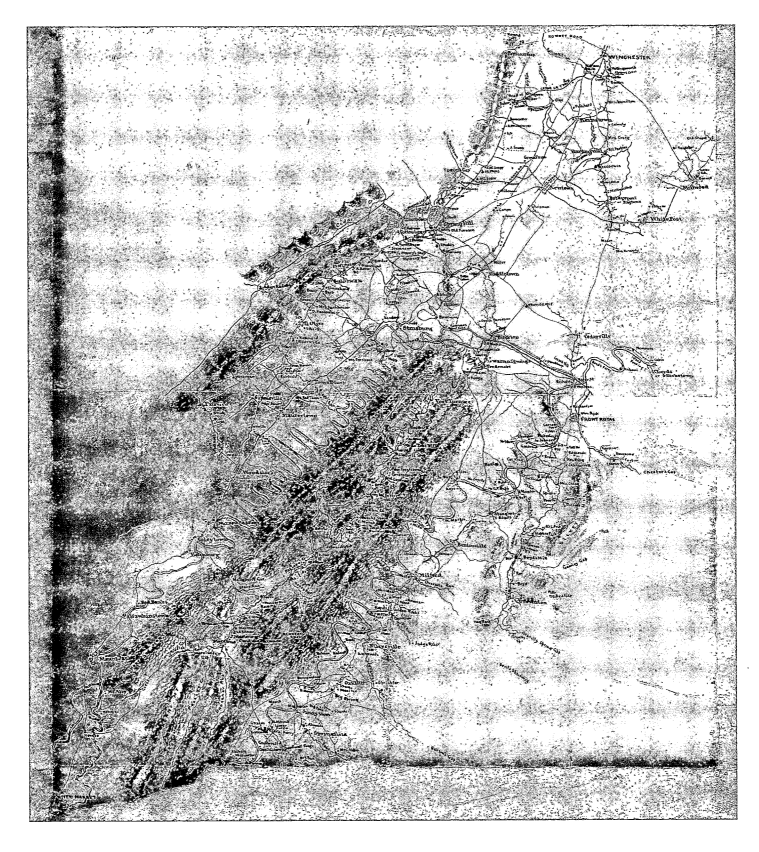

Jed Hotchkiss's Map of the Central Shenandoah Valley, 1862

A stunning map of the midsection of the Shenandoah Valley of Virginia by the Confederacy's and the Civil War's best-known topographical engineer. The map is done in pencil, color pencil, India ink, and watercolors on tracing paper mounted on a heavier paperboard (the bottom of the map is torn off and missing). Almost every square inch contains a significant Civil War site. The depiction of elevations is especially interesting; Jed Hotchkiss uses an impressionistic contourlike line with a hachured effect.

Topl. Engrs U.S.A.: Field Note Book of Wm. Luce, 1862

This printed field notebook was turned over to Confederate topographical engineer Jed Hotchkiss. William Luce was an assistant to Captain D. H. Strother. At the time of the Civil War, Strother was one of America's best-known sketch artists and travel writers. Hotchkiss, for example, was familiar with his work. In his published "Personal Recollections" (1866–68) and in his published diary (1961), Strother mentions William Luce repeatedly. He feared Luce would be captured or killed because he went so far afield in his mapping expeditions and was so absorbed in his measurements, compass reading, and notes. In March 1862, near Milnwood, Virginia, Luce was taken prisoner. Soon thereafter, Strother got the story from the Rebel cavalryman who had captured Luce but who was now, by the fortunes of war, a prisoner himself.

Field Note Book Notations of Wm. Luce, 1862

After William Luce's captured notebook was very properly turned over to Jed Hotchkiss, information it provided may have been valuable to Rebel mapping efforts. It contained notes of areas that were more or less under Union control, and this data would have been difficult and dangerous for Hotchkiss to gather personally. Luce's observation that the "New Cut Road" was fifty years old and his accurate little sketch of a house were probably meant to jog his memory when he got around to collating his information. D. H. Strother, Luce's superior, knew details of Luce's captivity. It can be presumed, therefore, that Luce was paroled and survived his captivity.

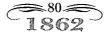

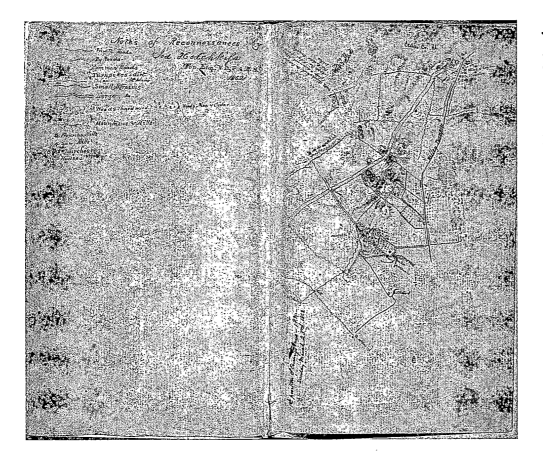

Jed Hotchkiss's Sketchbook, 1862–65, inside front cover and page 1

The initials following Jed Hotchkiss's name stand for Valley District Provisional Army, Confederate States. The inside front cover contains a carefully thought out and nicely rendered legend for the sketchbook maps. Page 1 contains sketches of the action on March 23, 1862, at Kernstown, Virginia, just south of Winchester. It shows Virginia rail fence lines, a stone and log house, orchards, a gate through which Federal troops advanced, cornfields, and meadows. A tactical defeat for the Confederates, Kernstown resulted in important strategical gains for them nonetheless. It would be regarded as the first action in Stonewall Jackson's great Valley campaign of 1862. Three days after this sketch was done, Jackson called Hotchkiss to headquarters at Miss Stover's stone house near Narrow Passage Creek and told the thirty-three-year-old engineer, "I want you to make me a map of the Valley. . . . "

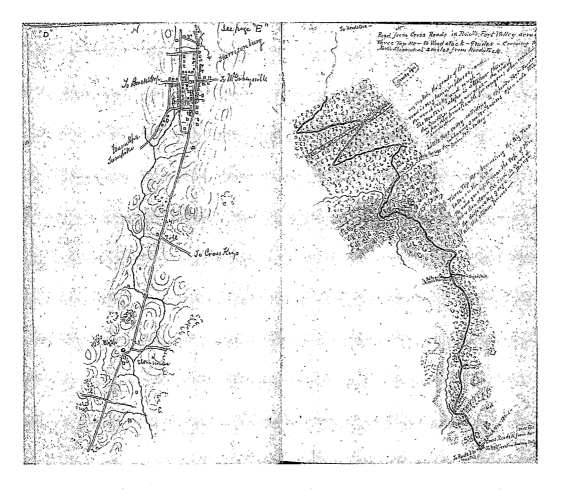

Jed Hotchkiss's Sketchbook, 1862–65, page 11

A beautiful sketch map. This mountain road over Three Top or Massanutten Mountain shows Jed Hotchkiss's expertise in mapping a gap or "heavy grade," a particularly valuable talent in the Shenandoah Valley when working with Confederate general Stonewall Jackson. Jackson's ability to make strategic use of the mountains and valleys bewildered his foes. Hotchkiss warns, "The road is hardly practicable for wagons when much loaded." This is one of Hotchkiss's most impressive representational sketches.

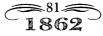

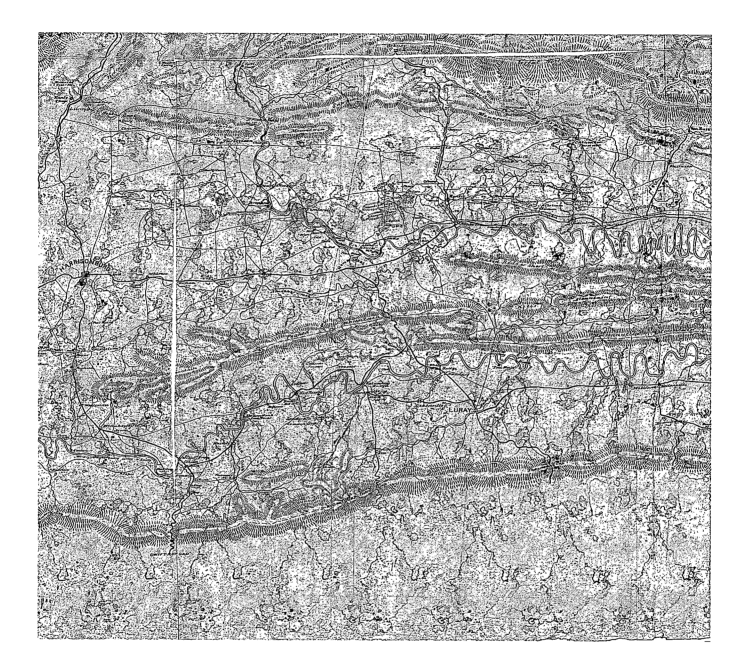

Detail of the Map of the Valley of Virginia from Staunton to the Potomac River, 1862

Taken from a very large map (approximately 6 x 4'), this section shows in detail the topography of the valley that Confederate general Stonewall Jackson took advantage of to baffle his enemies and defeat them. The gap in Massanutten Mountain between New Market and Luray brought Jackson unexpectedly to Front Royal to overwhelm the small force of Federals there. When Jed Hotchkiss received orders to prepare a map of the valley for Jackson on March 26, 1862, the Jefferson County surveyor, S. Howell Brown, had already been detailed to make a map of the lower (northern) portion of the valley. Hotchkiss, who came to be constantly associated with Brown, described him as a stout, accommodating, pleasant man but a "mass of facts; painfully matter-of-fact . . . fearfully exact." Despite the sobriquet "Sergeant," Brown, like Hotchkiss, was a civilian employed by the army. Mounted in eight separate squares on paper backed with muslin, this map was copied from the original in the Confederate Engineer Bureau, November 1863.

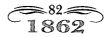

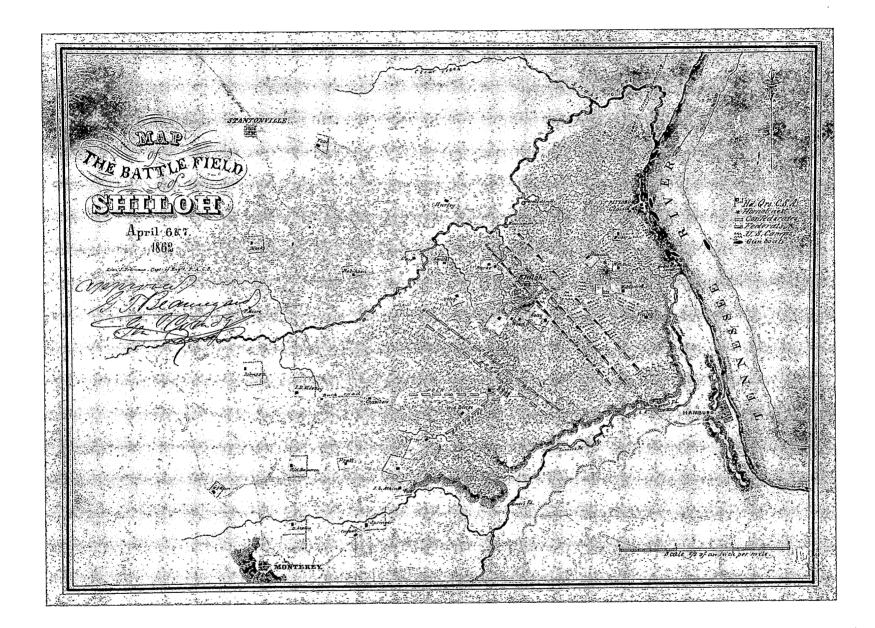

Léon J. Frémaux's Map of the Battlefield of Shiloh, Tennessee, April 6-7, 1862

This map by Confederate captain of engineers Léon J. Frémaux, a civil engineer by trade and a watercolorist by avocation, was likely drawn to illustrate Confederate general P. G. T. Beauregard's report of the battle. Done in ink, color pencil, and watercolor, the map exhibits very fine draftsmanship and lettering, and the embellishments are works of art. The misspelling of Confederate general and former United States vice president John C. Breckinridge's name is perhaps the only thing wrong with this map.

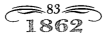

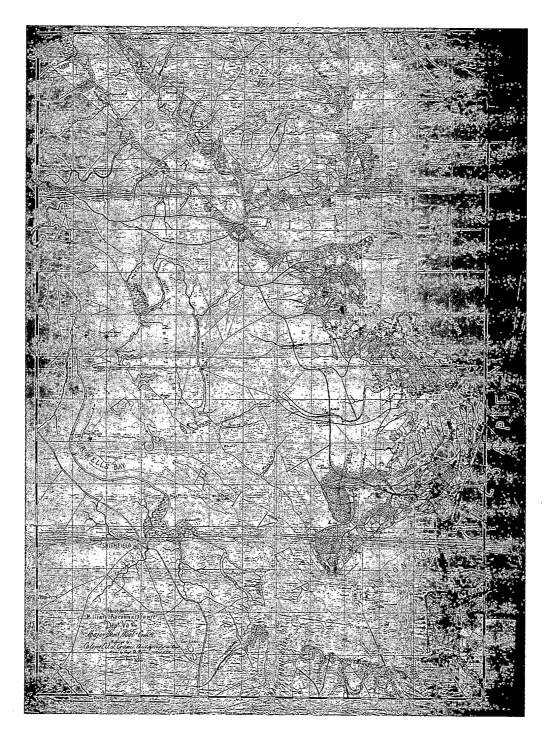

Thomas Jefferson Cram's Map of a Section of the Virginia Peninsula, 1862

This colorful ink-and-watercolor map, hand-copied on tracing paper, is a better cartographic effort than it is an accurate military map. Thomas Jefferson Cram, captain and veteran topographical engineer on the staff of General John Wool at the Union's Fort Monroe, prepared a map for General George B. McClellan to use in planning and launching his spring 1862 campaign up the Virginia Peninsula against the Confederate capital of Richmond. Cram got information from black contrabands, Rebel prisoners, and local residents in lieu of actual reconnaissance on the enemy-occupied portions of the peninsula. He also dug through War Department files for Revolutionary War–era material. The resulting map, not surprisingly and probably inevitably, was quite inaccurate. Unfortunately, McClellan took the mapped intelligence, as he took all his intelligence, at face value. McClellan planned to advance his forces up the peninsula behind the Confederate forces at Yorktown, isolating them there and cutting them off from Richmond. The Warwick River, shown on this map flowing parallel to the James River, would serve a handy strategic function by protecting the left flank of McClellan's columns as they moved up toward Yorktown. In reality, however, the Warwick River does not flow in the direction shown on this map. It flows across the peninsula. It was thus athwart the Union advance, and McClellan's plans were thrown into utter confusion. This was one of the first snags in a campaign that came completely unraveled. Cram's map of the area is therefore one of the Civil War's most momentous maps. It must be recalled, however, that the Confederates' knowledge of the same terrain was utterly lacking as well. A lithographed version of the map appears in the *Official Records Atlas* (plate 18, no. 1).

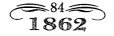

Plan of the Investment and Attack of York in Virginia, c. 1781

The decision by Union general George B. McClellan to move the huge and cumbersome Army of the Potomac by water to the Virginia Peninsula in the spring of 1862 and to advance on Richmond from that direction necessitated the preparation of a map of that unfamiliar region. Veteran topographical engineer Thomas Jefferson Cram began work on such a map. Among the more imaginative and resourceful ideas Cram had while he prepared the map was to pore over War Department files for Revolutionary War data on the siege of Yorktown. He also made use of Revolutionary War memoirs, histories, and biographies to glean relevant information, which was then incorporated into his own map. Revolutionary War mapping was a highly evolved, formalized discipline—more so than American Civil War mapping—especially with the French *ingénieurs géographes,* or topographical engineers associated with General Comte de Rochambeau's French expeditionary force in North America. This map was likely one of, or similar to, the resources Cram used.

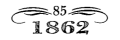

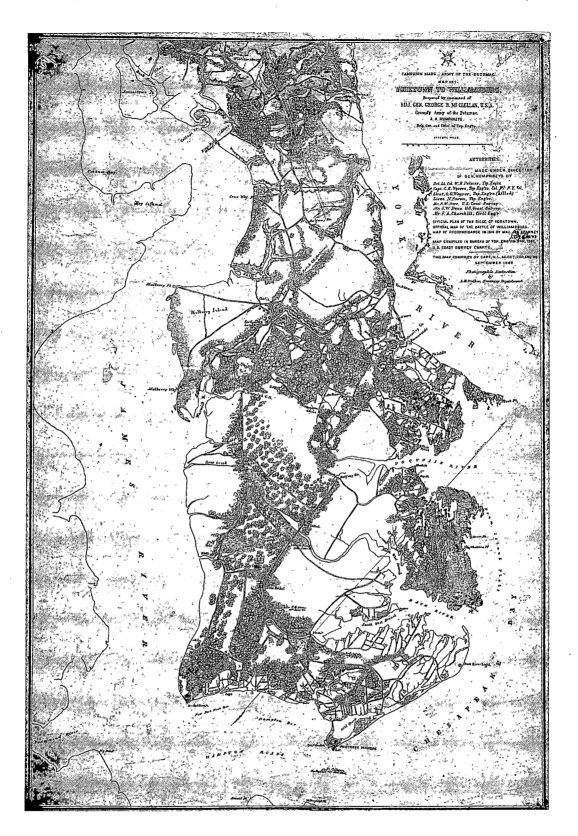

A. A. Humphreys's Map Number One, Yorktown to Williamsburg, 1862

A photographic print, this map, which covers the same ground as the Thomas Jefferson Cram map on page 84, provides a graphic example of the inaccuracies Union forces had to make allowances for as they campaigned on the Virginia Peninsula. A. A. Humphreys had the advantage of actual reconnaissance and physical presence on the ground being mapped and lists the authorities upon which the mapped information is based. In a subsequent report, Humphries apologized for not accompanying this map with a map memoir, explaining that the maps "can present comparatively little of the information required concerning a country that is the scene of military operations." Later in the same report, Humphreys acknowledges fourteen individuals who took part in the preparation of this map and others in the set that topographically detailed General George B. McClellan's operations. Standing out in bold relief is a name that would be seen much more in United States military history: "Lieut. George A. Custer, Fifth Cavalry." A lithographed version of this map appears in the *Official Records Atlas* (plate 18, no. 2).

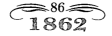

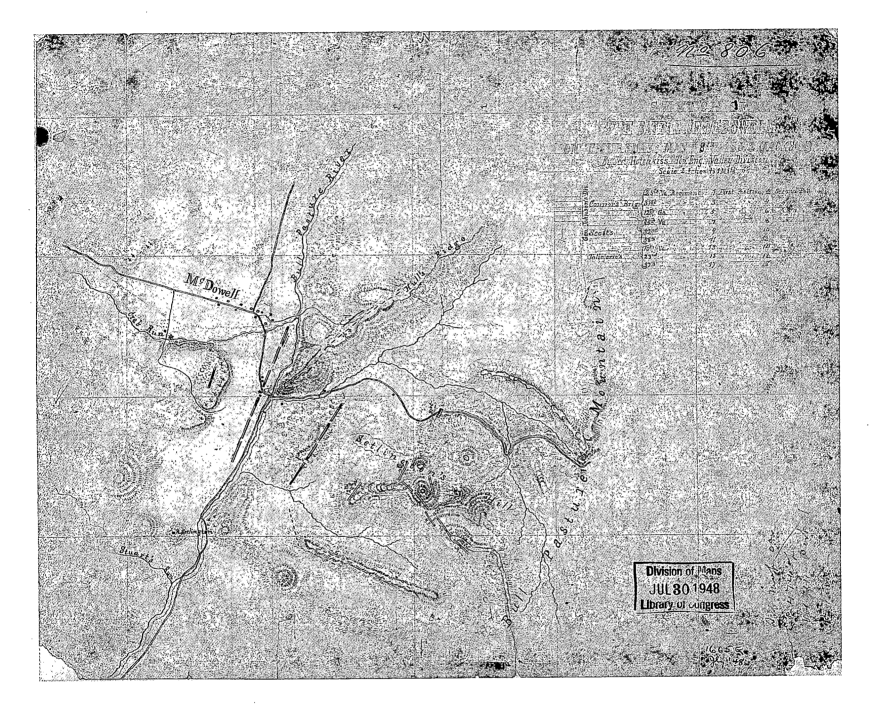

Jed Hotchkiss's Map of the Battle of McDowell, Virginia, May 8, 1862

Jed Hotchkiss took Confederate general Stonewall Jackson "to the end of a rocky spur overlooking the Bull-Pasture Valley [Sitlington's Hill]" and showed him the enemy in position near McDowell. At the same time, "[he] made him a map of McDowell and vicinity, showing the enemy's [Union generals Robert Milroy and Robert Schenck's] position, as in full view before us." Jackson had studied topographical engineering and drawing at West Point, but it was a subject he had difficulty with. He gratefully gave Hotchkiss free rein in topographical matters, willingly accepting his advice and insights. McDowell was a minor Confederate victory in the scheme of things—Jackson said, "God blessed our arms with victory at McDowell yesterday"—but it was the beginning of a string of victories known as the Valley campaign. The lithographed version is in the *Official Records Atlas* (plate 116, no. 1).

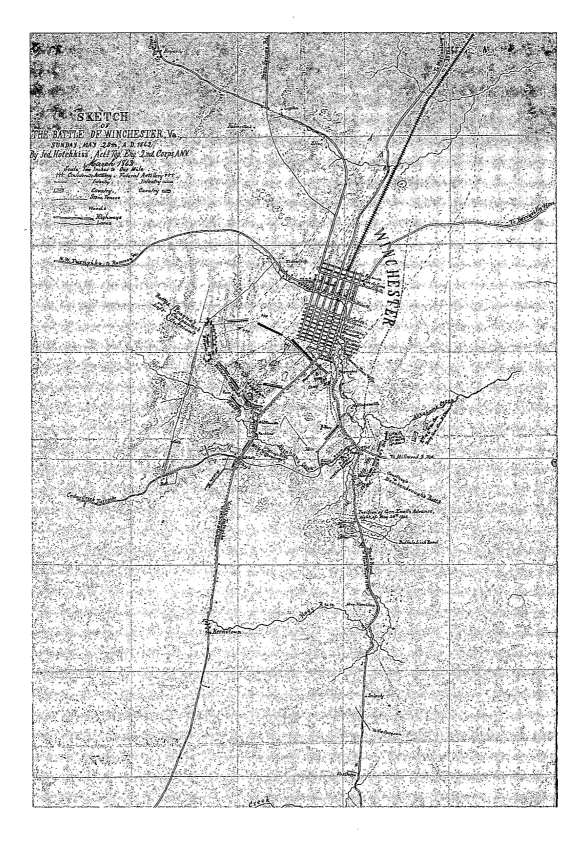

Jed Hotchkiss's Sketch of the Battle of Winchester, Virginia, May 25, 1862

Jed Hotchkiss spent five or six days in late March 1863 working on this sketch of Winchester. The town changed hands repeatedly throughout the war. Hotchkiss studied the reports of the battle to aid him in preparing the map. In one of the pitched battles of the legendary Valley campaign of General Stonewall Jackson in May and June 1862, the Federals were sent whirling in retreat back north. This large-scale tactical map shows details right down to stone fences and the deployment of individual Confederate regiments. The original manuscript was prepared in red-and-black ink and color pencils. A lithographed version appears in the *Official Records Atlas* (plate 85, no. 2).

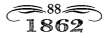

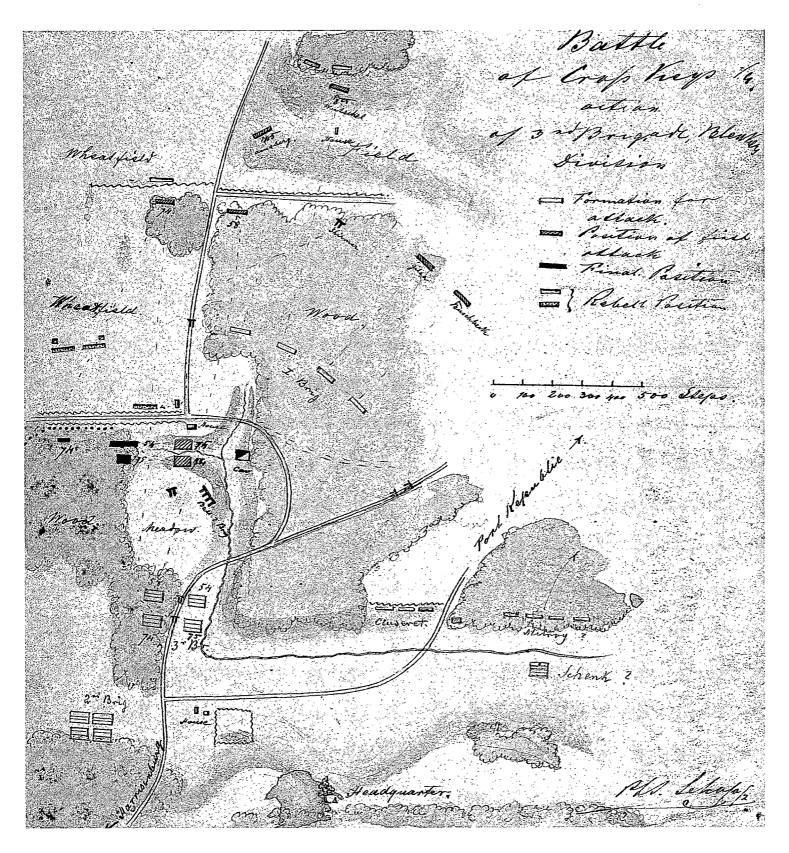

Ph. I. Schopp's Map of Cross Keys, Virginia, June 8, 1862

This very confusing map, with a road configuration that fails to correspond with either contemporaneous military maps or modern topographical maps, possibly explains the Union failure here. The map is strangely oriented; north is off by approximately 135 degrees, and the watercourses appear to have been drawn at random. Some of the cultural landmarks of the field are identified with unlabeled squares, including Union Church. However, the entire left wing of the Confederate forces on the field is overlooked. Note several nice touches nonetheless, including the precise, almost poignant rendering of a tent and tree indicating "Headquarters" at the foot of the map.

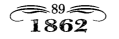

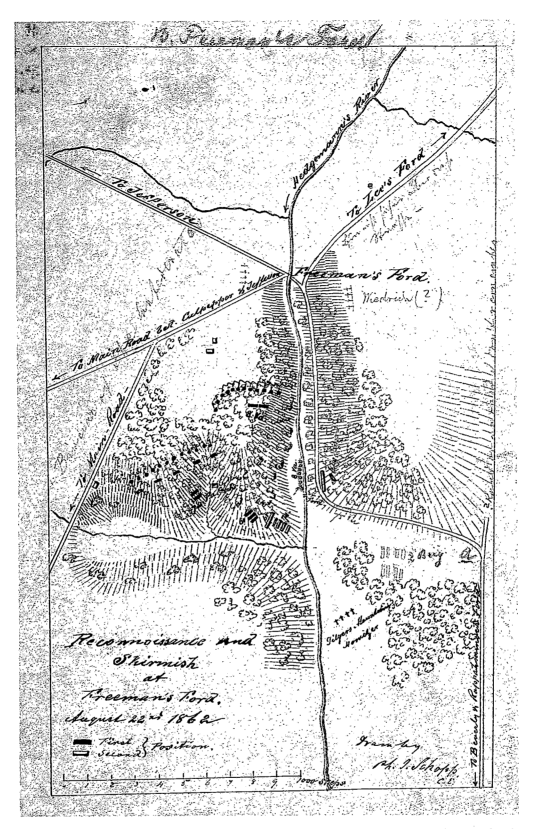

Ph. I. Schopp's Sketch Map of Freeman's Ford, Virginia, on the Rappahannock River

Located about five miles upstream from Rappahannock Station, Freeman's Ford was a frequent fording site throughout the course of the war. This sketch, as German native and civil engineer Ph. I. Schopp describes it, is drawn on heavy watercolor-type paper. He said, "The topography of the actual battlefield is correct enough. I completed it from a fleeting sketch out of my diary." Schopp made these comments in a lengthy description of the Battle of Freeman's Ford fought on August 22, 1862, as Confederates under Stonewall Jackson sidled upstream to locate an undefended crossing of the Rappahannock. Schopp's report, written in a High German military dialect (see examples on the sketch in the penciled notations), was translated for this atlas by Dr. Patrick J. Kelly. Schopp labels the Rappahannock as Hedgemann's River. Confederate maps do not. The scale is shown in "steps." The action here resulted in the death of Union general Henry Bohlen, a German immigrant. Interestingly enough, a requisition book kept by Lieutenant L. B. Norton, chief signal officer of the Army of the Potomac, describes numerous fords on the Rappahannock River. For Freeman's Ford he writes simply, "Best on the River."

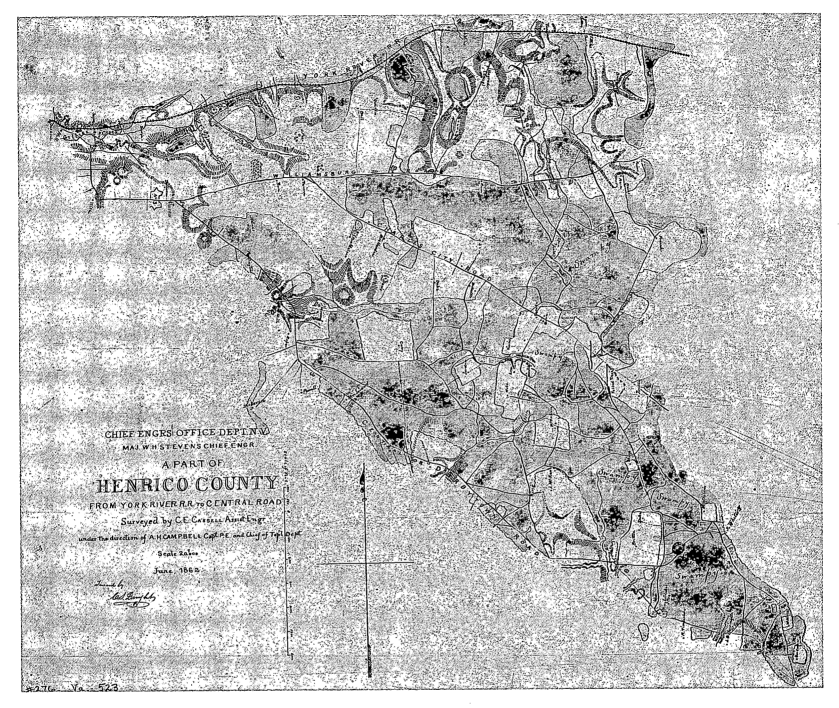

Charles E. Cassell's Map of Part of Henrico County, Virginia, June 1862

This map on tracing paper is copied from a map that was made at a pivotal moment in Confederate mapmaking. Captain Albert H. Campbell assumed control of the Confederate Bureau of Engineers on June 6, 1862. There was universal complaining, some of it quite eloquent, on the part of Rebel commanders such as D. H. Hill, Richard Taylor, and E. M. Law, about the lack of Confederate maps. Law sarcastically described the attempt to wage a campaign without maps as "a new principle in modern warfare." This map represents the beginning of a concerted effort to remedy the dearth of topographical information, but it also represents the results of a survey conducted, by necessity, within the Confederate lines. Any Confederate offensive—and with Robert E. Lee assuming command on June 1, 1862, an offensive would not be long in coming—would quickly march off this map. When the Union army retreated (or changed its base, as the case may be), among the tons of supplies left behind were two hundred yards of tracing paper, an immense boon to the Confederate map-reproduction effort. Though this map is dated June 1862, it could very likely have been traced on paper from the July 1862 capture. Charles E. Cassell was a lieutenant in the Confederate Bureau of Engineers.

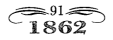

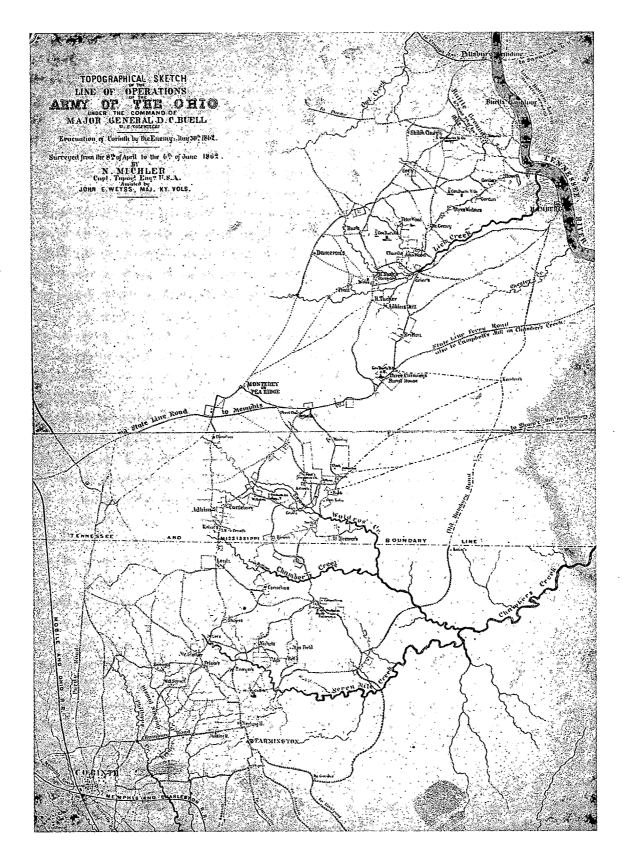

Topographical Sketch of the Line of Operations of the Army of the Ohio, 1862

Mounted on muslin, this map is a hand-colored, sun print–positive photocopy of an original map surveyed by Union topographical engineer Nathaniel Michler, assisted by John E. Weyss. Michler and Weyss went on to become premier Civil War mapmakers and immediately after the war were set to work surveying the major eastern battlefields with peacetime thoroughness. Union general Henry W. Halleck's advance on Corinth, Mississippi, from Pittsburgh Landing, Tennessee, was so slow and methodical that it afforded plenty of time for adequate surveys, a rare luxury for military mapmakers.

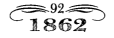

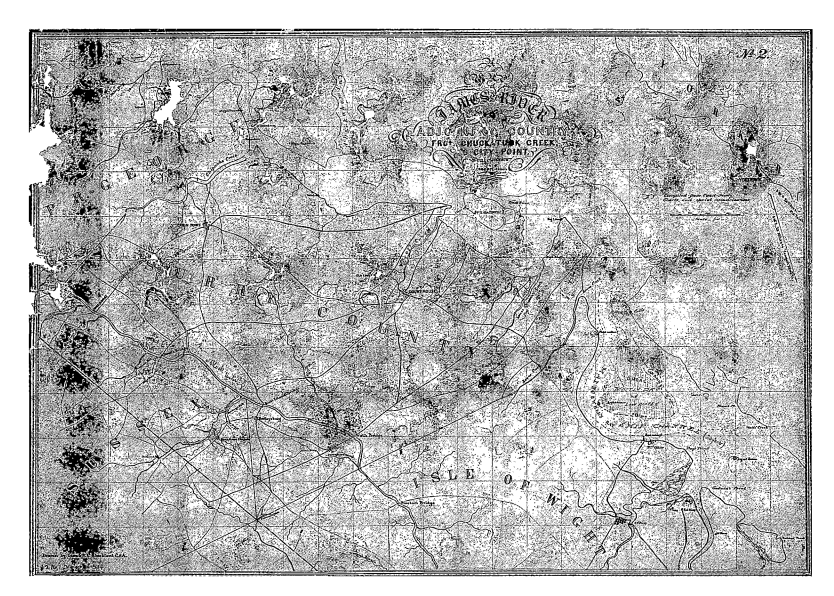

B. L. Blackford's Map of James River and the Adjoining Country, Virginia

The Blackfords constitute a unique triumvirate of Confederate brothers, analogous to the Union's Cushing brothers. William W. Blackford was Confederate cavalry general Jeb Stuart's staff engineer and a brillliant memorialist whose *War Years with Jeb Stuart* is one of the finest and most exciting first-person accounts of the Civil War ever published. A second brother, Charles Minor Blackford, wrote very descriptive and thoughtful letters, diary accounts, and a memoir of his experiences in "Lee's Army." And a third brother, B. Lewis Blackford, was, like his brother William, a topographical engineer. This map, with its pencilled grid lines faintly visible, is a very distinctively designed production. The border alone is noteworthy, and the title and lettering are particularly expressive and decorative for a military document. The map (or perhaps, more accurately, the chart) is left unfinished. The pencilled data detailing the depth of water in Burwell's Bay, for example, is more fully developed than the land information. Though a cluster of buildings indicates a small village in Isle of Wight, it is not identified as Smithfield. Blackford indicates that United States Coast Survey charts form the basis of the map. The Confederate takeover of the U.S Naval base at Norfolk, Virginia, would likely have yielded those valuable and accurate charts in abundance.

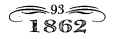

Lieutenant C. S. Dwight's Tracing of Chickahominy River, Mechanicsville, Virginia, c. 1862

A map traced by Dwight, a draftsman and engineer whose name appears on a number of other maps and map sketches of the vicinity of Richmond. This map is typical of the mapping agenda being followed by Confederate topographical engineers in the spring and early summer of 1862 as General George B. McClellan's massive Union army made its cautious way toward Richmond up the Virginia Peninsula. Captain Meade had been sent for from North Carolina to come to Richmond to help work on the defenses of that city as McClellan inched closer and closer. Captain Alfred L. Rives, acting chief of the Confederate Bureau of Engineers, wrote to Robert E. Lee on June 4, 1862, saying that if directives were given to the various officers to make tracings of any topographical data they gathered and then forward them to the bureau, a valuable general map could probably be prepared. This tracing may well have been one of those sent in. It shows several features that do not appear on subsequent, more general maps. These include a canal and a number of bridges on the Chickahominy River between Mechanicsville Bridge and New Bridge.

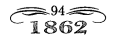

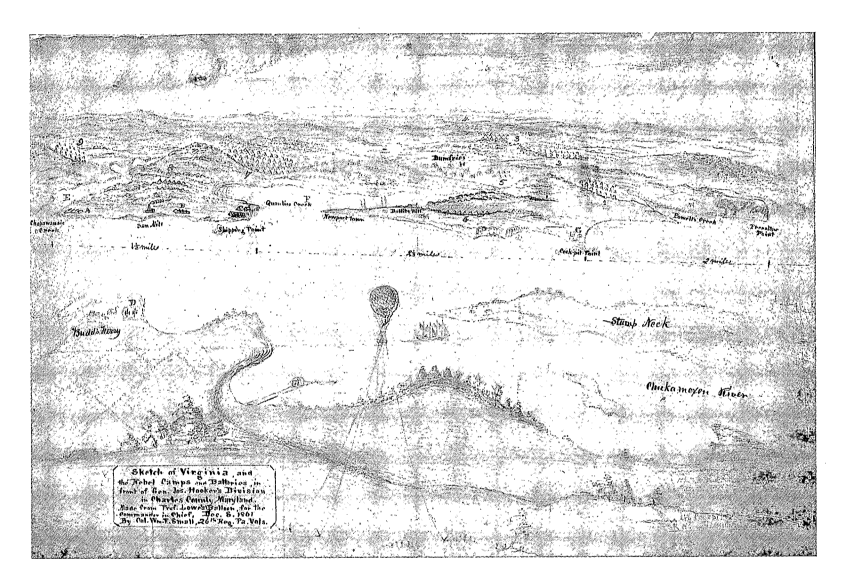

Small's Aerial Sketch Map of Virginia, 1862

This rare isometric map was drawn from a Federal observation balloon. Few of these sorts of maps exist, despite innumerable balloon ascensions, because the motion of the balloons made drawing very difficult. The Balloon Corps had a checkered career and faded out of existence after June 1863, shortly after the Battle of Chancellorsville. There was mutual resentment between the civilian aeronauts and military men. "They always see lions in our path" was one topographical engineer officer's opinion of them. At least one high-ranking Confederate, General Edward Porter Alexander, disagreed. He thought that the Federals never appreciated the amount of roundabout marching the presence of observation balloons high above the field forced on the Confederates. One of the most enjoyable accounts of reconnoitering the Rebel lines from a balloon was written by Lieutenant (and, in April and May 1862, topographical engineer) George Armstrong Custer, who also drew several aerial sketch maps.

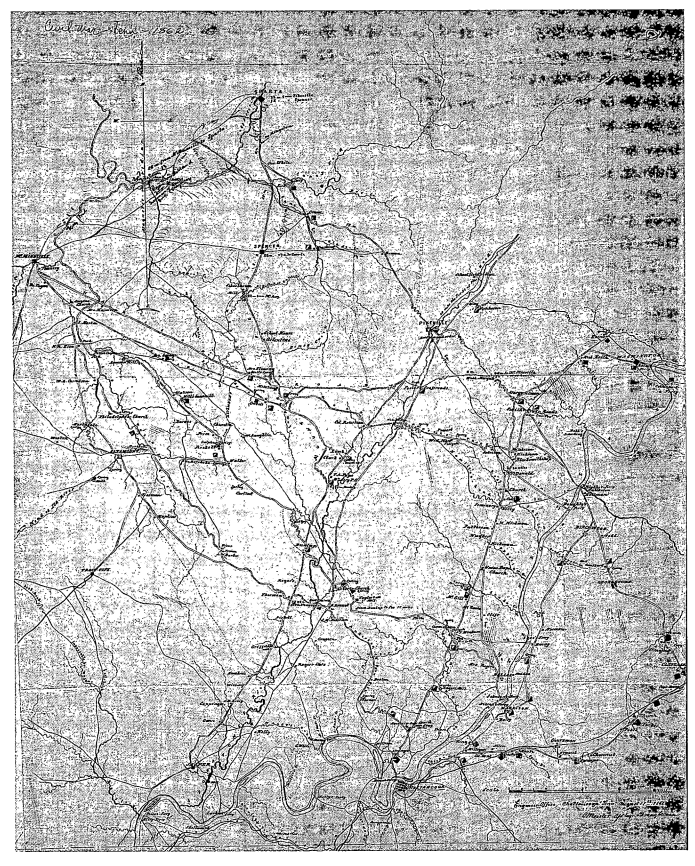

Chattanooga, Tennessee, to Sparta, Tennessee, August 1862

A theater map on muslin by C. Meister, "Apt. [Apprentice] Engineer" and draftsman in Braxton Bragg's Confederate Army of Tennessee. The map shows very few terrain features—Lookout Mountain, for example, is not shown—but cavalry outposts are marked with green boxes, and plenty of other cultural and physical details are noted. Springs are described as "good" or "constant"; some roads are marked "impassable for wagons." Mills, ferries, fords, blacksmith shops, wells, and other useful resources are located. An odd feature of Meister's original map is the red pencil outline and yellow surface of the principal roads, a unique embellishment.

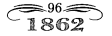

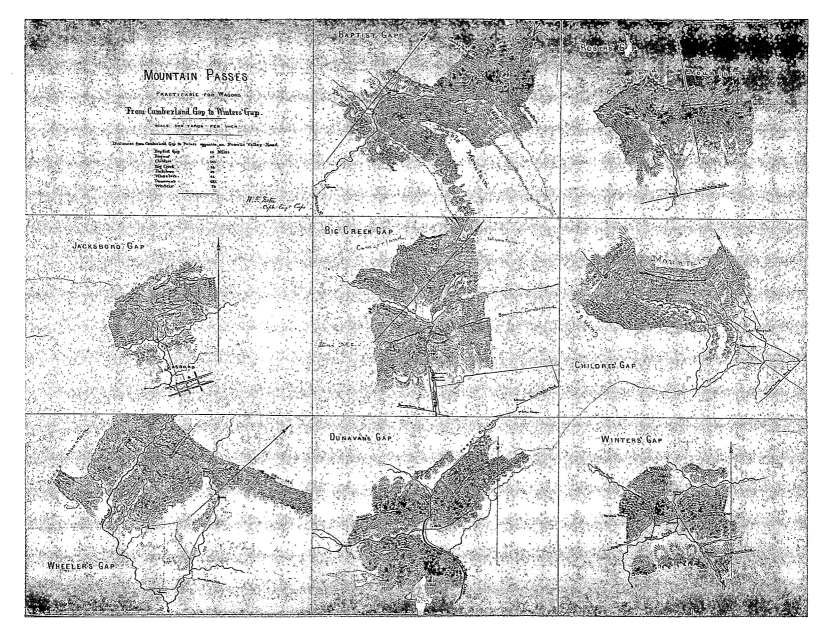

Cumberland Mountain Passes Practicable for Wagons

These map vignettes were meticulously drawn by Confederate engineer W. F. Foster. Knowledge of the practicability—a word in frequent use during the Civil War—of a mountain gap was critical in planning campaigns and maneuvers. A gap was relatively easy to defend and commensurately difficult to attack. Gaps could be of great strategic and tactical importance, as the care taken to get and record this information attests to. A map, clearly compiled from these sketches, appears in the *Official Records Atlas* (plate 95, no. 3). The Official Records Atlas map is heavily annotated by Foster with notes on the practicability of the gaps. In describing one gap, Foster writes: "This road is very rough in places and full of rocks. Four mules could pull 1000 lbs. over it."

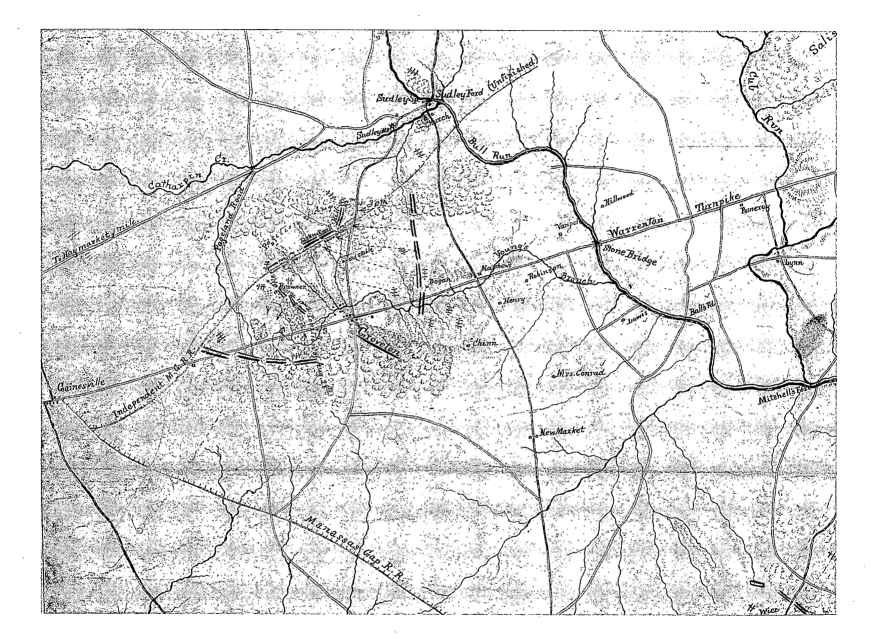

Detail Showing Engagement at Groveton or Brawner's Farm from Jed Hotchkiss's Sketch Showing Positions of Second Corps, Army of Northern Virginia, August 26 to September 2, 1862

A very detailed and accurate sketch of the vicinity of Groveton, where the opening fight of the Battle of Second Manassas took place in the early evening of August 28, 1862. Union troops moving eastward on the Warrenton Turnpike got into one of the sturdiest skirmishes of the war when Confederate general Stonewall Jackson opened fire on them from positions north of the turnpike, around the Brawner farmhouse and along the Independent Manassas Gap Railroad, or unfinished railroad, as it was called. Behind the Brawner house, the unfinished railroad can still be seen today to end in a squared-off dug gap exactly as Hotchkiss depicts it. A lithographed version of the overall map appears in the *Official Records Atlas* (plate 111, no. 1).

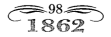

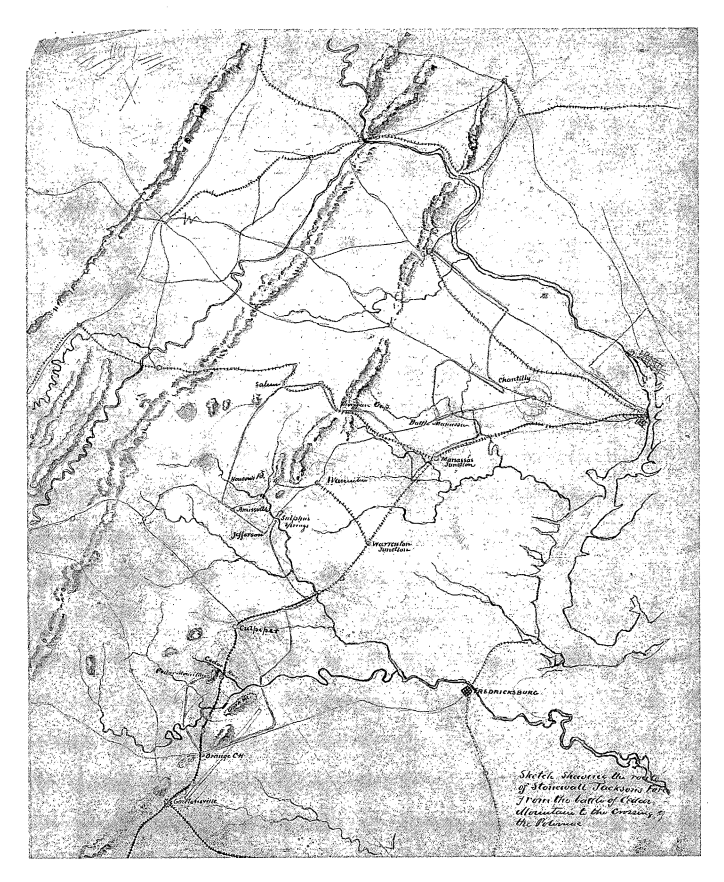

Stonewall Jackson's Second Manassas Campaign

This uncompleted map by John S. Clark reflects the operations of Stonewall Jackson through August and early September 1862. Civil War historian Douglas Southall Freeman describes this period as a time when Jackson "comes to life" after his catatonic performance on the Virginia Peninsula in June 1862. In fact, of the four separate actions included on this map, Stonewall Jackson turned in somewhat marginal performances on three of them: Cedar Mountain, Brawner's Farm (Second Manassas), and Chantilly. The map, on tracing paper, is in color pencil and India ink, with terrain represented with pencil shading.

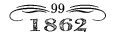

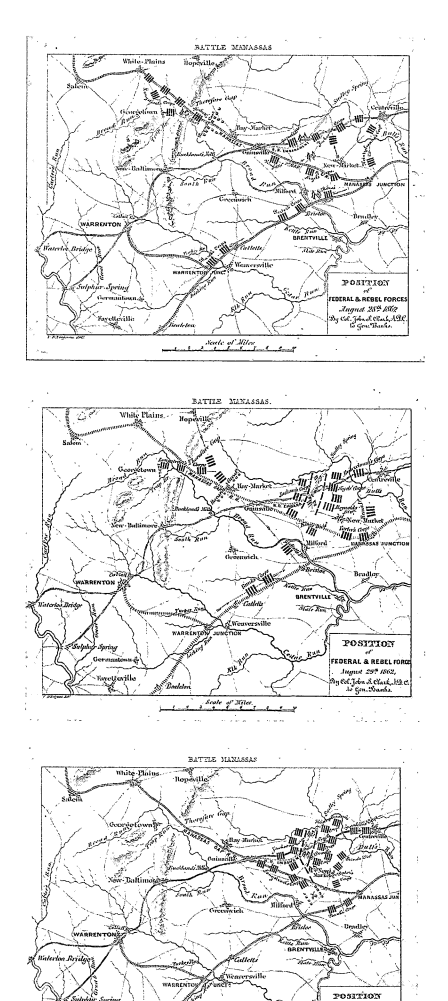

Second Manassas, August 28, 1862

Knowledge of the deployment of Confederate troops shown on this map would probably not have helped Union general John Pope. He drew consistently erroneous conclusions about the whereabouts of the various Rebel units, and even when he knew where they were, he completely misunderstood their intentions. A former topographical engineer, Pope always knew where he was. However, his maps failed to inform him where Confederate generals Robert E. Lee, James Longstreet, and Stonewall Jackson were and what they were doing. On this day, the well-fought battle of Brawner's Farm marked the beginning of the Second Manassas battle. The map is in watercolor and color pencil.

Second Manassas, August 29, 1862

Fitz-John Porter's inaction on this day would bring an end to his Civil War career. It was John S. Clark who first discerned and reported Confederate general Stonewall Jackson's flank march toward Thoroughfare Gap, which was the principal maneuver of the campaign and led to the second great Rebel victory at Manassas. The map is in watercolor and color pencil.

Second Manassas, August 30, 1862

A postbattle theater map of the Second Manassas campaign, this very finished production was prepared by Colonel John S. Clark. A draftsman, or delineator, as he styled himself, F. D'Avignon, copied the map. Although not the most accurate map—Clark was more of a cartographer and stylist than an engineer—it gives a good sense of the overall troop positions. In an odd reversal, Union troops are shown in red, Confederate troops in blue. The town of Hopeville should be Hopewell. The map is in ink and watercolor. Pencil tracings are evident.

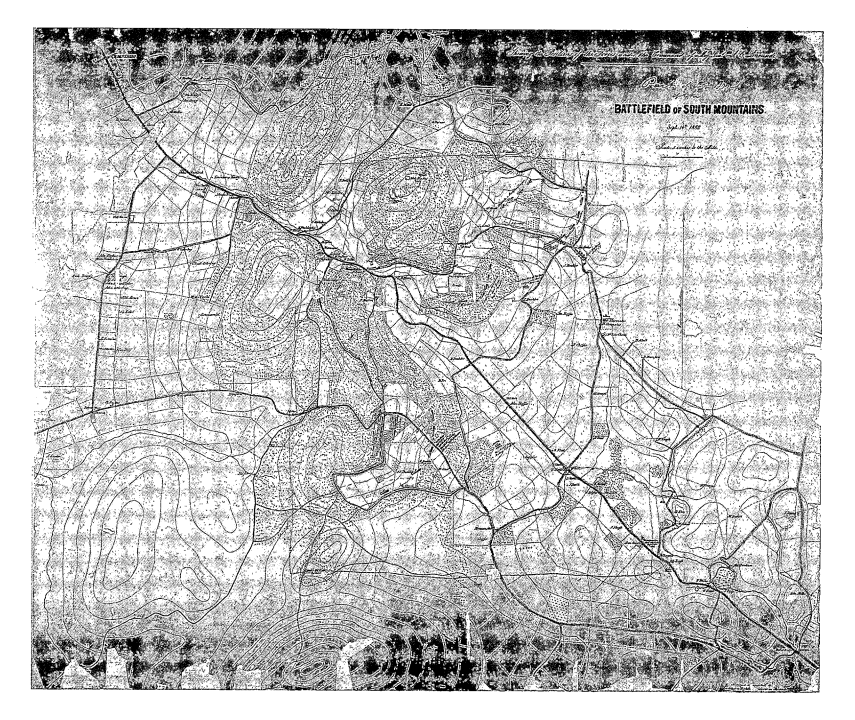

Arthur de Witzleben and Theodor von Kamecke's Map of South Mountain, Maryland, September 14, 1862

This map shows Turner's Gap in South Mountain where the National Road passes through. It is a very precise production, with contour lines instead of hachure marks and a thorough presentation of every fence, farm, and field, including Wise's field, death site of two generals: Confederate Samuel Garland and Federal Jesse Reno. The map is reproduced in a lithographed version, credited to the Bureau of Topographical Engineers, in the *Official Records Atlas* (plate 27, no. 3). It is one of the finest military maps from that early phase of the war. De Witzleben and von Kamecke were apparently draftsmen with the Bureau of Topographical Engineers in the Winder Building in Washington. Singly or jointly, they are credited with drawing, under the directions of others, a number of excellent maps throughout the course of the war.

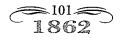

Washington Roebling et al.: Part of the Battle of the Antietam Fought September 16–17, 1862

This large-scale map, drawn on tracing cloth, is packed with contemporaneous information about one of the Civil War's great battle-fields. Rebel general James Longstreet said that the loss of this battle "sprung" the keystone of the arch upon which the Confederate cause rested.

Because the map was prepared so soon after the battle, it contains a number of intriguing perspectives. The famous sunken road known to history as "Bloody Lane" is here labeled "Hallow Road." The Dunker Church, now regarded as the symbol of the battle, is simply an unremarked building, but the cornfield on the Miller farm is carefully rendered, cornstalks and all. Fields are identified as corn, pastures, or plowed. Orchards, woods, and fence lines are exhaustively drawn. The configuration of roads and watercourses is not exact, but the overall layout of the field is quite accurate. The map is limited to the part of the ground associated with the Union First Corps, so the Rohrbach or Burnside's Bridge area is not shown. A remarkable wartime map.

Macomb and Paine's Map of Northeastern Virginia, November 13, 1862

Colonel John N. Macomb, seldom mentioned in the Civil War literature, was chief topographical engineer of the Union Army of the Potomac and head of its Office for Surveys and Maps from late 1861 until the spring of 1863. Colonel Macomb was more of a deskbound topographer, and while Captain William Henry Paine was performing reconnaissance in the field, Macomb was more likely to be evaluating information coming in from the field and searching libraries for map data. It is thought that Macomb was one of the first Civil War mapmakers to use information from the Census Bureau to create statistical maps that provided not only cultural and physical details but the population density of a given area—the sort of information Union general William Tecumseh Sherman used in November 1864 to plan the March to the Sea. This map shows the site of the Battle of Cedar Mountain (labeled here Cedar Run Mountain) and also depicts Clark's Mountain. A Confederate signal station there gave an extensive view of the surrounding countryside, which embraced some of the war's most noted rivers, towns, and fields; it was frequently used by Confederate generals Robert E. Lee and Stonewall Jackson.

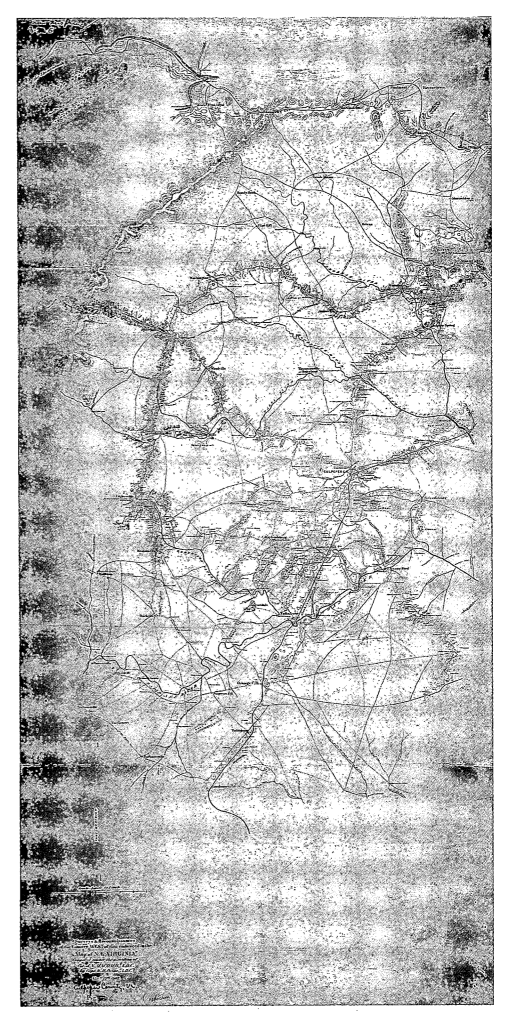

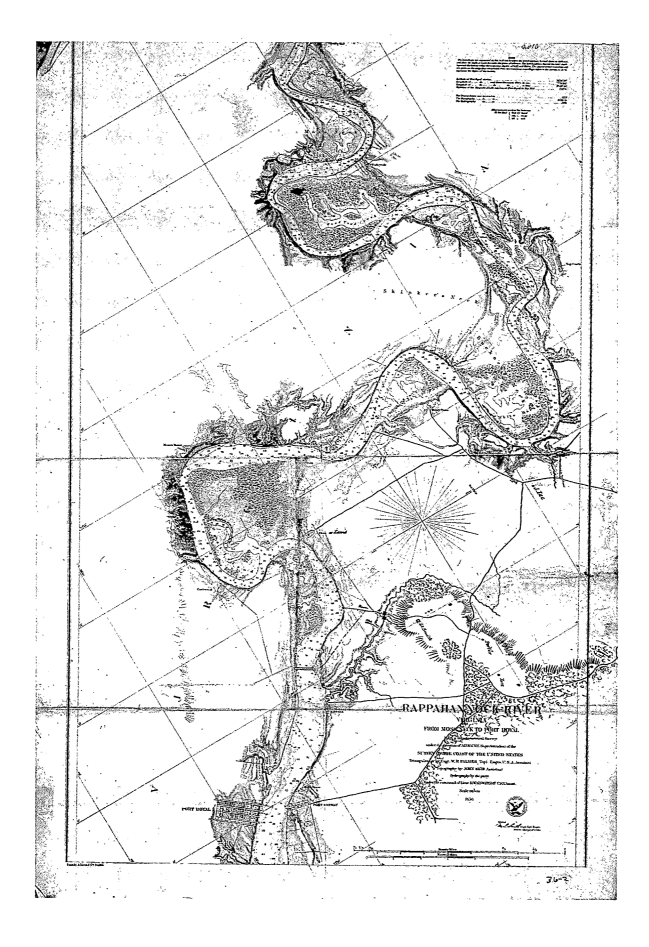

United States Coast Survey Map of the Rappahannock River, Virginia, 1856 (Additions, c. 1862)

The United States Coast Survey, a civilian agency of the United States government at the time of the Civil War, produced this map of the Rappahannock River downstream and southeast of Fredericksburg, Virginia. A thoroughly accurate map with a few wartime additions, including red ink on the roads to distinguish them from the watercourses.

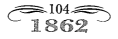

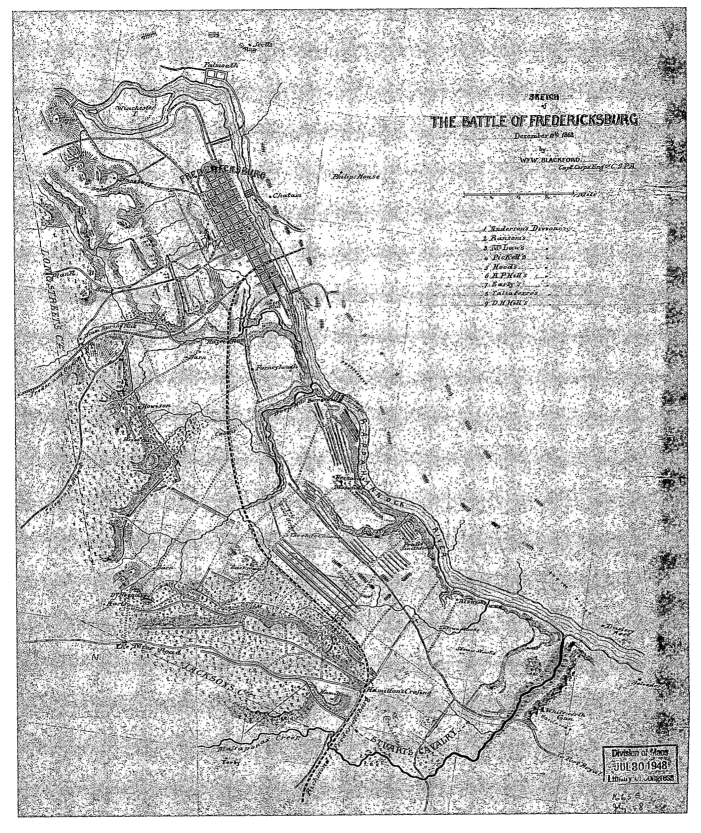

W. W. Blackford's Sketch of the Battle of Fredericksburg, Virginia, 1862

Captain W. W. Blackford, Confederate Corps of Engineers and cavalry general Jeb Stuart's topographical engineer, rode over the Fredericksburg battlefield with his brother, Charles M. Blackford, on January 10, 1863. His brother wrote home, "William knows more of the position of the different troops than any living man, not excepting General Lee, for he has surveyed the ground and gotten the position of each regiment on his map from the brigadier under whom it served." Confederate lines are in red, Union lines in blue. Note the triangular-shaped woods bordering the Richmond and Fredericksburg Railroad in front of Stonewall Jackson's corps. The tangled woods were deemed impassable and were left largely undefended. Union troops under General George Gordon Meade broke through here briefly but were not reinforced. This sketch formed the basis for the Fredericksburg section of Jed Hotchkiss's Chancellorsville map, reproduced on page 118.

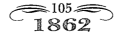

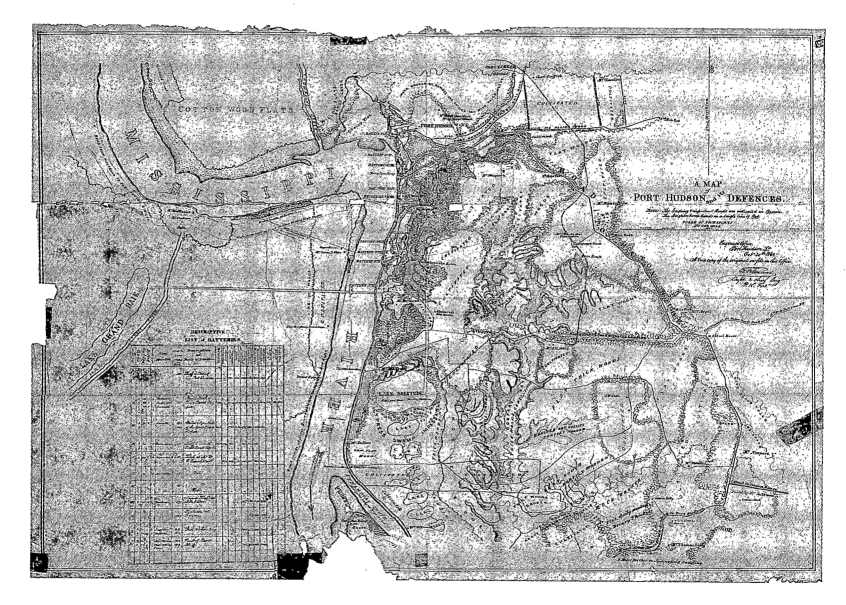

Léon J. Frémaux's Map of Port Hudson, Louisiana, and Its Defenses, 1862

Before the war, Léon J. Frémaux, a Frenchman by birth and largely educated there, was an assistant state engineer in Louisiana, where he was frequently engaged in surveying, mapping, and clearing watercourses, including the Mississippi River. His intimate knowledge of the region was utilized by Confederate general and former United States vice president John C. Breckinridge to select and fortify a point on the Mississippi where the Confederates could sink or intercept Union vessels steaming between New Orleans and Vicksburg, Mississippi. Frémaux was soon superintending the work of Port Hudson, Louisiana, a river port with a population of about three hundred in 1860 that shipped baled cotton and hogsheads of sugar. Frémaux contracted dysentery and was not at the fortifications to witness the lengthiest actual siege ever conducted in the United States, ending in the surrender of Port Hudson to Union general Nathaniel P. Banks on July 9, 1863. The fall of Vicksburg five days earlier had made Port Hudson a strategic nullity in any event. Beautifully executed, meticulously lettered, thoroughly detailed, and replete with range and other engineering data for the batteries, this is a quintessential military map by one of the war's great topographical engineers.

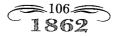

Drawer 125.

Sheet 18.

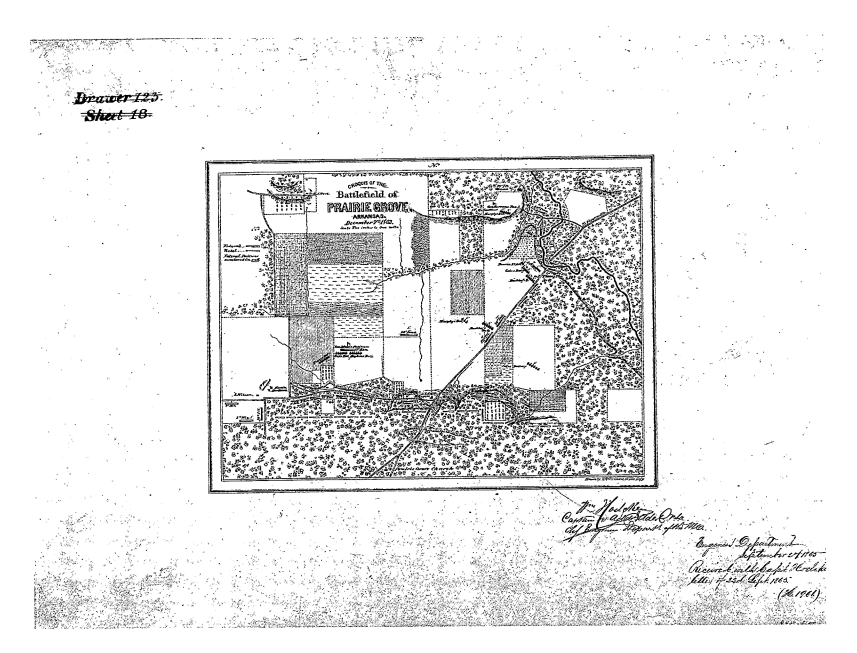

T. W. Williams's Croquis of the Battlefield of Prairie Grove, Arkansas, December 7, 1862

This map of the Prairie Grove battlefield is hardly the rough sketch that the term *croquis* implies. It is an evocative, beautifully detailed cartographic work of art. The square and rectangular fields reflect the planned divisions of the West, as contrasted with the irregular field patterns of the East. T. W. Williams of the Fifteenth Illinois Infantry later served as engineer officer of the Union Department and Army of the Tennessee. Prairie Grove, southwest of Fayetteville, Arkansas, was a victory for Union troops under Generals James G. Blunt and Francis J. Herron over Confederate generals Thomas C. Hindman and John S. Marmaduke.

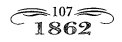

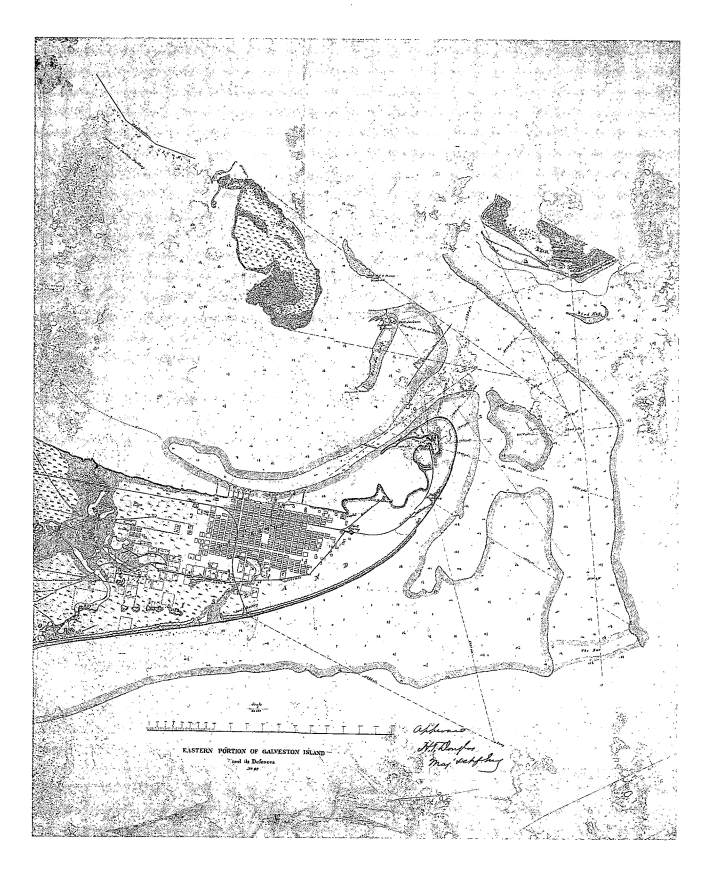

Detail of the Eastern Portion of Galveston Island, Texas

Finely drawn Confederate map, ink on linen. The topographical engineer responsible for preparing this map, or chart, is not known. It was approved by H. T. Douglas, major and chief, Engineer Department. The map very closely resembles, but is not the original of, the lithographed Galveston in the *Official Records Atlas* (plate 38, no. 1). Douglas commanded the First Battalion Engineer Troops (Mounted), a unique Confederate Trans-Mississippi organization. Two companies of the battalion were stationed on Galveston Island and were likely responsible for this map.

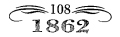

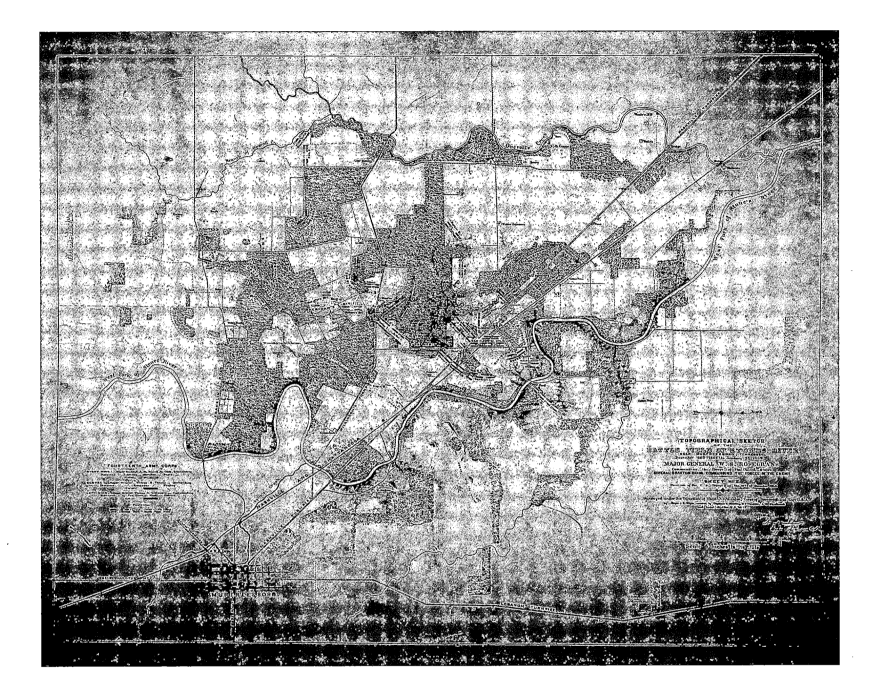

Nathaniel Michler's Topographical Sketch of the Battlefield of Stones River, Murfreesboro, Tennessee, 1863

Union general William S. Rosecrans, like William Tecumseh Sherman, was a great general on the march and maneuvering. Almost of necessity, he was reliant on good maps and worked hard and sensibly to build an effective mapping establishment in his army. Rosecrans, also like Sherman, was not a particularly good battlefield commander. Stones River was his important victory. Afterward, Abraham Lincoln wrote, "I can never forget whilst I remember anything, that about the end of last year and beginning of this, you gave us a hard earned victory, which, had there been a defeat instead, the nation could scarcely have lived over." Major John E. Weyss and Captain Nathaniel Michler produced a number of postwar maps of the major eastern battlefields. On this manuscript map, north is off by forty-five degrees. A detail of this map, not credited to Michler and in a slightly different version, lithographed, appears in the *Official Records Atlas* (plate 30, no. 1).

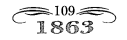

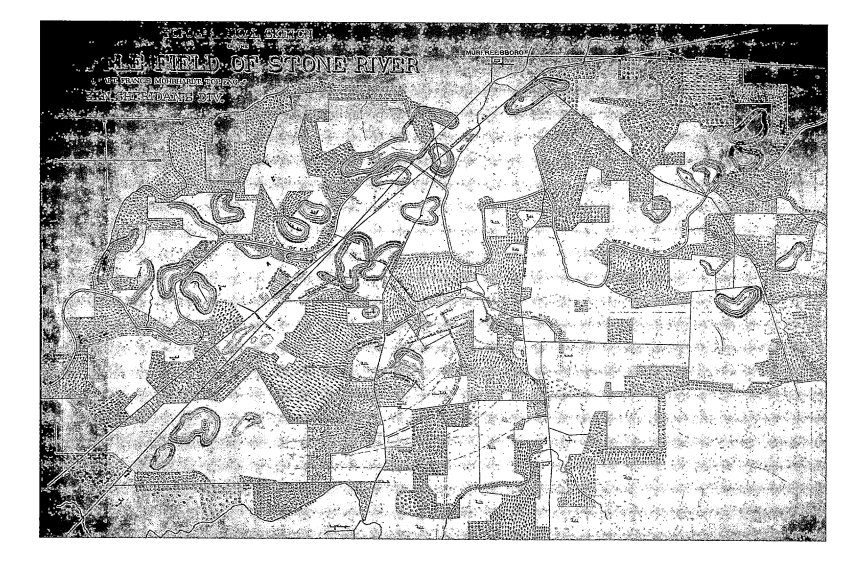

Mohrhardt's Topographical Sketch of the Battlefield of Stones River, 1863

This manuscript map by the topographical engineer of Union general Philip Sheridan, another western general who was topographically literate, is oriented from the Federal perspective with north off by forty-five degrees. More stylized than Michler/Weyss's map of Stones River, it is also more detailed and more encompassing.

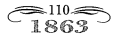

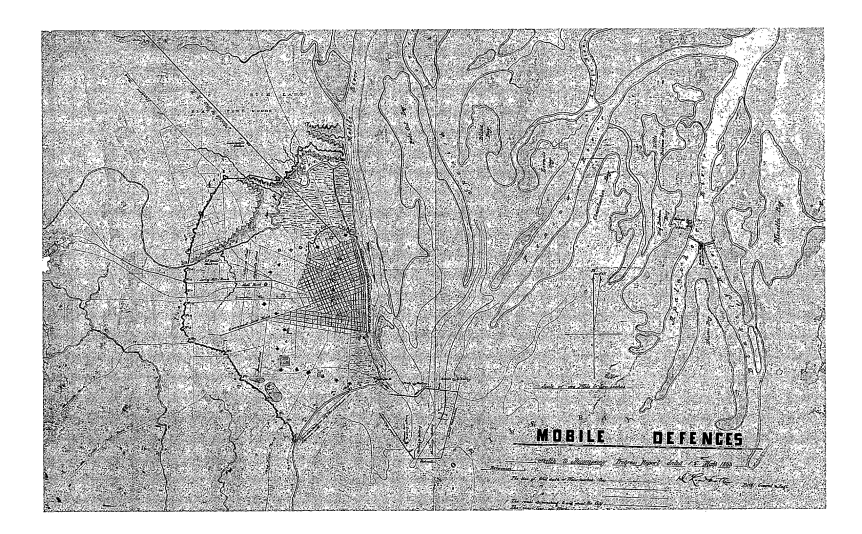

Sketch of Mobile Defenses to Accompany Progress Report, March 15, 1863

Danville Leadbetter, brigadier general and engineer, was the Confederate chief of engineers for Alabama and in charge of the defenses of Mobile. This was the same duty he had performed in antebellum days, after his graduation from West Point, as an engineer in the United States Army. There are some unusual features in this map. The use of blue, green, and yellow watercolors to graphically differentiate depth is one, and the employment of pencil-shaded hachure marks is another effective device to present topographical information in a readily understandable format. An interesting touch that underscores the steady hand and concentration of these wartime mapmakers is the error in the date of the map, that is, March, where the letter *c* is inserted, as indicated, by an arrow. None of the extensive, minute, and detailed work on this map could be erased and corrected.

Leadbetter was a fine mapmaker, but his engineering resumé included disasters such as the Confederate line on Missionary Ridge, Tennessee, and his intervention at Knoxville was not helpful to General James Longstreet's failed November 29, 1863, assault there.

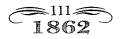

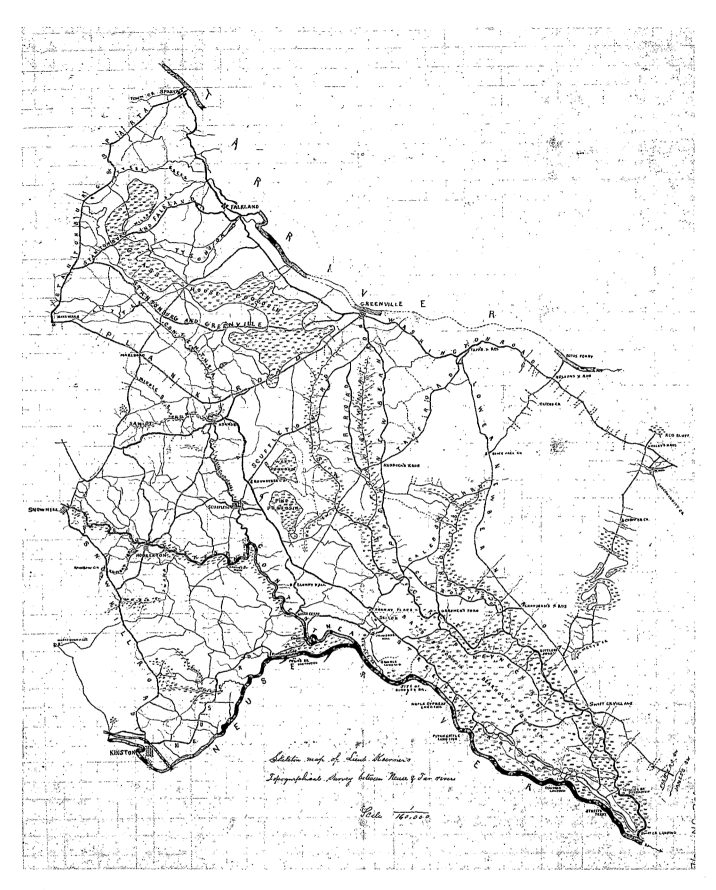

Skeleton Map of P. W. O. Koerner's Topographical Survey between Neuse and Tar Rivers, North Carolina

A skeleton map usually consists of a framework of fixed features, relatively easy to map accurately, such as railroad lines, rivers, and principal roads. Subsequent surveys and reconnaissance, other maps, and interrogations develop additional details that are added to fill out the skeleton map. Lieutenant P. W. O. Koerner's topographical map, drawn on tracing paper over a penciled grid in blue, red, and black ink, was probably prepared for Confederate operations under Generals James Longstreet and D. H. Hill in March and April 1863. Lieutenant Koerner of the Confederate Bureau of Engineers was a close associate of Jed Hotchkiss in the last months of 1863 and through much of 1864.

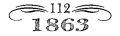

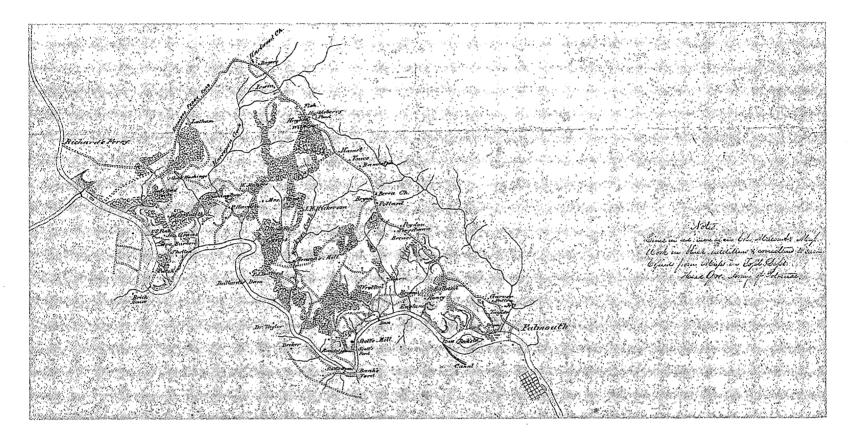

Sketch Map, Richards Ferry to Falmouth, Virginia, February 1863

Drawn on heavy tracing paper, this map is based on existing information with new details added. It was most likely incorporated into G. K. Warren's field operations map of Fredericksburg and Chancellorsville in the *Official Records Atlas* (plate 39, no. 3).

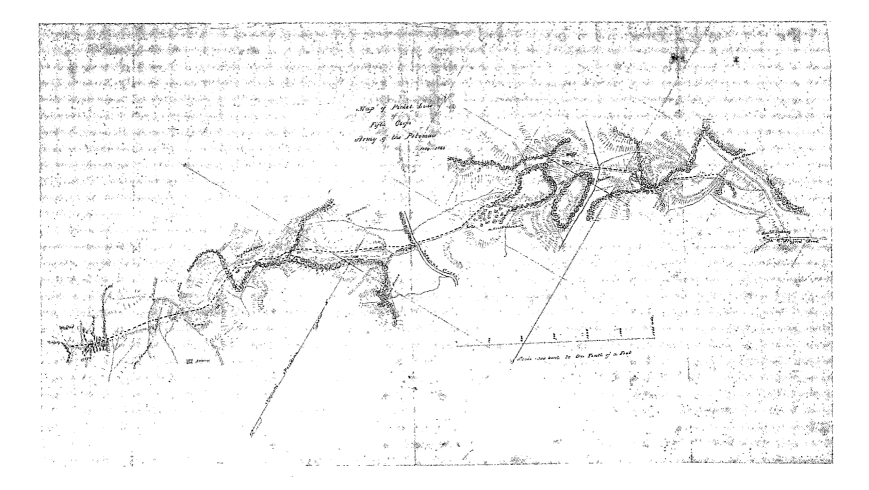

Map of the Picket Line of the Fifth Corps, Union Army of the Potomac

The picket line was an edgy, dangerous place, and no officer wanted to stumble around near it, especially in the dark. This very large scale, highly detailed map would help minimize the difficulties and dangers of checking the line. Drawn in ink on cloth, it was easily folded and hence convenient to use.

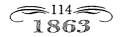

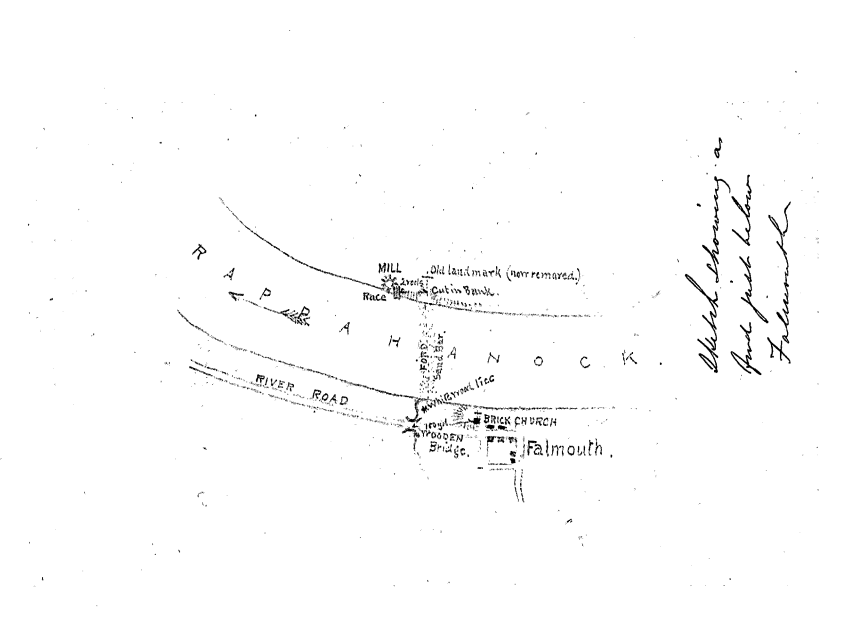

Falmouth, Virginia/Whitewood Tree

A map drawn from the perspective of the Union forces on the north bank of the Rappahannock River looking across at the Confederate lines just west of Fredericksburg, Virginia. The arrow indicates the direction of the river current. This minimal sketch in pencil and blue pencil provides numerous cultural and physical details: a whitewood tree marks the location of the sandbar ford; hachure marks realistically portray the cut in the bank that gives access to the ford; and a wooden bridge carries the River Road into Falmouth.

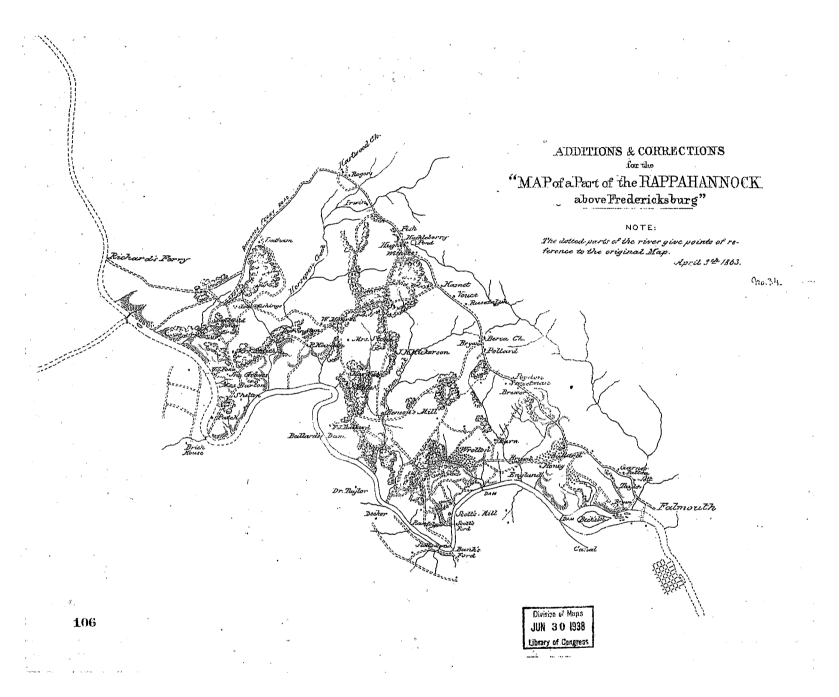

Division of Maps
JUN 30 1938
Library of Congress

106

William Henry Paine: Additions and Corrections—the Rappahannock River above Fredericksburg, Virginia, April 3, 1863

The Civil War catchphrase "All Quiet along the Potomac" might more appropriately have been "All Quiet along the Rappahannock." During much of the war in the East, the Union Army of the Potomac occupied the north side of the Rappahannock near Falmouth, while the Army of Northern Virginia held the opposite bank at Fredericksburg. U.S. topographical engineer William Henry Paine added new information and corrections to his hastily drawn, unadorned map of this section of the river. Within three weeks to the day, across this mapped ground, the Union commander, General Joseph Hooker, stole a march on Robert E. Lee. Hooker then crossed farther upstream at Kelly's Ford and came down on the flank of Lee's army, closer to Richmond than Lee was. Paine's plain, businesslike maps were accurate and timely productions.

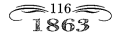

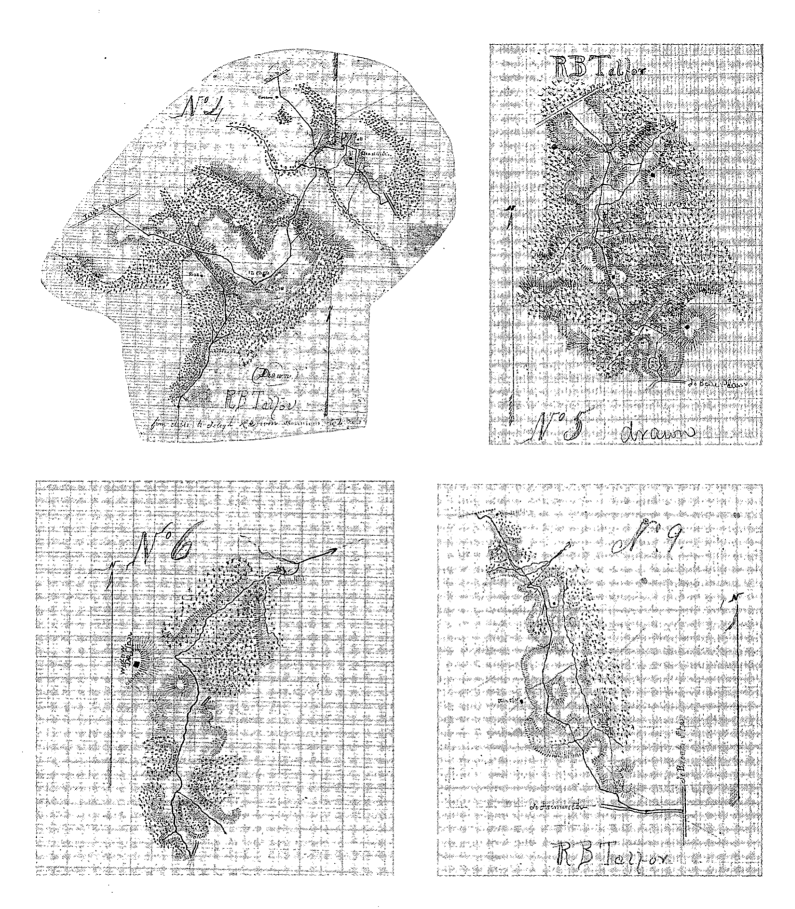

R. B. Talfor's Sketch Maps

The four small, oddly shaped sketch maps (approximately 4 x 6") of the relatively unrecorded area northeast of Falmouth, Virginia, are the work of civilian topographical engineer R. B. Talfor. An excellent draftsman, Talfor was a careless letterer. Such identifying names as Widow Daffom and Catlet are almost indecipherable. The sketches exhibit an excellent use of hachure marks to depict elevations and ridgelines as well as a good representation of Virginia's difficult woodlands.

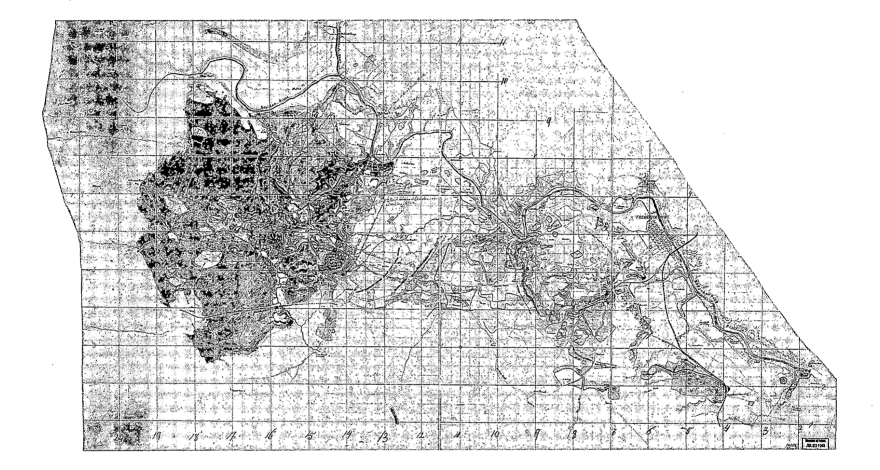

Jed Hotchkiss's Map of Chancellorsville, Virginia, Battlefield, 1863

Almost as soon as the smoke had cleared, and well before the battlefield had been cleaned up, Robert E. Lee set Jed Hotchkiss to work mapping this great but flawed Confederate victory. The loss of Stonewall Jackson was an irreplaceable one for the South, whereas the Union had lost nothing at Chancellorsville that could not be replaced easily. Hotchkiss spent most of three weeks arduously mapping the field. The map encompasses an area of approximately 250 square miles, much of it densely wooded, cut by ravines and watercourses. It was an emotionally and physically exhausting assignment for Hotchkiss. His revered chief was gone, and his best friend and fellow topographical engineer, Captain James Keith Boswell, was dead, killed in the same mistakenly fired Rebel volley that mortally wounded Jackson. A visually beautiful map, it contains major errors, especially in the courses of the streams and rivers. Upstream from Ely's Ford, for example, the Rapidan River flows from the northwest, not the southwest. The fact that Hotchkiss was preparing this map on the eve of the invasion of Pennsylvania indicates a strange sense of priorities on the part of General Lee. After all, the Chancellorsville battlefield was not going anywhere, while the Army of Northern Virginia certainly was; and into unfamiliar, unfriendly territory at that. There were assuredly more appropriate uses for Hotchkiss's talents at this critical juncture of Confederate affairs. A lithographed version of the map is in the *Official Records Atlas* (plate 41, no. 1).

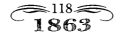

Lafayette McLaws's Map of the Battles of Fredericksburg and Chancellorsville, 1862, 1863

This very attractive manuscript map, watercolor and ink on tracing linen, is not credited or dated. The map is informative in terms of cultural details—tollgates, barns, mills, houses, and the like—but the courses of the Rapidan and Rappahannock rivers are badly oriented, so that the rest of the map is pulled far out of kilter. To give just one example: Fredericksburg is shown slightly northeast of Ely's Ford. In fact, it is southeast of the ford. The route of the Richmond, Fredericksburg, and Potomac Railroad is also incorrect. The line runs due south out of Fredericksburg. The way it swings to the southeast in this map again reflects the distorted representation of the Rappahannock River.

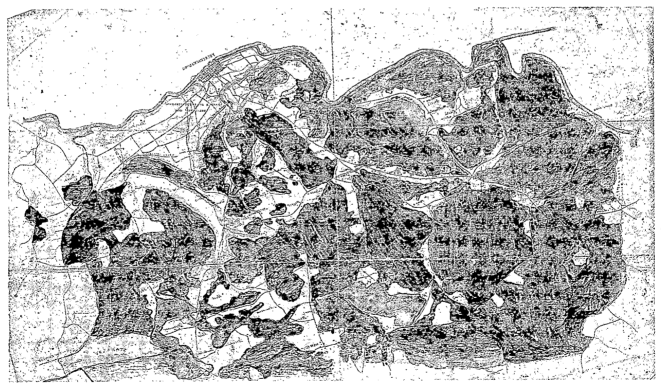

Reverse Side of Lafayette McLaws's Map of the Battles of Fredericksburg and Chancellorsville, 1862, 1863

The reverse side of this map reveals the technique used to create the delicate green tints, or washes, that made the vast woods on the map look so attractive and also transparent for legibility when it came to labeling the features of the map. Heavy, dark green was simply daubed on the reverse side of the tracing linen, to show faintly through to the image side of the map.

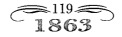

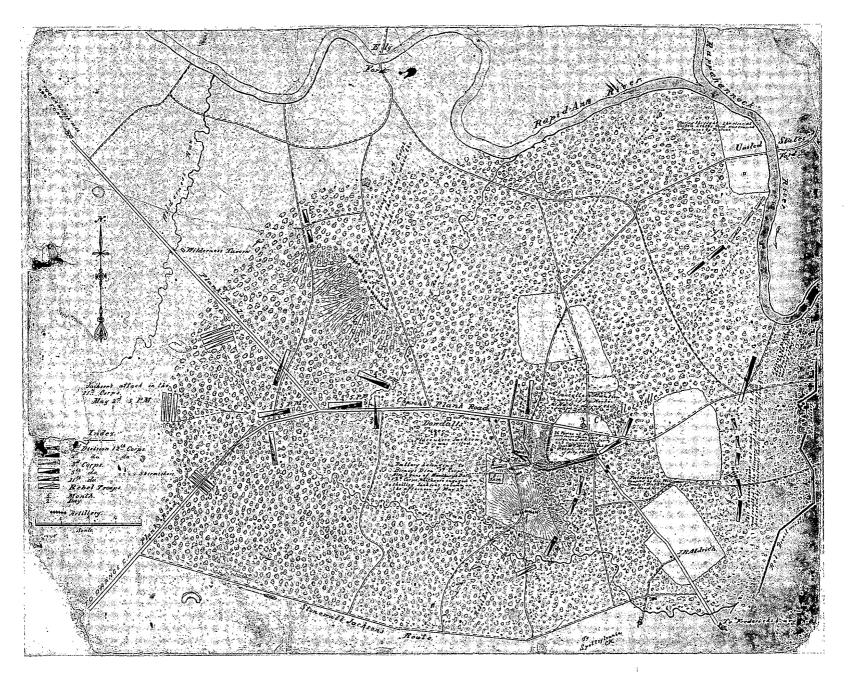

Lucius T. Stanley's Chancellorsville

Lucius T. Stanley's regiment, the 107th New York, was at Chancellorsville as part of Union major general Henry W. Slocum's Twelfth Corps. The map is not dated nor is it credited to Stanley, but it is undoubtedly his work. Very detailed, it is essentially accurate, though the course of the Rapidan (or Rapid-Ann, as Stanley has it) River is off substantially and many minor roads are missing. The map notes are interesting, particularly the one identifying "Place where Jackson was killed." Confederate general Stonewall Jackson, of course, was mortally wounded at Chancellorsville on May 2, 1863, dying eight days later. The site of Jackson's wounding is a subject of debate, but no one had ever suggested that it was south of Dowdall's Tavern. There are some fanciful touches to the map—for example, the hand pointing directions at the limit of the principal roads and the elaborate directional arrow.

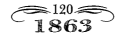

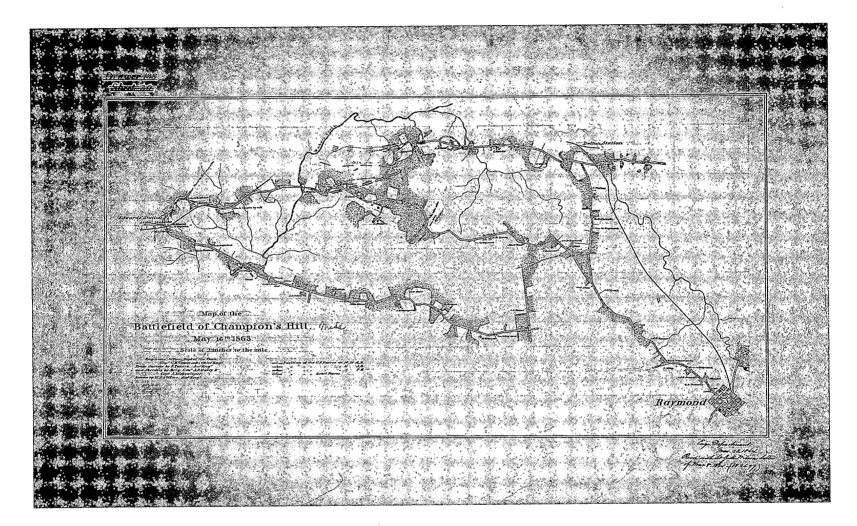

Cyrus Ballou Comstock's Map of Champion's Hill, Mississippi, May 16, 1863

There were fewer large-scale, highly detailed military maps drawn by the Union topographical engineers in the West than by their counterparts in the East. The Confederates in the West did not generally fight for every square inch of ground, as Robert E. Lee tended to do. Nor was the same ground fought over again and again, as was often the case in the East. This manuscript map gives an overview of a crucial field battle in the midst of Ulysses S. Grant's brilliant Vicksburg campaign. The woods, carefully delineated here, played a vital obstructive role, from the Federal point of view, in the results of the battle. The name Hickenlooper is misspelled on the map.

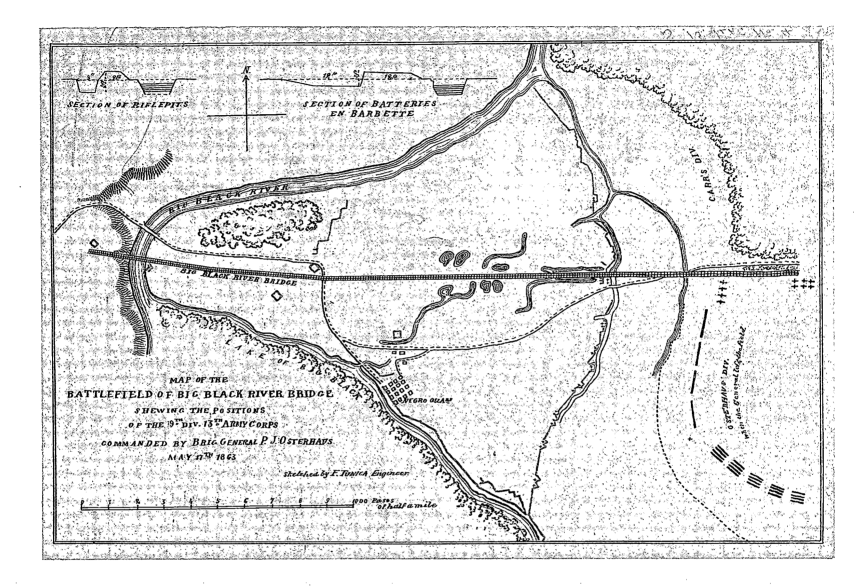

Francis Tunica's Map of the Battlefield of Big Black River Bridge, Mississippi, May 17, 1863

This sturdy, unembellished, competent map reflects Union general Ulysses S. Grant's command style. The Confederates, after their defeat and retreat at Champion's Hill, Mississippi, the day before, tried to halt or slow the Union advance by defending the crossing at Big Black River. The Confederates were routed, and by May 19, Union forces closed in around Vicksburg. Grant distributed his few engineer officers to the staffs of his various corps. Lieutenant Francis Tunica was one of them. Tunica sketched the routes of march as he himself proceeded along them. He copied them himself at night and forwarded them to corps and army headquarters. If Grant required additional information, Tunica went back to confer personally with his commander. He also prepared maps such as this one to record troop positions and ground-cover data for use in subsequent, more elaborate maps. It is interesting to examine the map of this engagement by Tunica in the *Official Records Atlas* (plate 37, no. 6).

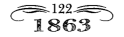

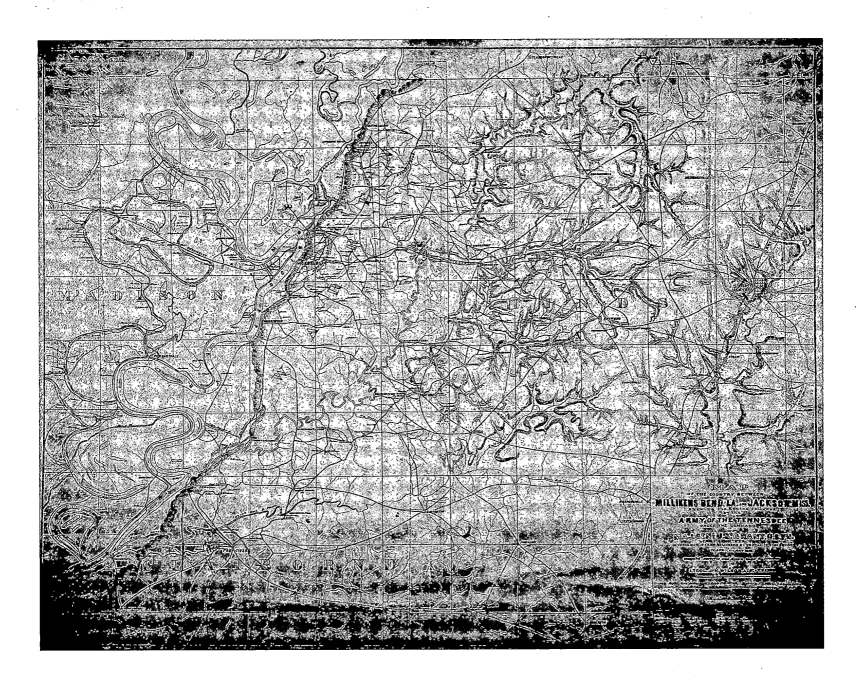

James H. Wilson's Vicksburg Campaign Map, Louisiana and Mississippi, April and May 1863

This manuscript map of arguably the greatest campaign of the Civil War provides a good sense of the lay of the land. Its creator, James H. Wilson, was soon to emerge as one of the finest Union cavalrymen in the climactic closing days of the war. Wilson complained that General Ulysses S. Grant did not display much interest in the business of topographical engineers, but the general had a way of getting what he needed out of what he had. He also was reputed to have an unerring sense of direction, a feel for terrain, and a photographic memory for graphic presentations of topography. The note section is an inventory of the cultural and physical details of the region.

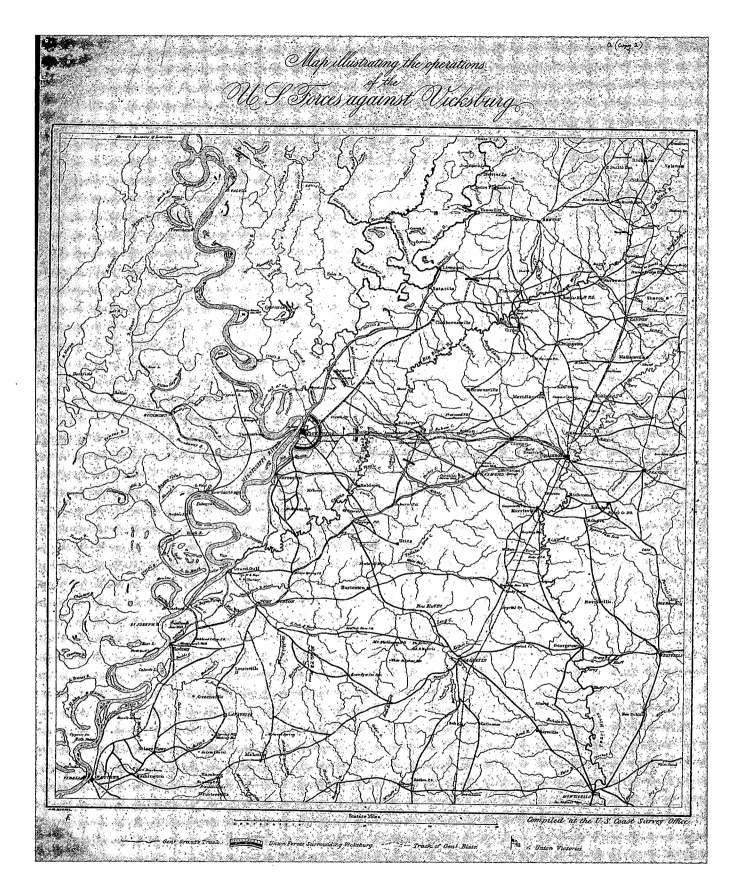

United States Coast Survey—Operations against Vicksburg, Mississippi, 1863

A map of Union general Ulysses S. Grant's innovative and brilliantly successful campaign against the Confederate stronghold on the Mississippi River. Though the map is not credited to them, two United States Coast Survey civilians, "Messrs. Fendall and Strausz," were on the scene surveying. A medical disability prevented Fendall from completing a survey of the approaches to Vicksburg prepared by Grant's troops.

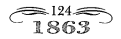

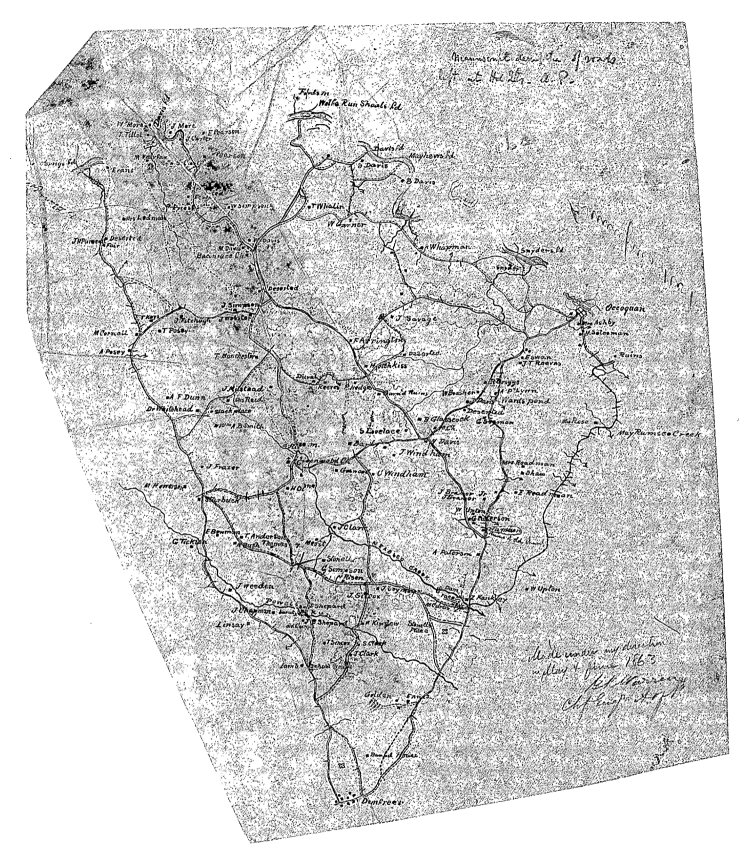

Manuscript Depiction of Roads, Left at Headquarters, Army of the Potomac, May and June 1863

The Telegraph Road leads from Dumfries (Dumfrees on this map), Virginia, to Occoquan, roughly inland from and paralleling the Potomac River. Two pieces of tracing paper pasted together make up this map, whose original data would be used to update and supplement the existing maps of the region. After Chancellorsville, Union mapping concentrated on the areas between Fredericksburg, Virginia, and Washington.

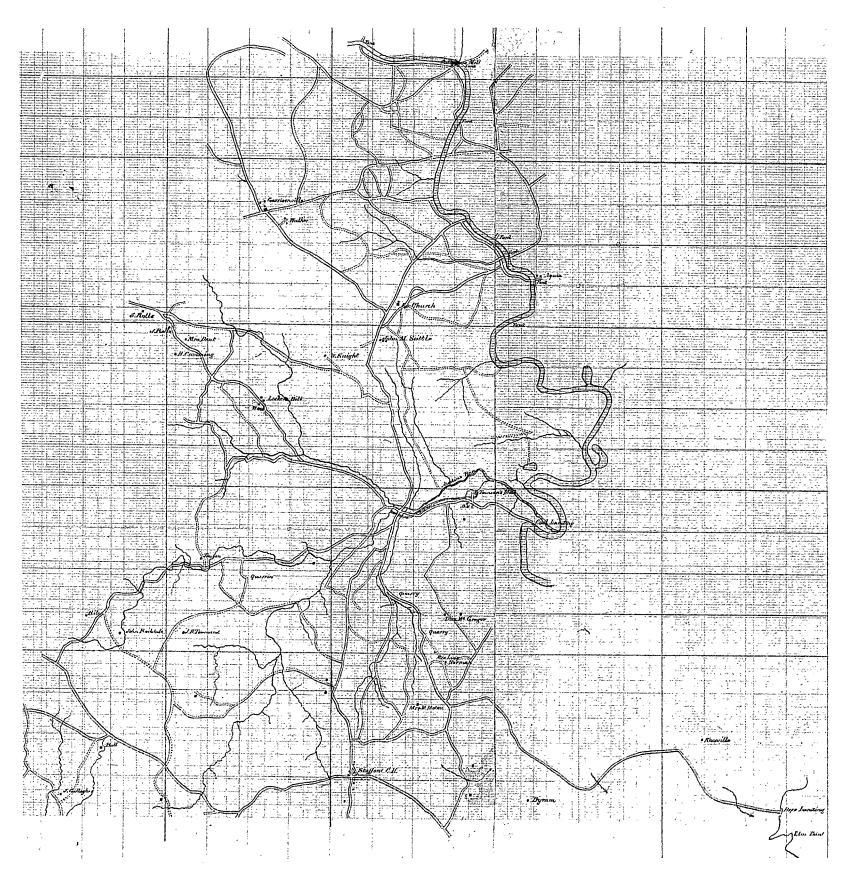

Field Sketch, Area of Stafford Courthouse, Virginia

Original map data. Two pieces of graph paper pasted together make up this meticulous field sketch of the area north of Stafford Courthouse. Note the careful rendition of Townson's Mill and the millrace through which water was diverted to power the mill. The precision of this sketch and the delicate blue watercoloring give some notion of the aesthetic sensibilities topographical engineers brought to their duties, especially considering that this was simply a study, not a final map.

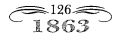

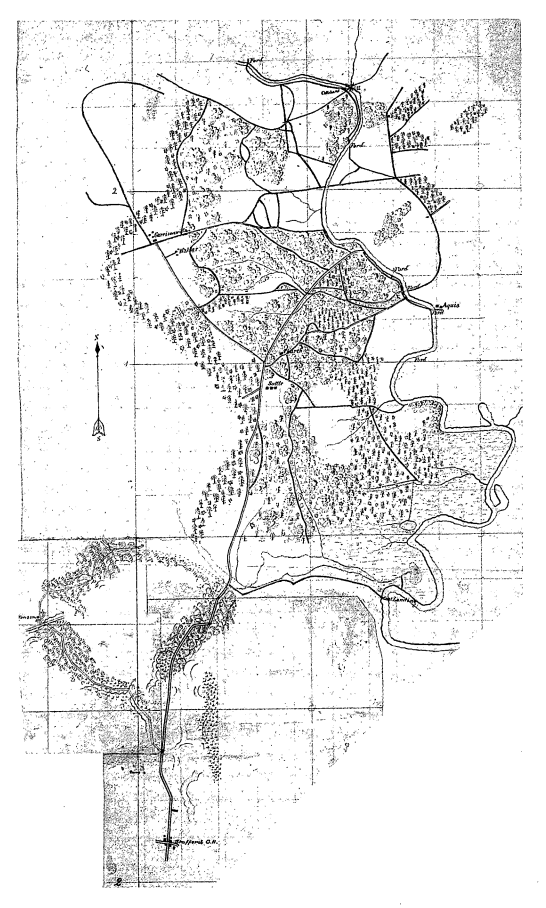

Field Sketch, North of Fredericksburg

A second map study of the region immediately north of Stafford Courthouse, Virginia, due north of Fredericksburg and just inland of Acquia Landing on the Potomac River. The topographical engineer distinguishes between pine trees and deciduous trees because pines form darker, distinctively shaped natural landmarks. The care used in the execution of the map contrasts with the ad hoc quality of the pasted-together watercolor paper.

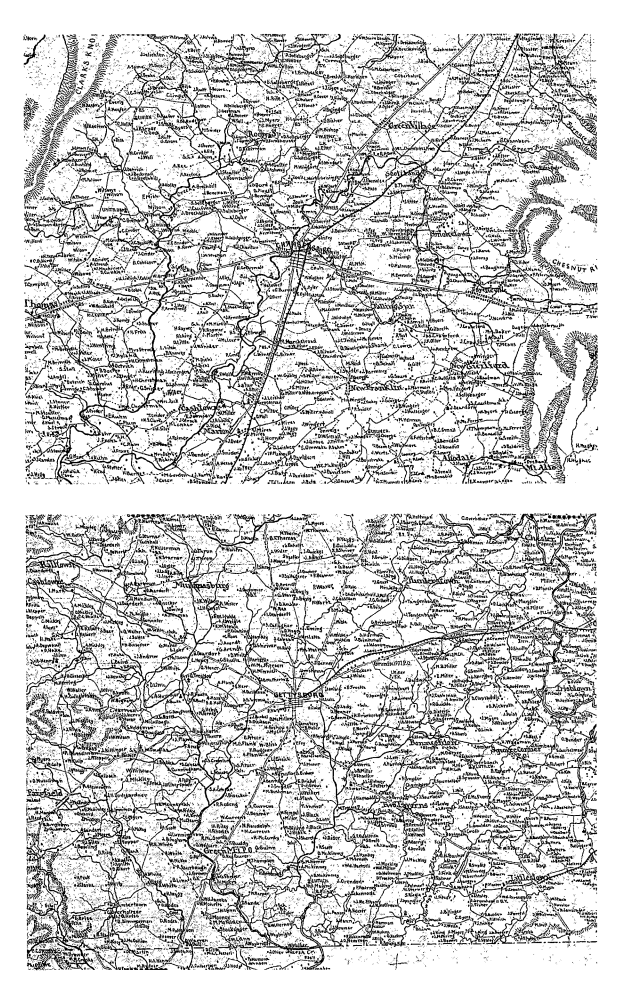

Detail of Jed Hotchkiss's Map of the Valley of Virginia Extended to Harrisburg, Pennsylvania, 1863— Chambersburg

A large 32 x 52" map of what became the Gettysburg campaign was begun on February 23, 1863, on the secret orders of Stonewall Jackson. This is one of the most significant maps produced during the Civil War, and it has never before been published. Its lithographic reproduction in the *Official Records Atlas* (plate 115A, no. 2) leaves off the names of the occupants of the farmhouses, one of the most striking features of the map. A pencil grid in square centimeters is visible. A 9"-wide strip of paper is pasted across the top of the overall map to complete the northern section of the map. This detail from the map shows Chambersburg, where Robert E. Lee waited uncertainly for news of the whereabouts of his own cavalry as well as of the Union army.

Detail of Jed Hotchkiss's Map of the Valley of Virginia Extended to Harrisburg, Pennsylvania, 1863— Gettysburg

By chance, the town of Gettysburg was the epicenter of Jed Hotchkiss's theater map. Based largely on the 1858 G. M. Hopkins/Converse map of Adams County, Pennsylvania, this detail of the map contains names that would reverberate in military history, including McPherson, Culp, Spangler, Trostle, and Bliss. The map concentrates on residents' names and on blacksmiths, mills, and landmarks, as well as watercourses and significant elevations. The map does not show, and Hotchkiss would have had no way of knowing about, Seminary Ridge, Cemetery Ridge, Little Round Top, and other terrain features that dominated the Gettysburg battlefield.

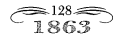

Detail of Jed Hotchkiss's Map of the Valley of Virginia Extended to Harrisburg, Pennsylvania, 1863— Harrisburg

This detail shows where the farthest confirmed penetration by armed, uniformed Confederate soldiers occurred during the Gettysburg campaign. Near Yellow Breeches Creek and just upstream from Milltown (due west of New Cumberland) is a house identified on the map as belonging to one W. H. Smith. William Henry Harrison Smith recounted that he was dozing on his back porch late in June when he was startled to find himself in "the rebels' hands. They said, 'Yank, what are you doing here?'" In Carlisle, Jed Hotchkiss drew a map of Adams County for his chief, General Richard S. Ewell.

Detail of Jed Hotchkiss's Map of the Valley of Virginia Extended to Harrisburg, Pennsylvania, 1863— Williamsport, Maryland.

Jed Hotchkiss, with General Richard S. Ewell and his Second Corps of the Army of Northern Virginia, advanced into Maryland at Shepherdstown and retreated at Williamsport and Falling Waters upstream on the Potomac River. Ewell's men left eight thousand pairs of shoes on the swollen Potomac River bottom as they retreated across the ford on July 14, 1863. The phrase "Slackwater Navigation" is written in the unpainted section of the Potomac River. It refers to slack current behind Dam 4. This esoteric bit of information was likely supplied to Hotchkiss by his fellow topographical engineer, S. Howell Brown. Brown had surveyed this region of western Virginia before the war and was intimately familiar with it.

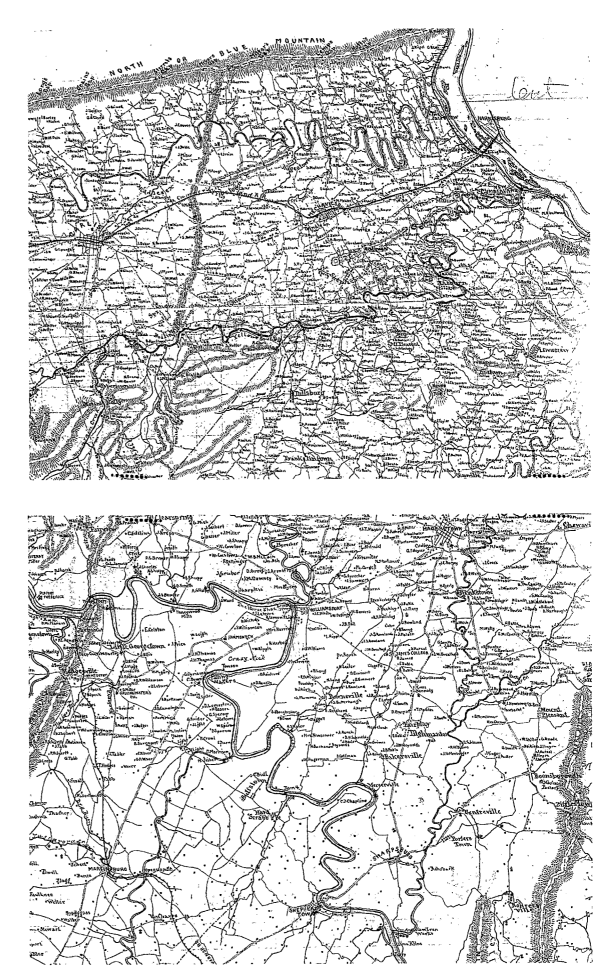

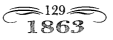

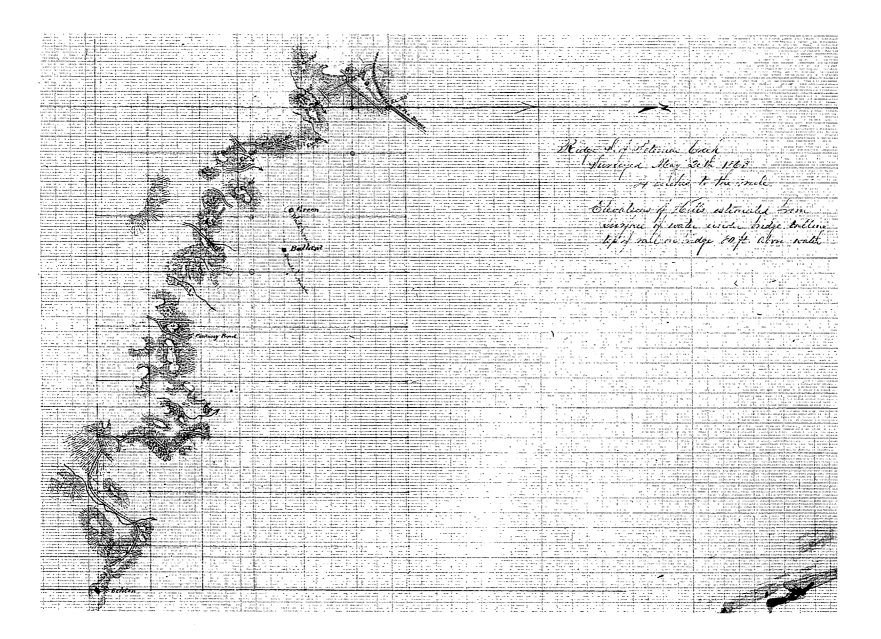

Sketch Map, Ridge South of Potomac Creek

Original map data prepared on graph paper. This is the work that was actually done in the field at the moment of passing the mapped details. General G. K. Warren ordered surveys of the region between Falmouth and Centreville, Virginia, after the Union defeat at Chancellorsville in early May 1863. Warren was caught flat-footed when the Confederates moved north and Union headquarters scrambled to procure even the most general published maps of Maryland and Pennsylvania.

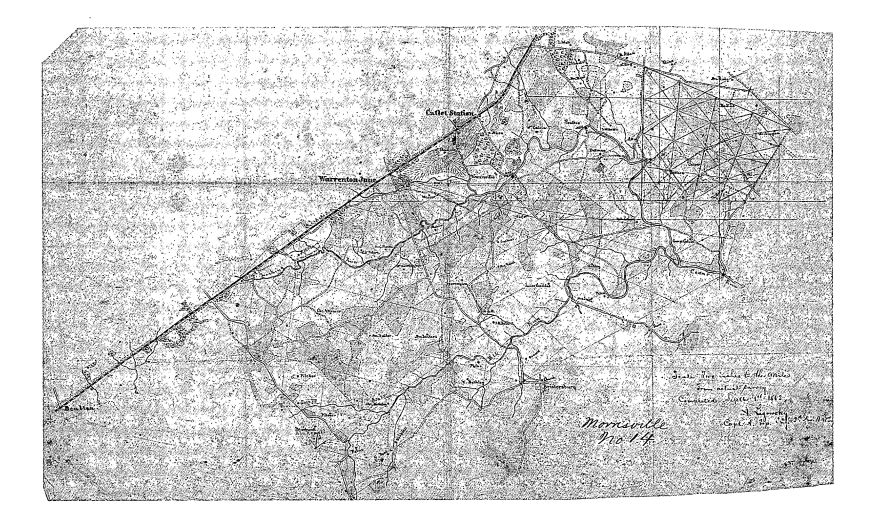

Morrisville, Virginia, No. 14

This carefully executed map was based on actual surveys (a survey line is clearly visible on the map). The names of residents have been added or corrected, and the ground-cover information is much more extensive than that shown on earlier maps of the same area. The mapmaker, Captain A. Ligowsky, was probably a civil engineer in antebellum days. His use of a distinctive alternating black and white line to represent the Orange and Alexandria Railroad through Bealton and Warrenton junction reflects the style of antebellum county maps.

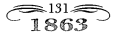

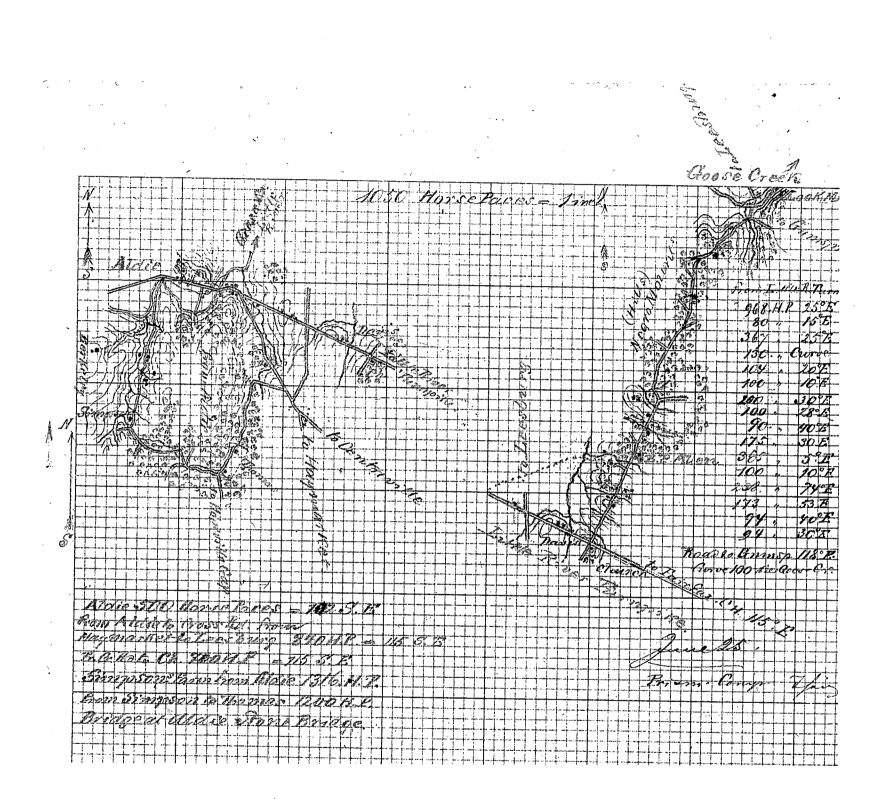

Little River Turnpike/Aldie and Negro Mountains, Virginia

A wonderfully informative piece of graph paper, almost certainly prepared en route, recording information at the moment of passing. The use of horse paces as a scale is combined with the use of a prismatic compass to provide an extremely detailed and exact depiction of twisting roads and complex terrain. The commentary on the map amounts to a map memoir.

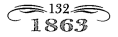

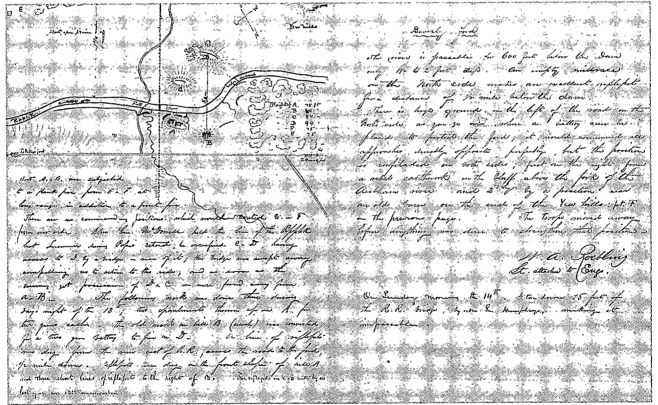

Washington Roebling's Map Memoir, June 1863

A map memoir by Washington Roebling, oriented with south at the top to reflect the Union's viewpoint of Rappahannock Station and Beverly Ford on Virginia's Rappahannock River. This sketch and memoir were prepared as the Gettysburg campaign was getting under way. Roebling provides information about the practicability of the fords as well as detailed tactical instructions for defending them.

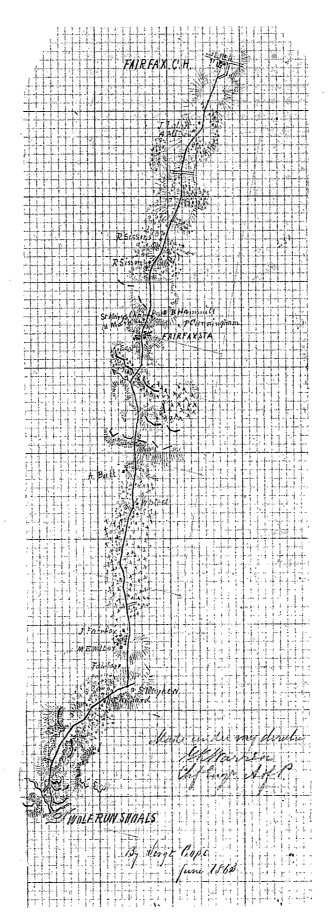

Route Map Prepared on the Eve of the Gettysburg Campaign, June 1863

A route map from Wolf Run Shoals north of Falmouth, Virginia, to Fairfax Courthouse in the vicinity of Centreville. This map is part of the reconnaissance ordered by the Army of the Potomac's chief topographical engineer, G. K. Warren, as the army awaited developments in the wake of the Chancellorsville defeat. The names of residents served as route markers along the way and were one of the tirelessly acquired, carefully denoted components of a military map.

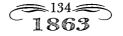

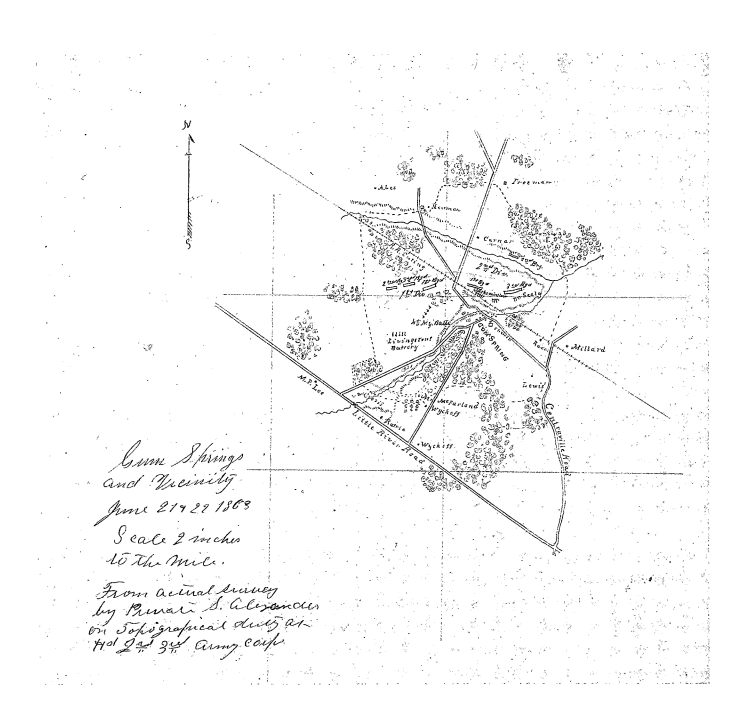

Little River Road/Turnpike at Gum Springs, Virginia

The Second, Third, and Fifth Corps of the Army of the Potomac followed this exact route through Gum Springs, Virginia, crossed the Potomac at nearby Edwards Ferry, and marched on to Gettysburg in mid-June 1863. This handsome map by Private S. Alexander, a soldier in the Corps of Topographical Engineers, records their temporary appearance on the important route between Manassas and Leesburg, Virginia.

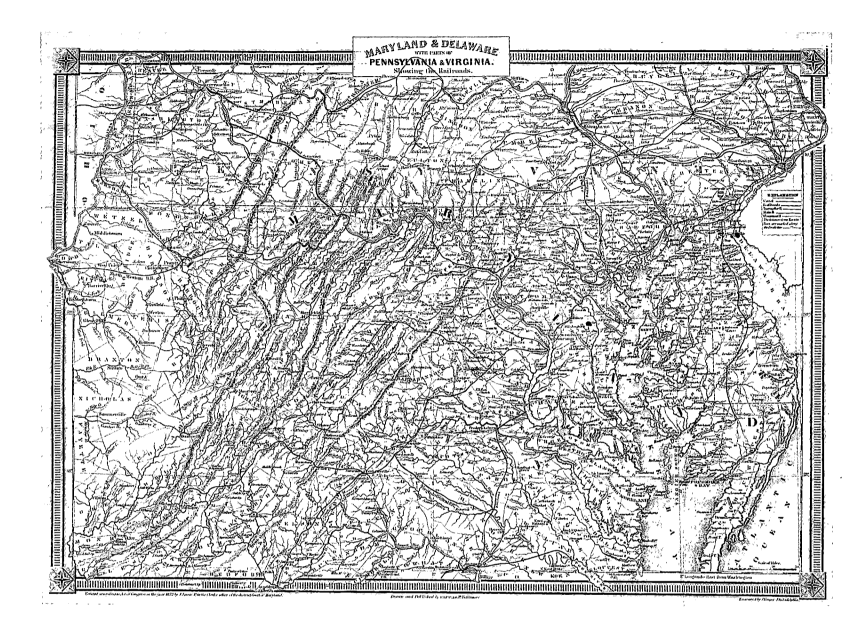

Fielding Lucas, Jr., Map: Maryland and Delaware, 1852

During the Gettysburg campaign, the Union Army of the Potomac scrambled to find maps of the new area of operations. A request was sent by the army's chief of staff, Major General Daniel Butterfield, for copies of this Lucas map of Maryland. The map was out of print, but General Robert Schenck in Baltimore assured Butterfield that the publisher had telegraphed to Philadelphia for the plates and would immediately print some. A cartographer and publisher, Lucas, a Virginia native, was on the Baltimore City Commission and was also a member of the board of directors of the Baltimore and Ohio Railroad—hence, the prominence of rail lines in this map. A small-scale map, it would be of marginal military use in tactical terms but would serve on a strategic level to coordinate movements of the different corps and divisions.

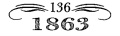

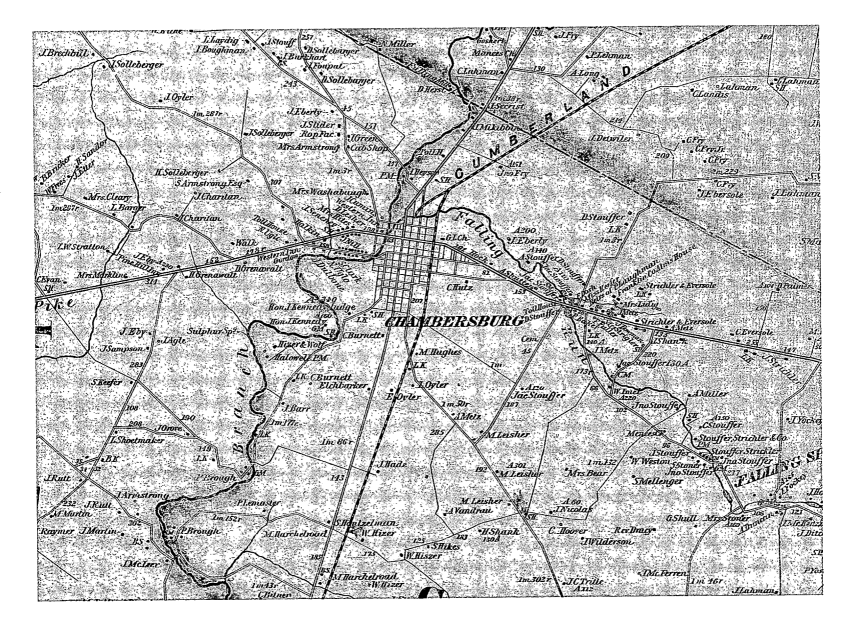

Detail of Map of Franklin County, Pennsylvania, 1858

This lithographed county wall map, surveyed by D. H. Davison and published by Riley and Hoffman, Greencastle, Pennsylvania, was used by Confederate mapmaker Jed Hotchkiss as one basis for the southern Pennsylvania section of the theater map he prepared for what became the Gettysburg campaign. A copy of this map, hanging on a sitting-room wall in a private home in Mercersburg, Pennsylvania, caught the attention of Captain W. W. Blackford, topographical engineer on the staff of Confederate cavalry chief Jeb Stuart. According to Blackford, "these maps had every road laid down and would be of the greatest service to us." Blackford had to push his way into the room and "coolly cut the map out of its rollers and put it in my haversack." To a Confederate levy on Greencastle for 2,000 pounds of lead, 1,000 pounds of leather, 100 pistols, and 12 boxes of tin, Hotchkiss added his mite: 2 maps of Franklin County.

Not everybody wanted a copy of this map. A Mercersburg constable carried one of these maps to a company of one hundred Union cavalrymen stationed at Cove Gap outside of town. He told the troopers that Rebel cavalry was approaching the town and proffered the map. Without glancing at the map, the cavalrymen all leaped into their saddles and fled over the mountain.

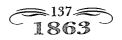

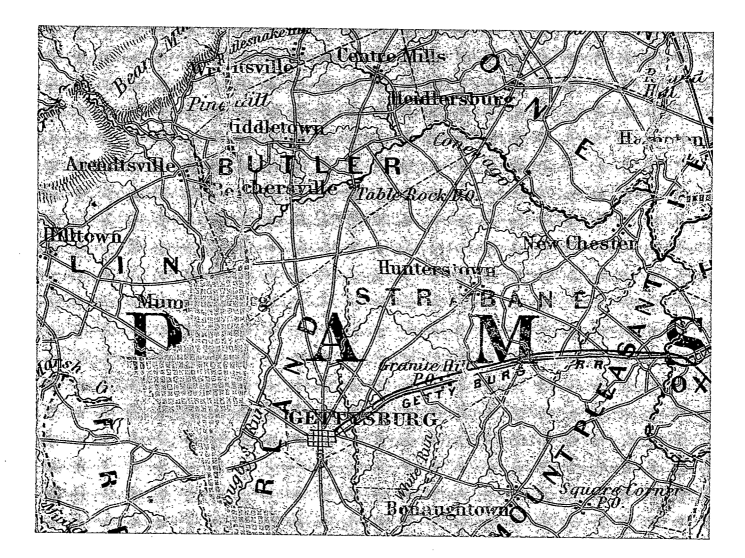

Topographical Map of Pennsylvania, 1862

Surveyed by H. F. Walling and published by Smith, Palmer and Co., New York. Walling was noted as one of the greatest county mappers of his time. This detail of the map shows Gettysburg, which was where a copy of it is believed to have been carried by Washington Roebling on July 2, 1863. Roebling was dispatched by the chief topographical engineer of the Union Army of the Potomac, General G. K. Warren, to Washington, Baltimore, and Philadelphia to procure maps of Maryland and Pennsylvania when it became evident that the Confederates were mounting what was to be the Gettysburg campaign. Roebling recollected that there was a new topographical map of Pennsylvania at his father's home in Trenton, New Jersey, and he rode there to get it. He then rode back through "an eerily deserted" Maryland to rejoin the army at Gettysburg, where fighting had already commenced.

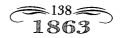

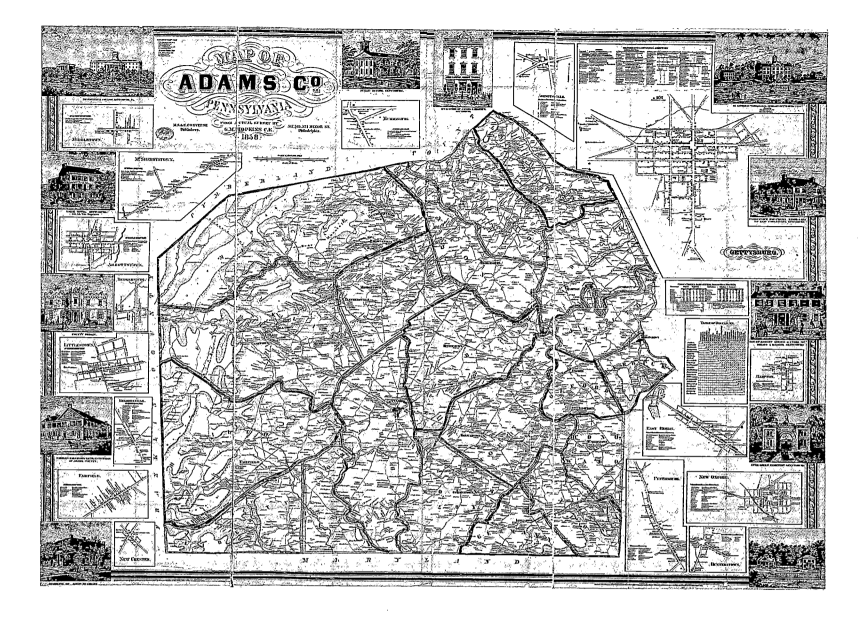

Map of Adams County, Pennsylvania, 1858

A classic period (1845–65) wall map that became the Civil War's most famous county map. The county seat of Adams County is Gettysburg. The town is nestled in the lower center of the county and also is featured on a much larger, more detailed scale on the upper right of the map. The lithographed map is quite accurate. It was made by G. M. Hopkins and published by M. S. and E. Converse. The map was sold by subscription at five dollars per copy. This was the most detailed map available of Adams County at the time of the Gettysburg campaign in June and July 1863. The scale of the map was much too small for tactical military purposes, but it did show a great deal of valuable information. It was one of the base maps incorporated by Jed Hotchkiss into his Gettysburg theater map, and Hotchkiss made traced copies of the map available. Confederate generals Robert E. Lee and Jubal A. Early are believed to have used this map, and a copy of it is among Confederate general Richard S. Ewell's wartime papers. Union commander George Gordon Meade is described as poring over a large map of Adams County, which could only mean this one.

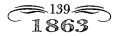

The New York Herald, Monday, June 29, 1863

Confederate cavalry general Albert G. Jenkins led the advance of the Confederate Army of Northern Virginia into Pennsylvania. At Mechanicsburg, Pennsylvania, Jenkins and his aides "elevated their heels" at the Ashland House Hotel and perused the New York, Philadelphia, and local newspapers. Jenkins undoubtedly saw and benefited from this conveniently provided map. Mechanicsburg, between Carlisle and Harrisburg, is inexplicably overlooked on the map. The Rebel scouts and cavalry mentioned at Sterrett's Gap in the newspaper's columns are, in fact, Jenkins's men. Jenkins's picaresque but valuable role at the forefront of the invading Confederates is often overlooked.

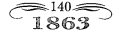

Emmor B. Cope's Map of Gettysburg, Pennsylvania, 1863

This is a photograph of a map based on the 1858 Adams County, Pennsylvania, map by G. M. Hopkins. Corrections and additions have been added by hand. Union general G. K. Warren, chief engineer of the Army of the Potomac during the Gettysburg battle, notes on the reverse of this map that Sergeant Emmor B. Cope made the original map from sketches made while on horseback reconnaissance of the field. This pieced-together map is oriented with East on the bottom. This is a perfect example of the transformation of a published, small-scale civilian map into a large-scale military map. A lithographed version of the map appears in the *Official Records Atlas* (plate 40, no. 2).

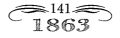

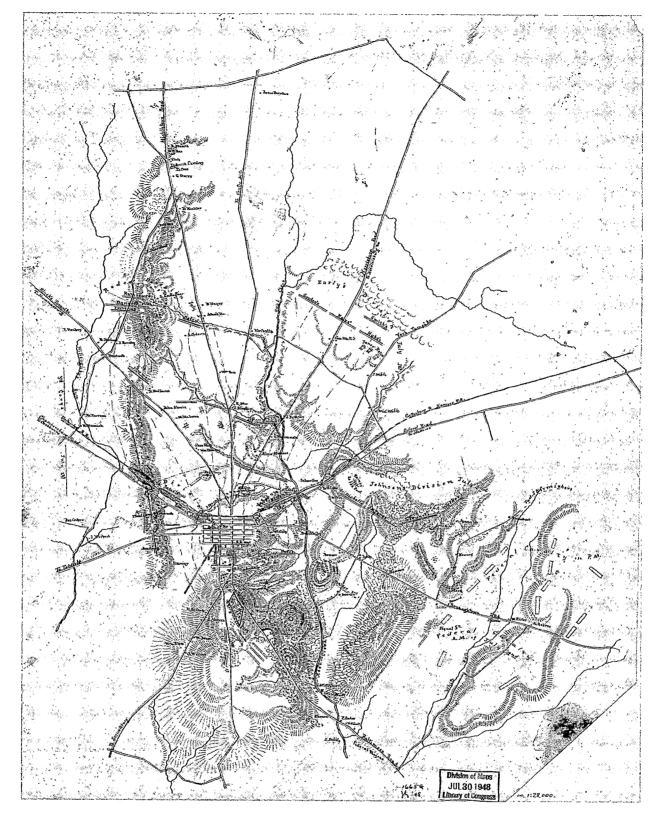

Jed Hotchkiss's Map of Gettysburg, 1863

Copies of this map illustrated the written reports of the battle by Generals Robert E. Lee and Richard S. Ewell. The configuration of the map is apparently based on the 1858 Adams County map, though some features of Jed Hotchkiss's map reflect his vantage point on the ground north of the battlefield. The terrain added by Hotchkiss, for example, is good for the left of the Confederate line but is very vague over to the right, including the Round Tops. Hotchkiss's contemporaneous view was that the Battle of Gettysburg was a "draw." A lithographed version of this map appears in the *Official Records Atlas* (plate 43, no. 2).

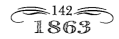

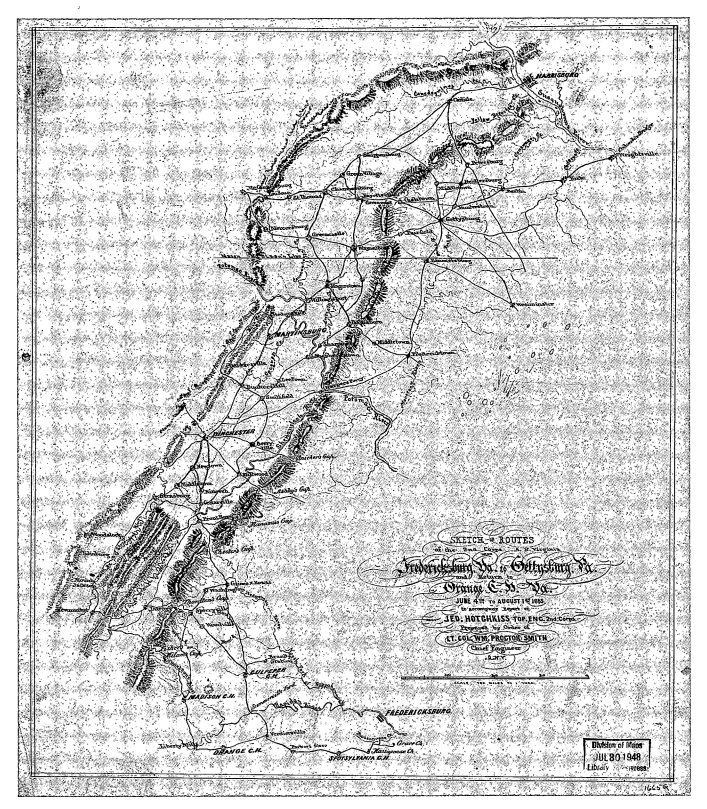

Jed Hotchkiss's Sketch of the Routes of the Second Corps, Army of Northern Virginia, from Fredericksburg, Virginia, to Gettysburg, Pennsylvania, and Return, June to August 1863

This theater map of the Gettysburg campaign clearly shows Robert E. Lee's strategy. He advanced down the Shenandoah Valley and up the Cumberland Valley, his troops hidden and protected by the Blue Ridge and South mountains. Lee emerged from behind the screen of the mountains at Cashtown Gap primarily because of all the major wind gaps (Harpers Ferry was the one water gap), Cashtown Gap was the widest, straightest, and most level. Through this gap Lee could safely and certainly move his large, cumbersome army, with all its bags and baggage, and at the same time prevent Union forces from doing the same thing in reverse. Gettysburg became a battleground in large part because Cashtown Gap was the most practicable gap at Lee's disposal. A lithographed version of this map appears in the *Official Records Atlas* (plate 43, no. 7).

has surface water in several streams crossed by the road line. The isolated rounded materials forming the masses of Knobs near Bowlinggreen from a contin- -uous line of hills, reaching from Russellville Eastwardly for several miles toward Bowlinggreen. The Sand stone ridge which extends along the line of the road from Cave City to Barren River, lies farther to the north, and at no point approaching the R.R. nearer than 8 or 9 miles; after leaving it at Bowl- inggreen.

Cave City to Glasgow.

The country between Cave City and the Glasgow and Louisville turnpike road is gently undulating Sink Hole Country. The sinks becoming less deep and abrupt as the margin of this peculiar territory is approached. A number of good roads traverse the belt of country lying between the R.R. and Glasgow and Louisville turnpike, surface water is found in branches. Dirt road to pike 7 miles, along turnpike road 5 miles — whole distance 12 miles. The country along the turnpike is more hilly.

The hills have a direction controled by the line of the branches, It is worthy of remark that the drainage of the Sink Country has no surface drainage. Water falling on the surface

A Page of Sidney S. Lyon's Map Memoir to Orlando M. Poe, Lexington, Kentucky, June 29, 1863

Captain Sidney S. Lyon, "Actg. Asst. Engr. 23rd A.C.," accompanied Union general George Lucas Hartsuff, commanding the Twenty-third Army Corps, on a tour of inspection through Kentucky that encompassed Louisville, Bowling Green, Russellville—and also Cave City, Columbia, Campbellsville, Lebanon, and Lebanon Junction. Captain Lyon wrote to Captain Orlando M. Poe, chief engineer of the Twenty-third Corps, "I have the honor to report such observations as I have been able to make of the topography of the country recently traversed." Reproduced here is one page of a report of twenty pages of analysis and observation by a man plainly trained to some extent as a geographer, geologist, and topographical engineer. He refers to the area between Cave City and Glasgow as "this peculiar territory . . . sink holes abound where there is no surface drainage."

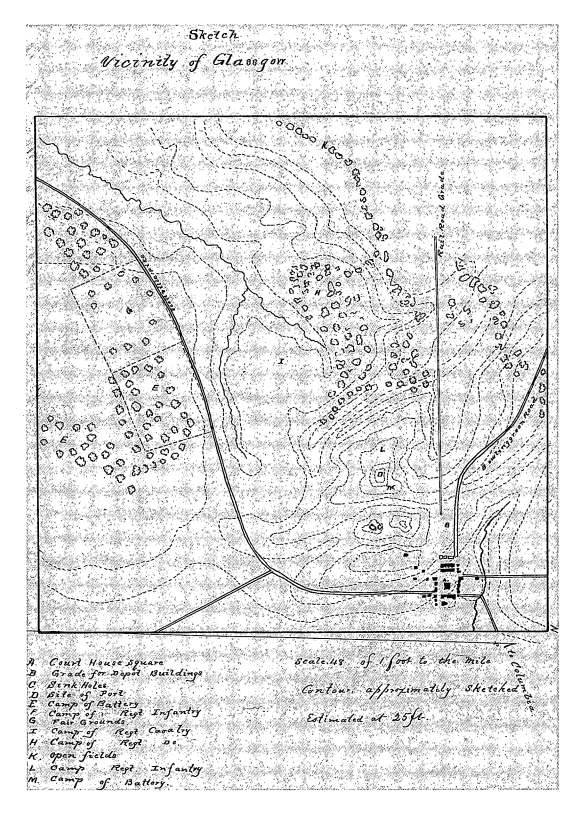

Inside the map:

Sketch
Vicinity of Glasgow.

A Court House Square
B Grade for Depot Buildings
C Sink Holes
D Site of Fort
E Camp of Battery
F Camp of   Reg.t Infantry
G Fair Grounds
I Camp of   Reg.t Cavalry
H Camp of   Reg.t   Do.
K Open fields
L Camp   Reg.t Infantry
M Camp   of Battery.

Scale.48. of 1 foot to the mile

Contour. approximately Sketched

Estimated at 25 ft.

Sidney S. Lyon's Map of Vicinity of Glasgow, Kentucky, c. June 1863

Drawn with black and red ink and blue pencil on tracing paper, this map is glued to the extensive map memoir that Captain Sidney S. Lyon, acting assistant engineer, Twenty-third Union Army Corps, sent to Captain Orlando M. Poe, his superior and chief engineer. Glasgow is the seat of justice for Barren County, Kentucky. The engineers planned to site a defensive work on one of the low knolls near the railroad grade, marked "D" on the map. Lyon qualified the map's accuracy, indicating that "no measures were made . . . all quantities estimated." The use of broken lines as contours is unusual, and Lyon would probably have used hachures if this map were intended for anyone but Captain Poe. Lyon consistently misspells Glasgow.

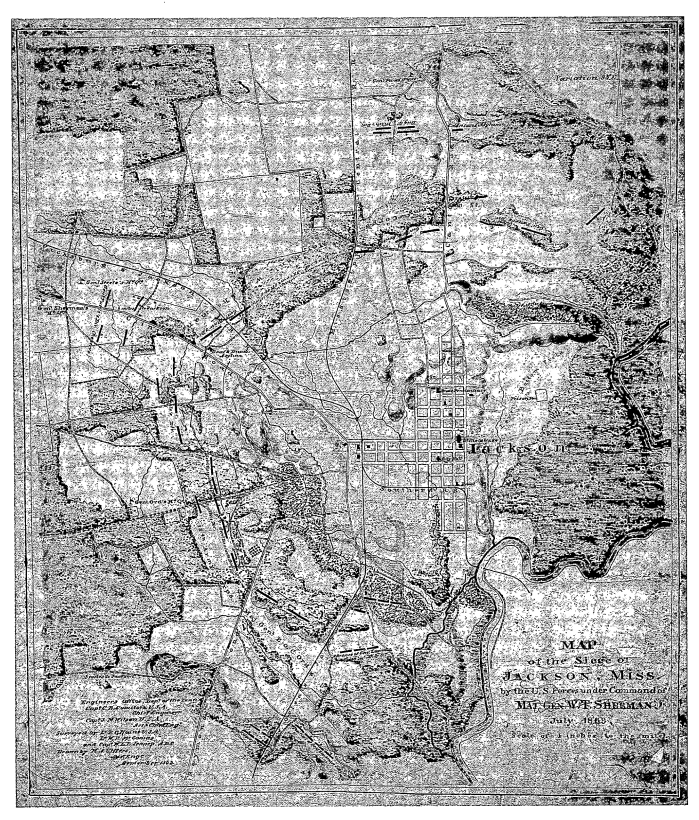

Hains's Map of Jackson, Mississippi, July 1863

Immediately after the capitulation of the Confederate stronghold of Vicksburg, Mississippi, on July 4, 1864, Union general William Tecumseh Sherman was ordered to turn about and attack the Confederate forces under General Joseph E. Johnston. These Rebel troops had hovered in the background during the operations around Vicksburg. Sherman quickly drove Johnston into the safety of the defenses of Jackson. The capital of Mississippi was evacuated on July 17. This map is one of the crowning achievements of military cartography in the West. Captain J. H. Wilson (the middle initial is mistakenly given as "M" on the map) became an eminent cavalry chief. Peter C. Hains was in command of a famous 30-pound Parrott gun at First Manassas and H. A. Ulffers was an assistant adjutant general and assistant engineer who was also a superlative cartographer. A lithographed version of this map appears in the *Official Records Atlas* (plate 37, no. 5).

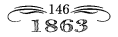

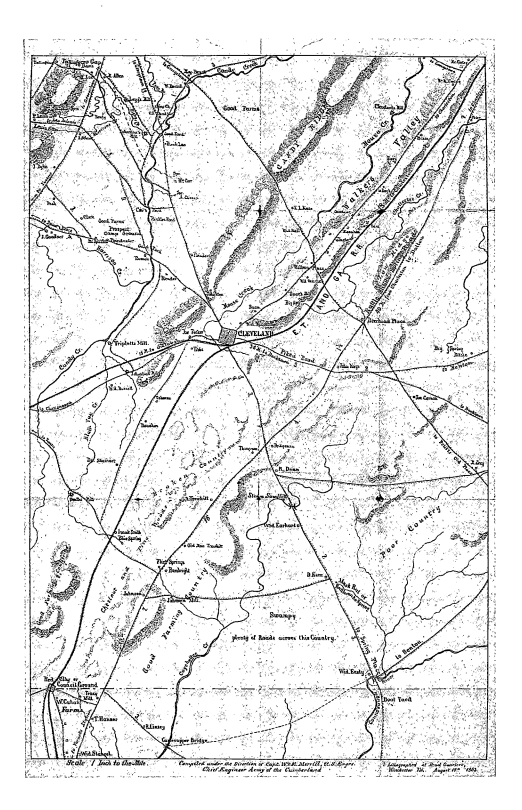

William E. Merrill's Lithographed Map of Vicinity of Cleveland, Tennessee, August 14, 1863

Major General William S. Rosecrans led a brilliant, bloodless campaign of maneuver against Confederate Braxton Bragg in Tennessee, prizing him out of Chattanooga, a crucial railroad junction. In his reports, Rosecrans singled out William E. Merrill for praise, one of the few instances when an engineer got some of the acknowledgment he was richly entitled to. Rosecrans claimed credit for constructing information maps, of which this is a good example. By utilizing already existing maps, increasingly more detailed, definitive maps were prepared. Rosecrans's real contribution was to give Merrill autonomy and to brook no interference in the duties of topographical engineers. They were not to serve as couriers, to plan fortifications, or to pick out campsites. They were to prepare and produce military maps. Note the militarily significant comments: "Swampy," "Good Ford," "Plenty of Roads." Cleveland is a railroad junction some twenty-five miles northeast of Chattanooga.

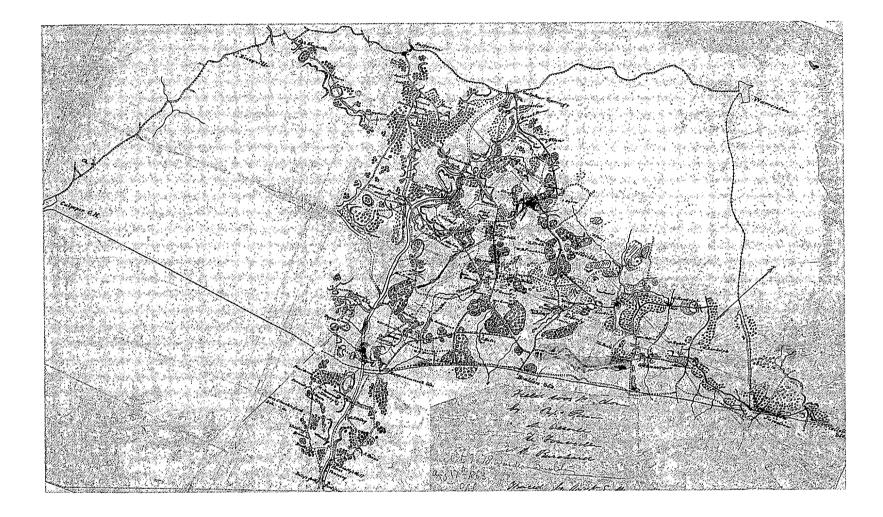

Field Work Map of Area East of Culpeper Court House, Virginia

This post-Gettysburg sketch is by Captain William Henry Paine. It is not oriented to the north; Kelly's Ford is actually due east of Culpeper. Copied on tracing paper and mounted, it examines areas of likely operations for the Army of the Potomac. A penciled notation reads, "Rec'd at No. 78 Winder Building, Aug. 17, 1863." The offices of the U. S. Army Bureau of Topographical Engineers were in the Winder Building in Washington.

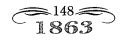

BRIDGES ON EAST TENN. & GA. R.R.

Corrected from better information. August 24th 1863.
to replace same list in "List of Bridges & Tunnels,,

Chattanooga to Cleveland 29 5/8 miles

Dist. from Chattan.	at or near what Place	over what	Length	Height above water	No. of Spans	Kind of Bridges.	Remarks
1 1/4.		Citico	120.'	38.'	1.	Stone arch	38' span, 19' rise
1 3/4.		W.&A.RR.	14.'	14.'		Wooden beam	(with struts)
7 3/4		....	14.'	14.'	1.	Stone arch	7' rise
8.		Chicamauga	150.'	24.'	2.	Stone arches	Segmental_each 50' span, 10' rise
10.	Tyners Station	Br. of	12.'	20.'	1.	" "	6' rise, 8' of earth on top
15.		Ooltawah	112.'	45.'	1.	" "	30' span, 15' rise
16 1/4.		Br. of "	12.'	20.'	1.	" "	6' rise
16 1/2.		" "	12.'	15.'	1.	" "	6' rise
16 3/4.		" "	28.'	12.'	2.	" "	each 6.' span, 3' rise
23.		Candys Cr.	15'	22.'	1.	" "	7.' 6" rise
24.		Black Fox	25.'	23.'	1.	" "	12.' 6" rise
25.		Mill Pond	12.'			Wooden beam	Track crosses diagonally

4. Tunnel. 830.' long_Stone side walls, 4' at base, 3' at top (poorly built) Brick arch from 3.' to 2' 6" thick; 200' at West end through rock, rest through earth._No shafts_The part between 200' and 350' from West end fell in_Was repaired with stone side walls 7' at base, 3' to 4' at top_half of arching repaired with small rough flat stones, making arch 3' thick_Key of brick; Rest of repaired arch of brick, 3' thick_

Additional Corrections

sent in by Col. Scribner Comdg. 1st Brig. 1st Div. 14th A.C.

Bridgeport to Chattanooga

1. 3rd Bridge_1/4 mile from Bridgeport_over low land_said to be filled in with earth_

2. 10th & 11th Bridges_over Lookout & Chattanooga Creeks_said to be Howe Truss Bridges on stone piers.

Hd. Qr. Topl. Engr. Office.

Ambrose Bierce Papers: Bridges on East Tennessee and Georgia Railroad, August 24, 1863

Union army operations in the West relied on rivers and railroads to bring in the enormous quantity of supplies that the men and animals required. Knowledge of the location, size, and construction of railroad bridges was important because, for example, a stone arch bridge was much more likely to survive the raids and depredations of an enemy cavalry raid than was a wooden beam bridge. This information was probably compiled for use by Union general William S. Rosecrans, a map connoisseur, as he maneuvered about Chattanooga in August 1863. His initial success in turning Confederate Braxton Bragg out of Chattanooga without firing a shot or losing a man made him overconfident, which led to disaster for Rosecrans at Chickamauga, Georgia, in September 1863.

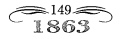

Ambrose Bierce's "Encampment" Sketch, c. Summer 1863

This sketch of an encampment along the Western & Atlantic Railroad north of Atlanta is a meticulously rendered map. It demonstrates Ambrose Bierce's technical competence on the use of hachures to depict elevations and the railroad embankment. Another fine feature is the use of shadows for the trees and buildings. Note the heavy dark lines on the sides of the structures, "away from the light." The light is supposed to enter the drawing at the upper left, in parallel rays forty-five degrees to the horizon. The shadows thus fall to the lower right of the trees. This technique helps establish, among other things, relative degrees of inclination for those interested in a very minute reading of a map. Ambrose Bierce was the one major American writer to see significant active service in the Civil War.

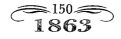

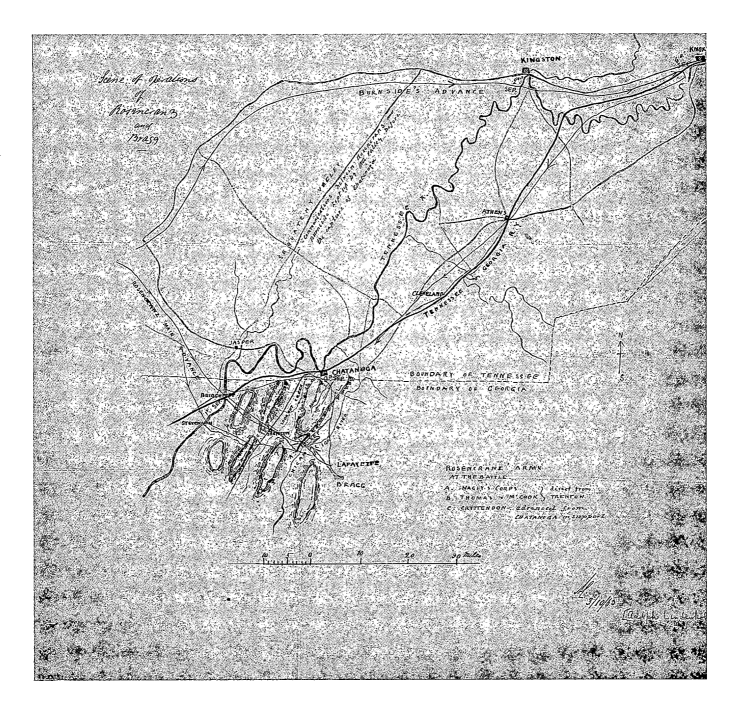

Scene of Operations of Rosecrans and Bragg

This eye-catching manuscript map, very bright and brisk, is by Adolph Lindenkohl, a civil engineer with the United States Coast Survey. Lindenkohl misspells nearly every proper name and place, but the deft map is a small cartographic treasure. The paper used for the map is much like that used for modern grocery bags.

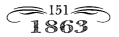

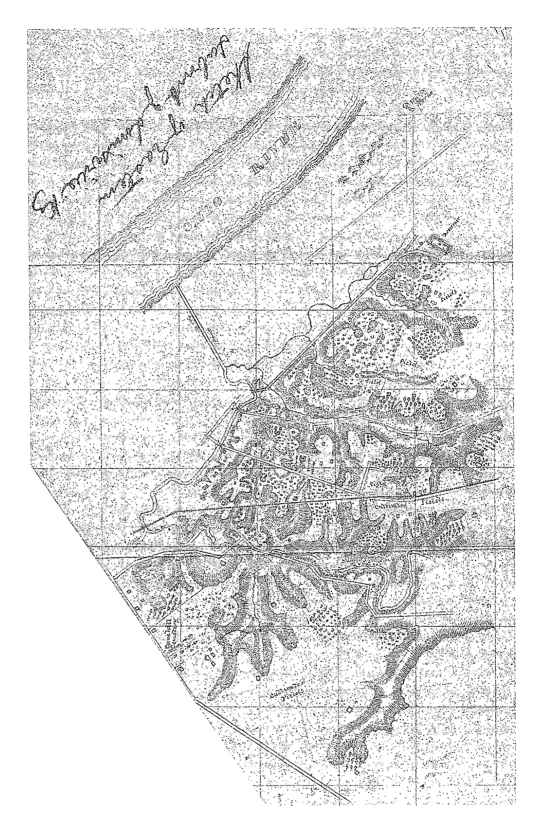

Sketch of Eastern Suburb of Louisville, Kentucky, 1863

This beautifully drafted, heavily detailed sketch map, drawn in pencil on paper prepared with a penciled-in one-square-inch grid pattern, was almost certainly the work of Orlando M. Poe, at the time chief engineer of the Twenty-third Union Army Corps of the Army of the Ohio. Poe created a most convincing rendition of terrain features, with an effortless, uncrowded presentation of numerous cultural details. This frail scrap of a masterpiece by one of the Civil War's great topographical engineers may have been used, without acknowledgments, as the partial basis of an overall map of Louisville in the *Official Records Atlas* (plate 102, no. 3). The atlas map identifies some of the unlabeled structures in this pencil sketch and is worth consulting for an interesting comparison.

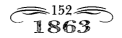

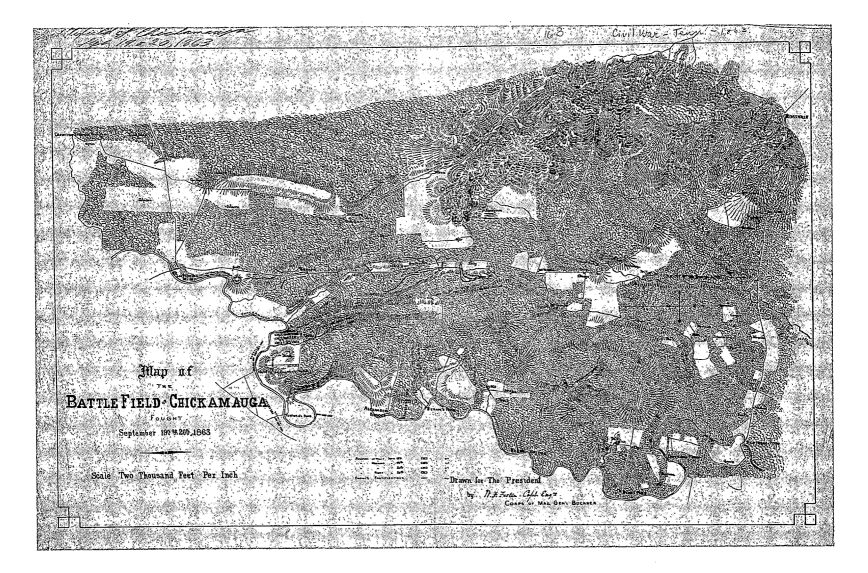

The Battlefield of Chickamauga, September 19 and 20, 1863

This striking map, on tracing linen, is executed with classic Civil War–era map features. Very precise hachure marks define the elevations with crisp, even lines, and wooded areas are painstakingly indicated by countless representations of trees. North is off by ninety degrees. Confederate engineer Captain W. F. Foster prepared a smaller-scale theater map of the Chickamauga battlefield and vicinity that appears in the *Official Records Atlas* (plate 111, no. 9). This map "drawn for the president" may have been the result of the controversy that swirled about the Confederate army after this flawed victory. Foster wrote a very interesting account of Confederate mapping in the West entitled "Battle Field Maps in Georgia" in *Confederate Veteran* magazine in 1912.

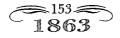

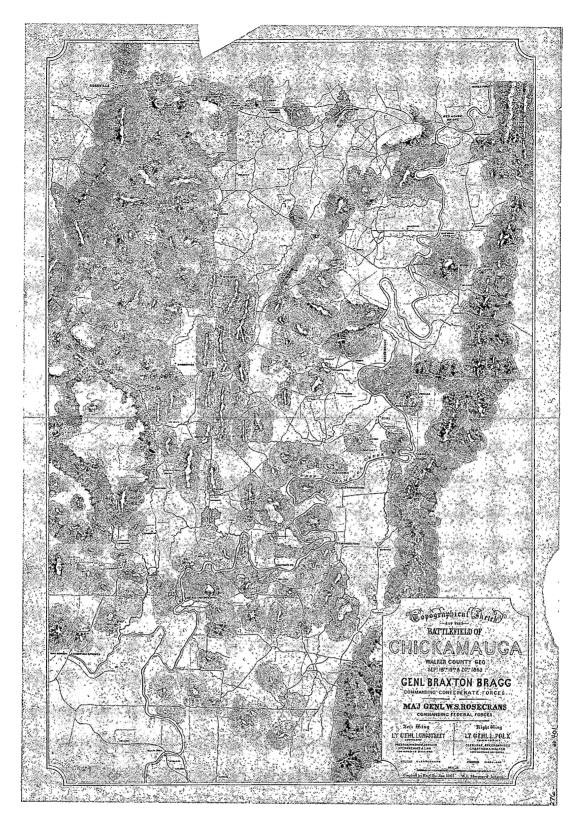

Topographical Sketch of the Battlefield of Chickamauga, Georgia, 1863

William L. Sheppard was a draftsman in the Confederate Bureau of Engineers in Richmond, Virginia. His source for this attractive and intriguing map of the Battle of Chickamauga, fought in September 1863, is not known. He depicts numerous cultural details—churches, residences, mills, and bridges. The physical aspects of the field, for example, West Chickamauga Creek, are carefully embellished with dark blue watercolor tint on the left bank of the creek. The course of the creek is, however, depicted inaccurately. The hachure work on the terrain is very expertly done, and the overall look of the map makes it one of the more attractive wartime map efforts. Chickamauga is one of the few battles called by the same name by both the Federal and Confederate armies.

The pay for a draftsman such as Sheppard was about ninety dollars per month. Most of the supplies he used, including instruments, pencils, paper, and watercolors, had been purchased for the Confederacy in England by Captain John M. Robinson of the Provisional Confederate Engineers early in 1863.

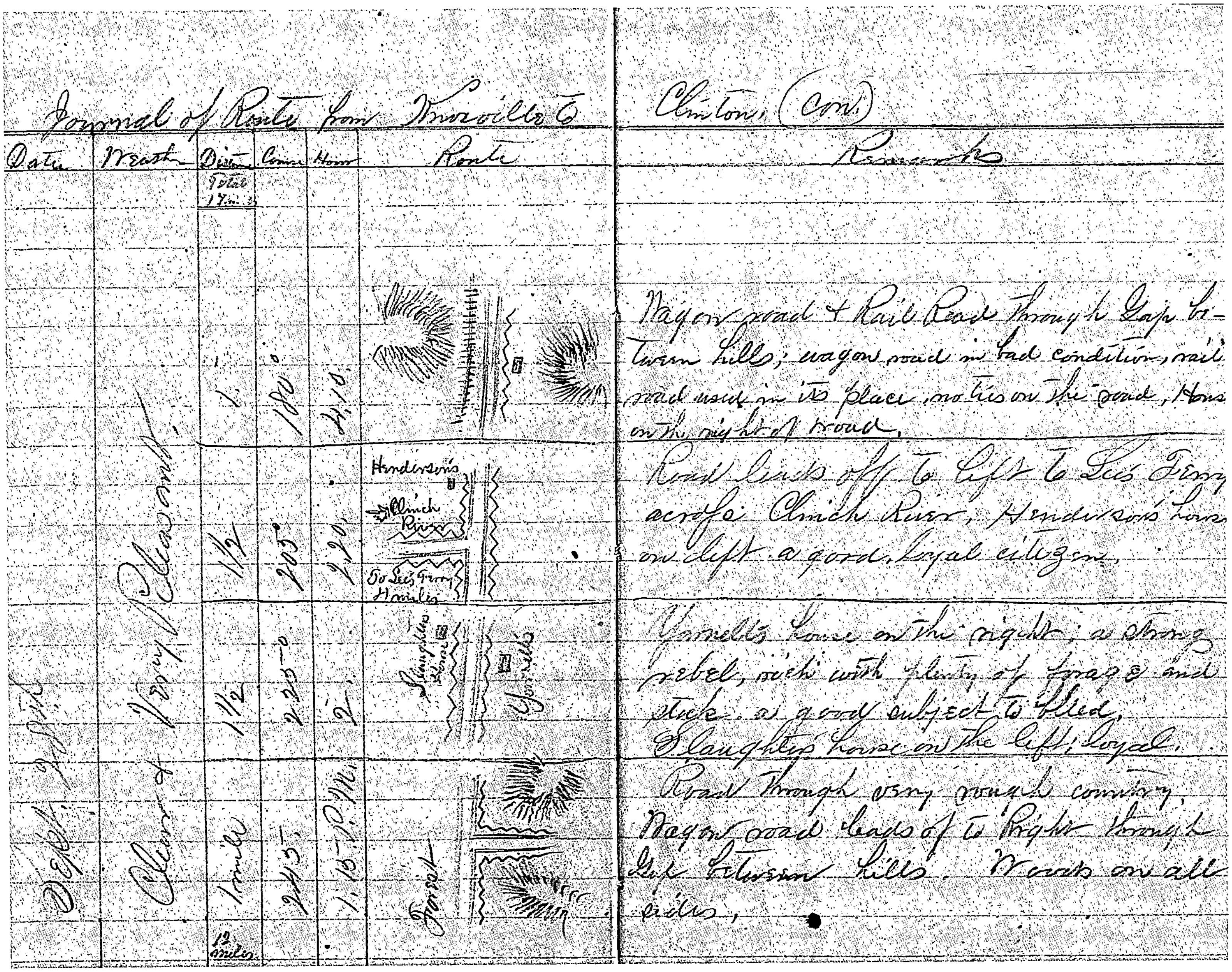

Journal of Route from Knoxville to Clinton (Con.)

Date	Weath	Dist	Comp	Hour	Route	Remarks
		Total 17				
Sept 28th	Clear & very Pleasant	1	180°	11.10		Wagon road & Rail Road through Gap between hills; wagon road in bad condition, rail road used in its place, no ties on the road, Hous on the right of road.
		1 1/2	205°	11.20	Henderson's; Clinch River; To Lee's Ferry 4 miles	Road leads off to left to Lee's Ferry across Clinch River, Henderson's house on left a good loyal citizen.
		1 1/2	225°	12.	Slaughter's House; Yarnell's	Yarnell's house on the right; a strong rebel, rich with plenty of forage and stock, a good subject to bleed. Slaughter's house on the left, loyal.
		1 mile	245°	12.10 M.	Forest	Road through very rough country. Wagon road leads off to Right through Gap between hills. Woods on all sides.
		12 miles				

Sidney S. Lyon's Journal of the Route from Knoxville, Tennessee, to Clinton, Tennessee, September 28, 1863

Captain Sidney S. Lyon of the Fourth Kentucky Cavalry was on detached duty as an assistant engineer in the Twenty-third Union Army Corps. (On another map, of an earlier date, Lyon signs himself "Acting Topog. Engineer.") In a 4½ x 7", soft, brown-covered notebook, Lyon keeps meticulous, detailed notes and sketches salient topographical features as he reconnoiters a route from Knoxville, Tennessee, to London, Kentucky, approximately ninety miles apart as the crow flies. Lyon made the trip in four days. An almost palpable sigh of relief comes with the final entry, "London at Last." Lyon records the date, weather, distance, compass course, and the hour, in the left columns, with the explanatory sketches. On the right, in his "remarks," he describes such facts as the wagon road "in bad condition, railroad used in its place, no ties on the road"; he comments that Henderson is "a good, loyal citizen," meaning that he could be relied upon for accurate directions and was to be left undisturbed. Yarnell, on the other hand, was "a strong rebel . . . a good subject to bleed," that is, to take advantage of for forage and supplies. "Woods on all sides" is the last comment on the page. This route guide would be heavily relied upon in planning or making a march on the roads it follows.

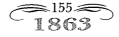

Field Sketches, Lookout Mountain, Tennessee

These rapid field sketches, in pencil, of Lookout Mountain, Tennessee, were drawn by an unknown topographical engineer. Union general Joseph Hooker used them in his successful assault of Lookout Mountain on November 24, 1863, during the Chattanooga campaign. Compare these to the "View of Lookout Mountain" from a wartime photograph in Robert Underwood Johnson and Clarence Clough Buel's *Battles and Leaders of the Civil War* (volume 3, page 694). Lookout Mountain was taken with relative ease by Hooker's three divisions, serving as a lesson on what were supposed to be "impregnable positions." Rapid field sketching was an interesting genre of pictorial military mapmaking, with careful rules for perspective drawing and relevant subject matter.

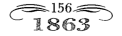

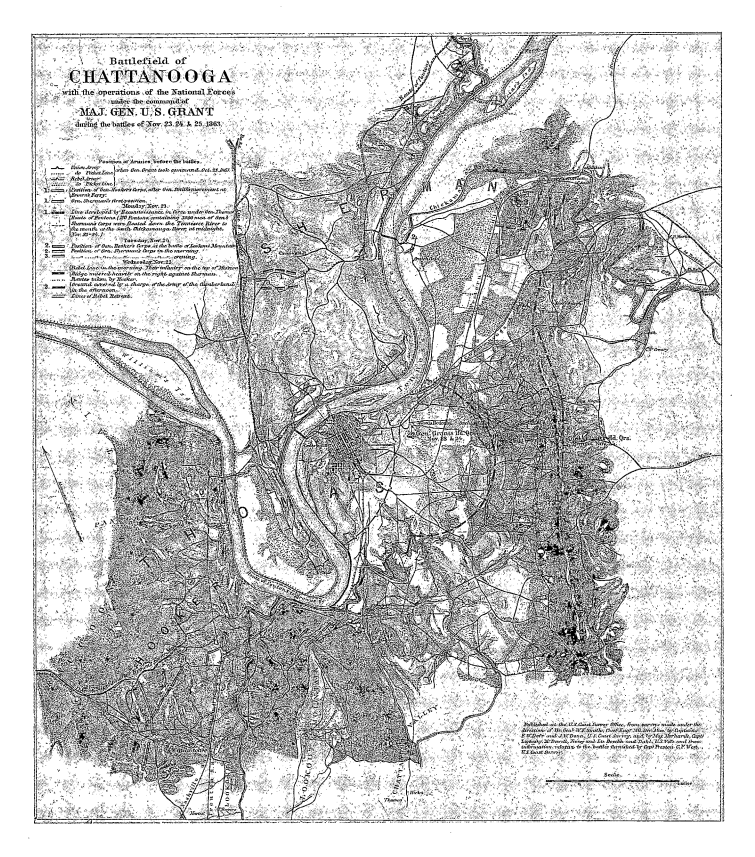

United States Coast Survey Map of the Battlefield of Chattanooga, Tennessee, November 23–25, 1863

An impressively rendered postbattle map that vividly documents and illustrates this dramatic engagement, with its "Battle Above the Clouds" on Lookout Mountain and the storming of the seemingly impregnable "Mission" or Missionary Ridge. The valuable contribution of the United States Coast Survey topographical engineers before the battle was not their maps but their establishment of very accurate artillery ranges for Federal guns against the fixed Confederate positions that overlooked them. The expertise of the civilian Coast Survey mapmakers is self-evident in this map.

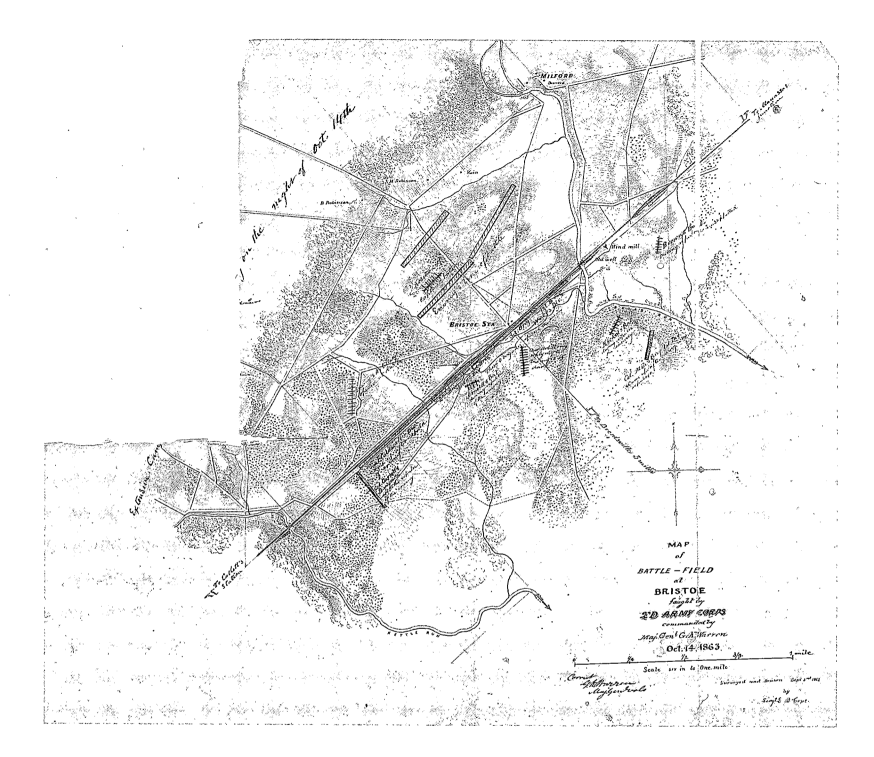

Emmor B. Cope's Map of the Battlefield of Bristoe Station, Virginia

Surveyed and drawn by E. B. Cope, whose postwar maps of the Antietam and Gettysburg battlefields are minor legends. The map depicts the October 14, 1863, Battle of Bristoe Station from the Union perspective. Compare the original manuscript map with its lithographed version in the *Official Records Atlas* (plate 45, no. 7) and with the Confederate map of the same battlefield on the facing page.

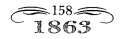

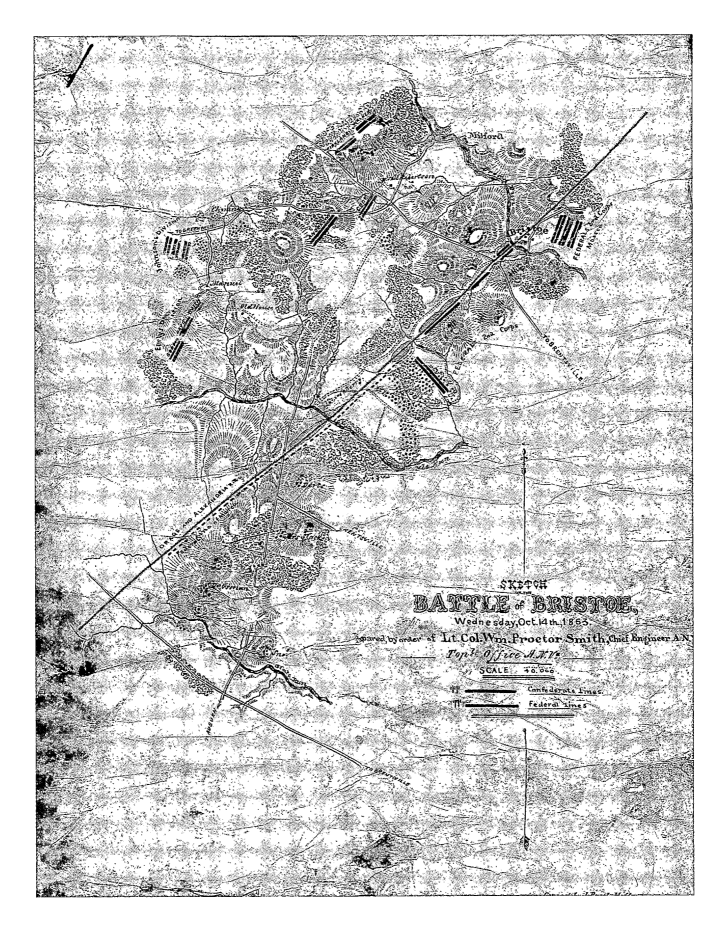

Confederate Sketch of the Battle of Bristoe Station

A copy on tracing paper, this Confederate map differs significantly from the Cope map (opposite). See, for example, the relative positions of Broad Run and Bristoe Station on the two maps. The Cope map is correct. This map is pasted and preserved in a letter book of Union general G. K. Warren's.

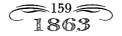

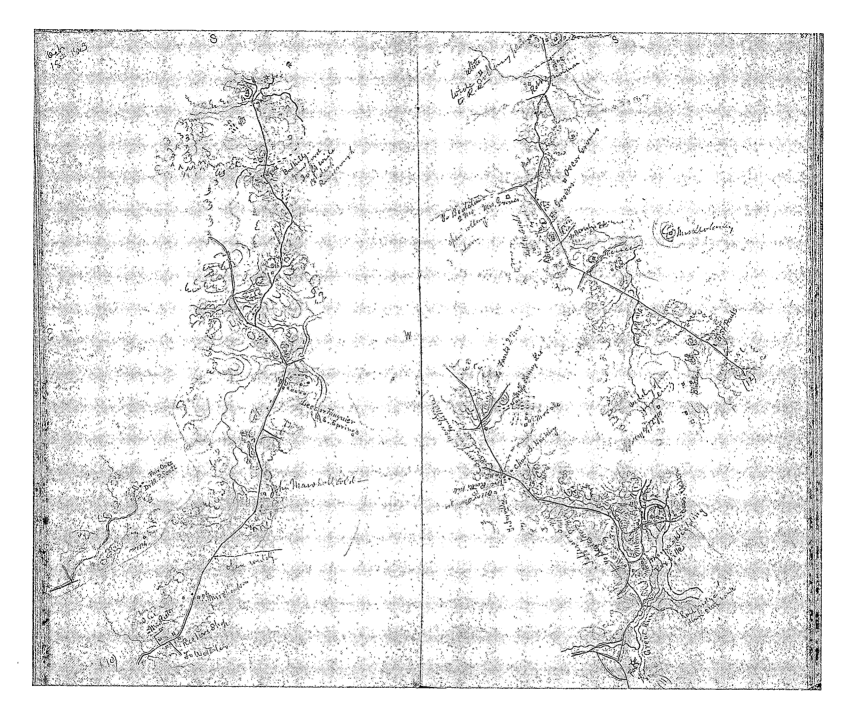

Jed Hotchkiss's Sketchbook, 1862–65, page 87

The raw material for Jed Hotchkiss's maps came from his fieldbook sketches. They preserve what he saw and drew from the saddle as he made his way along the roads and byways he wanted mapped. This data would be added to an existing base map that did not contain such highly detailed, militarily relevant information. Dated October 18, 1863, this mapping activity is inexplicably not noted in Hotchkiss's otherwise exhaustive journal. The sketching was done in Fauquier County, Virginia, south of Warrenton. The separate sketches connect: "10" to "11" to "12."

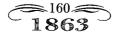

Union General George H. Thomas's Personal Map Journal

General George H. Thomas was probably the most completely professional soldier of the American Civil War. He gathered about him a staff of experts and specialists and he had an impressively equipped wagon that served as a mobile command post. When opened, the sides of the wagon folded down into traylike desks, and awnings folded out to protect the staff as they worked and to keep their papers dry. Thomas was as innovative and thorough in setting up a topographical engineer department. He gave complete authority over the topographical engineers to Captain William E. Merrill, who proceeded to assemble and put into operation the most efficient organization of its kind. General William Tecumseh Sherman seems to have regarded Thomas with muffled dislike and impatience, but he was quick to take advantage of Thomas's topographical engineer department and relied heavily upon both its mapping arrangements and its maps. This page indicates Thomas's own map-drawing expertise and his intense awareness of the kind of arcane information crucial to a marching army's well-being: terrain, ground cover, distances, road surfaces, fording sites, status of bridges, availability of drinking water, the names and reliability of residents. Of the other principal generals of either side, probably only Sherman was capable of drafting a map of this caliber. The written comments on the right constitute a map memoir, detailing information that could not be graphically presented.

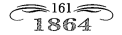

Overview of Red River Campaign Map Series

Red River Campaign

John S. Clark participated in Union general Nathaniel P. Banks's ill-fated Red River campaign of March, April, and May 1864. The complicated nature of this campaign—there were political, economic, and diplomatic factors at work—made it a dubious venture from the beginning. A mapping failure early on probably signaled its ultimate demise. An attempt to capture Shreveport, Louisiana, was at the heart of the expedition. The army, under Banks, was to move along the Red River, supported by a United States Navy squadron under the command of Admiral David Dixon Porter. At Grand Ecore, Banks followed an inland route that led him away from the protection of the navy. He appears to have overlooked a road that hugged the river on the opposite, eastern shore. Banks and Porter planned to rendezvous at Springfield Landing, thirty miles short of Shreveport.

Confederate forces were well aware of Union operations. Command differences were causing real frictions within the Rebel hierarchy, but they were forgotten temporarily in the face of the advance of Banks's troops. Confederate general Richard Taylor, son of the late former president Zachary Taylor, fell back before Banks's march. Finally, at Mansfield, Louisiana, Taylor turned and deployed his troops in a line across the route of the Union advance.

The following series of eight maps, or plans, as Colonel Clark called them, is from a set of eighteen maps that chronicles the campaign. Clark prepared them in late April 1864, intending to have them lithographed. At the same time, he was engaged in "collecting details and writing an account of the campaign," according to a letter written to his son, Charlie, on April 27.

The original maps were collected and bound between two heavy paperboard covers. They were tied together by looping a shoelacelike pink ribbon through three spaced punch holes. A lined piece of foolscap paper is pasted on the cover. In hand-lettered ink is written the title of the pamphlet, "Red River Campaign—Maps," followed by the names and numbers of the eighteen maps.

Each of the reproduced maps is on tracing paper. Secured beneath it and separated from it by a heavy piece of opaque paper is a rough outline version of the map upon which the finished map is based. The maps are watercolors, color pencil, and ink. The watercolors are applied on the reverse side of the tracing paper. The shaded billowy gray representing elevations is a trademark of Clark's and an attractive and effective technique. Banks was not a trained soldier—behind his back he was referred to as Mister Banks— and he was not adept at topographical matters. He probably appreciated and grasped Clark's rendition of terrain better than if it had consisted of contour lines or hachure marks.

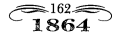

Red River Campaign, Map No. 2, Henderson's Hill, Louisiana, March 21, 1864

This map depicts one of the two Union successes of the Red River campaign. Fighting in rain, hail, and darkness, twenty miles northwest of Alexandria, Louisiana, six regiments of Federal infantry, cooperating with a brigade of General Nathaniel P. Banks's cavalry, surprised and defeated a regiment of Rebels, capturing a four-gun battery and many prisoners.

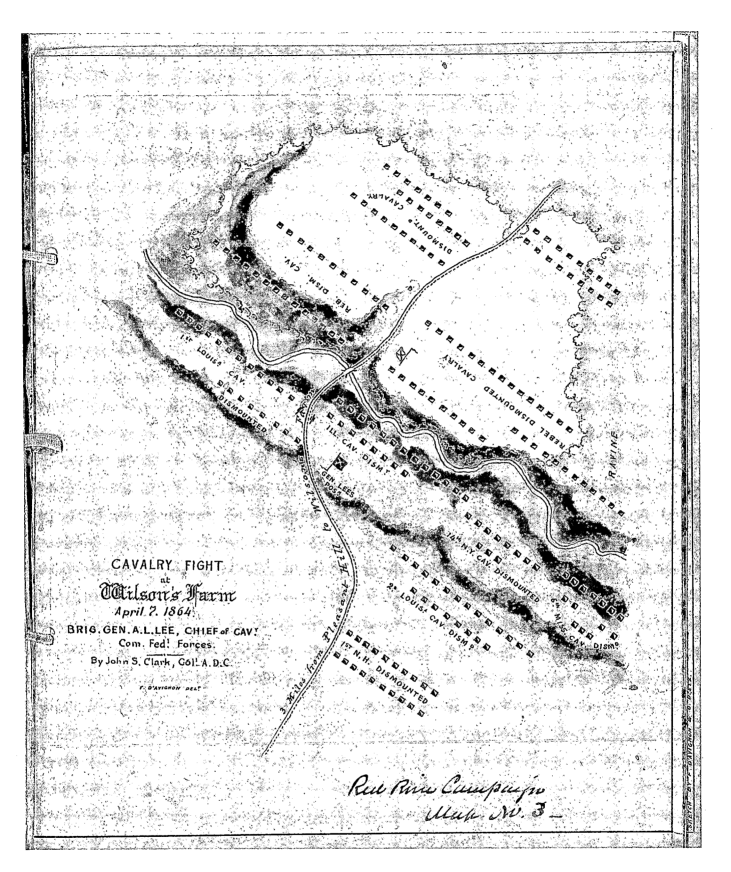

Red River Campaign, Map No. 3, Cavalry Fight at Wilson's Farm, April 7, 1864

The clash of Nathaniel P. Banks's and Richard Taylor's cavalry at Wilson's Farm or Wilson's Plantation was inconclusive, but for the first time in the campaign, the Confederates were showing an inclination to stand and fight. Here they fell back after a wild charge that convinced the advancing Yankees that there was likely to be fighting ahead. Consequently, John S. Clark, as one of Banks's staff officers, arranged for a brigade of infantry to move to the front to support the cavalry advance.

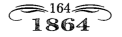

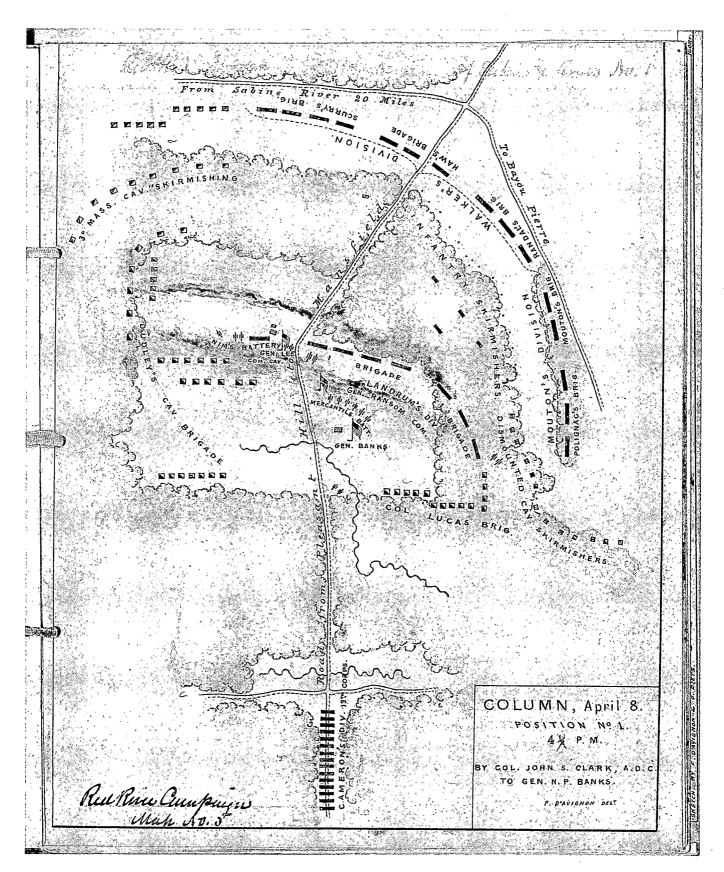

Red River Campaign, Map No. 5, Column, April 8, 1864, Position No. 1

The fighting came the day after Wilson's Farm. Banks's army moved beyond the support of David Dixon Porter's flotilla and marched on narrow roads over low hills through thick pine woods. Emerging from the pines into a large clearing, Rebel skirmishers were evident on the ridge shown bisecting the clearing. The Union cavalry commander, General Albert Lee, displaced the skirmishers and looked down upon a heavy Confederate line of battle. Just ahead was an intersection known as Sabine Cross Roads.

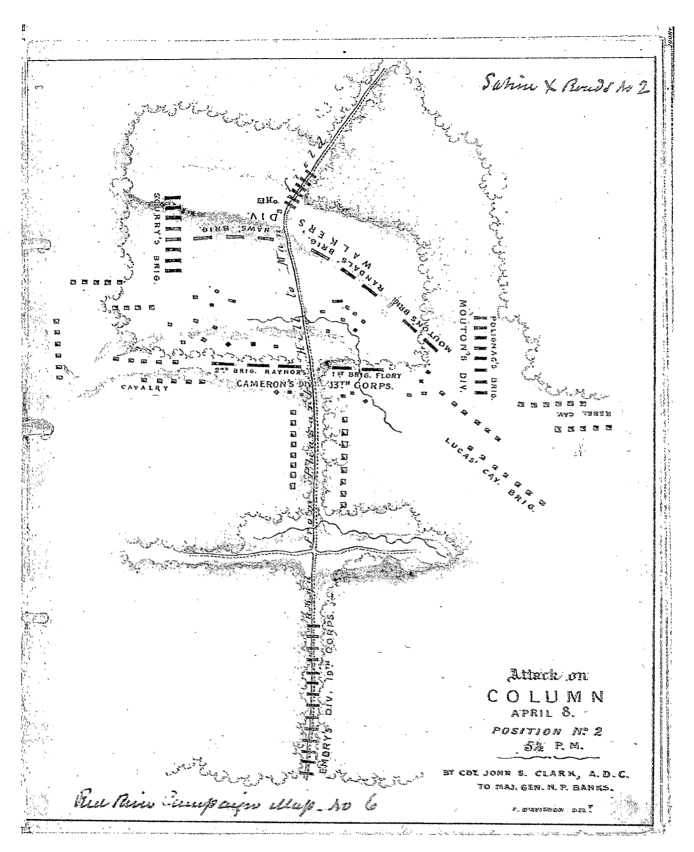

Red River Campaign, Map No. 6, Sabine Cross Roads, April 8, 1864, Position No. 2

An apprehensive General Albert Lee was ordered by Banks to advance against the Confederates in his front. The narrow road hemmed in by heavy woods hampered Lee's deployment. Before he could do so, the Rebels emerged from the pine woods to attack him, landing on the Union lines with a crash and the Rebel yell. The Union retreat turned into a rout, the Union troops into a mob. As he was pressing his victory, Taylor received a letter from his superior, General Edmund Kirby Smith, urging delay and caution. "Too late, sir," said Taylor, "the battle is won." The battle was fought by some twenty thousand men.

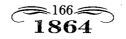

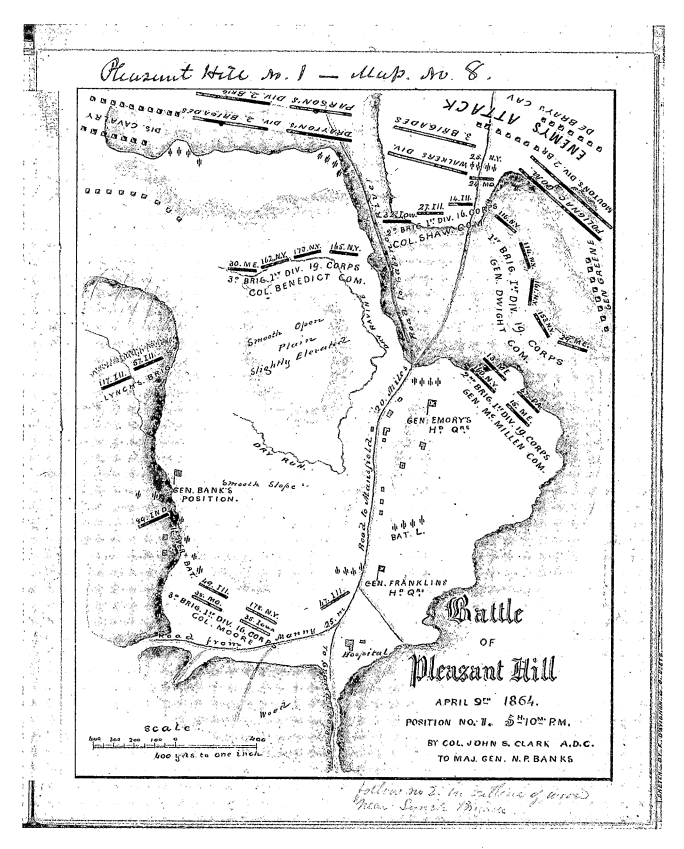

Red River Campaign, Map No. 8, Battle of Pleasant Hill, April 9, 1864

The Rebels, delighted by their easy, smashing success of the day before, followed the retreating army of Banks fifteen miles back toward Grand Ecore. Here, at Pleasant Hill, Taylor threw his army against the hastily devised Union line. Union general A. J. Smith's contingent, after serving as a transport guard for the navy, joined its frazzled Union comrades and was instrumental in repulsing the Rebels decisively. Once the fighting was over, both armies beat a retreat from the field. Dry Run was indeed dry, and the Union position was without a source of water.

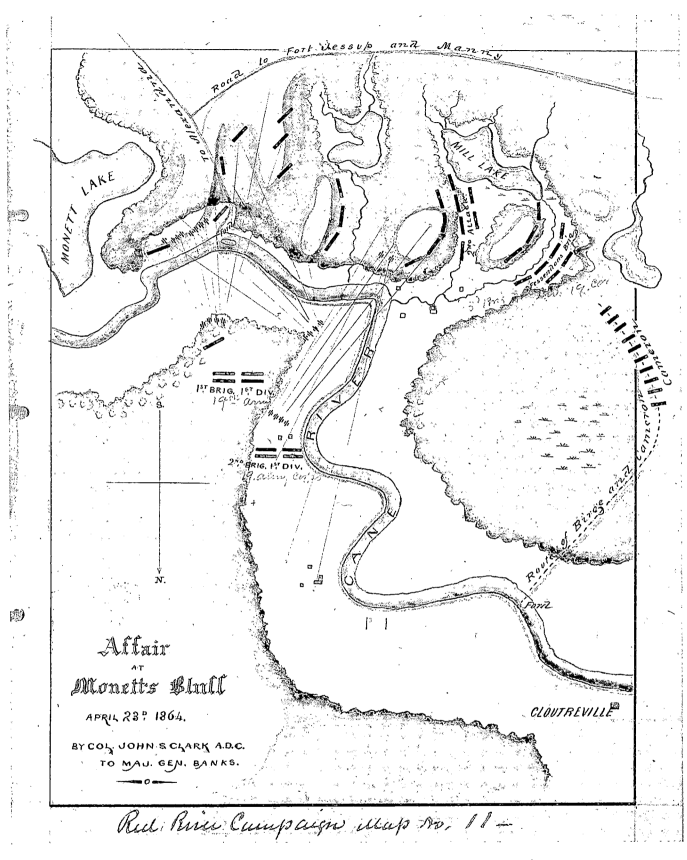

Red River Campaign, Map No. 11, Affair at Monett's Bluff, April 23, 1864

Also known as the affair at Monett's Ferry or Cloutiersville, this action was a futile attempt by Confederate general Taylor to impede Banks's retreat to Alexandria, Louisiana. The map is turned around 180 degrees, south being at the top. Compare this map of Clark's with the following far more detailed map of the same action by Frank W. Loring. The elegant simplicity of Clark's maps reflects the fact that they are to be lithographed.

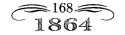

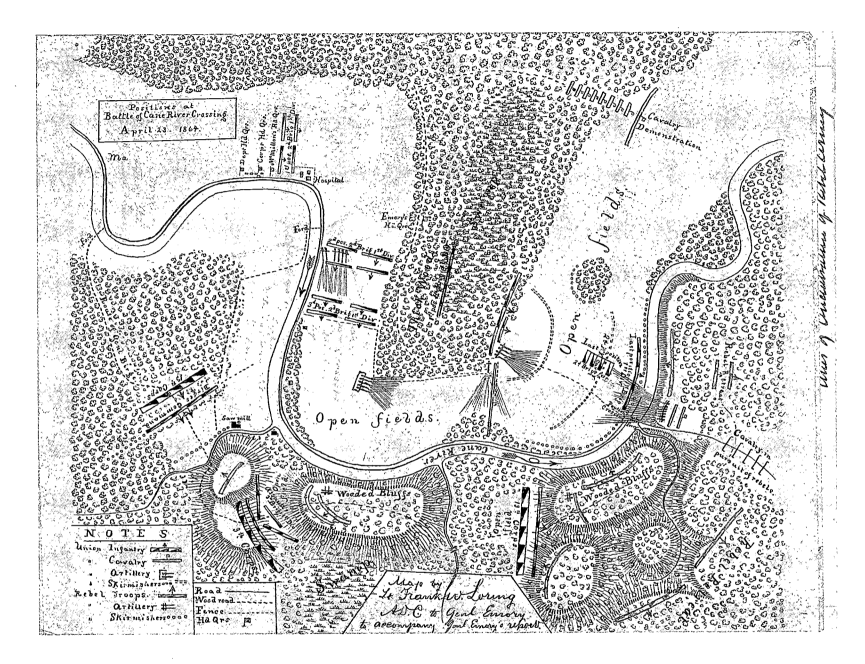

Frank W. Loring's Map of Positions at Battle of Cane River Crossing, April 23, 1864

This map is by Frank W. Loring, aide-de-camp to Union general William H. Emory, division commander in the Nineteenth Corps. The map is oriented traditionally, with north at the top. Though not as artistic as Clark's map of Monett's Bluff, Loring's is much more accurate and informative.

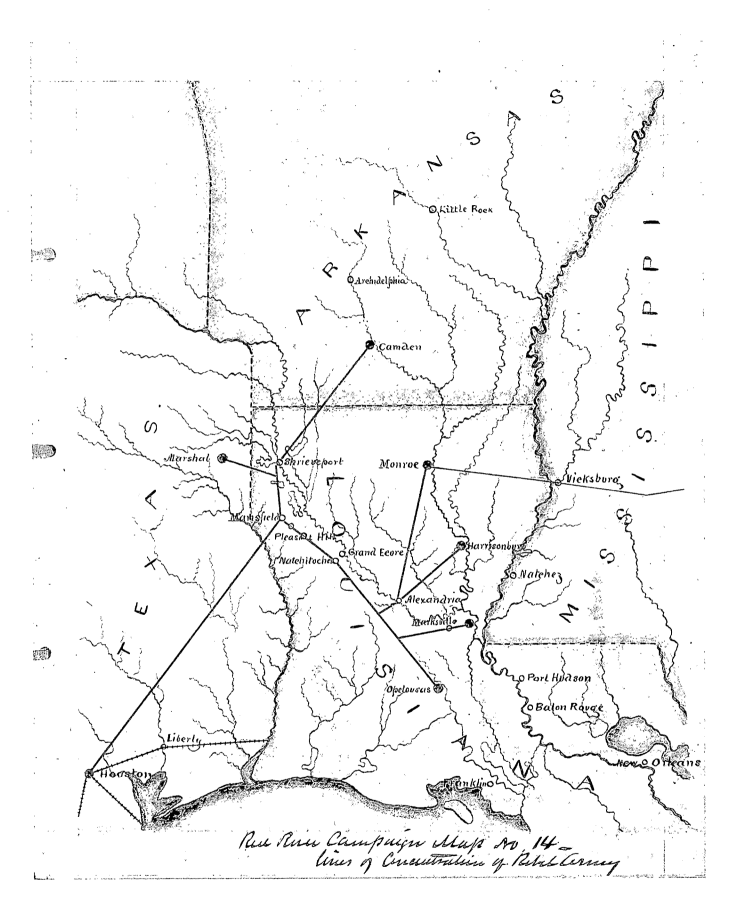

Red River Campaign, Map No. 14, Lines of Concentration of Rebel Army

John S. Clark's elegant theater map of an inelegant, incoherent Federal campaign. Sabine Cross Roads is located between Mansfield and Pleasant Hill, Louisiana. Monett's Bluff is halfway between Alexandria and Natchitoches. Henderson's Hill is halfway between Alexandria and Monett's Bluff.

TOPOGRAPHICAL ENGINEERS:

Prepared under the direction
of

U. S. ENGINEERS

Capt. Wm. C. Margedant 10th O.V.I.
SUPT. TOPOGRAPHICAL DEPT.

Pike	Burnt House	Stone Bridge	Signal Station
Good Wagon Road	Salt Works	Wooden Bridge	Fortifications & Intrenchments
Bad Wagon Road	Church	Suspension Bridge	Head Quarters
Bridle-Path	Post Office	Bridge over small stream	Camp
Foot Path	Foundry	Trestle Bridge	
Corduroy Road	Distillery	Ferry	
Rail-Road	Cotton Factory	Ford	Road
River	Saltpeter Works	Woods	River
Creek	Saw Mill	Pine Cedar	Creek
Run	Flour Mill	Wheat Field	Federal
Spring	Cotton Gin	Corn	Confederate
Marsh	Tanyard	Cotton	Federal
Town	Stone Fence	Orchard	Confederate
House	Rail Fence	Hills vertical	Artillery
Public Building	Stake Fence	Hills horizontal	

Printed and Issued Field Sketchbook of Ambrose Bierce

The extent to which the Union armies in the West organized and coordinated their mapping operations is evidenced by this printed field sketchbook used by Ambrose Bierce. A formalized, systematic legend is printed on the inside front cover so that the notes, sketches, memoirs, and maps of one topographical engineer made sense to all of them. If one engineer's "bad wagon road" looked like another engineer's "good wagon road," the consequences could be dire. An especially poignant symbol that appears regularly on contemporaneous maps is the bare chimney and hearth indicative of a burned-down house. On the inside page of the back cover, Bierce notes that he covers fifty feet with eighteen paces. Engineers estimated measurements with whatever instruments they had handy—including their hands, their arm-spread, or their horses' length.

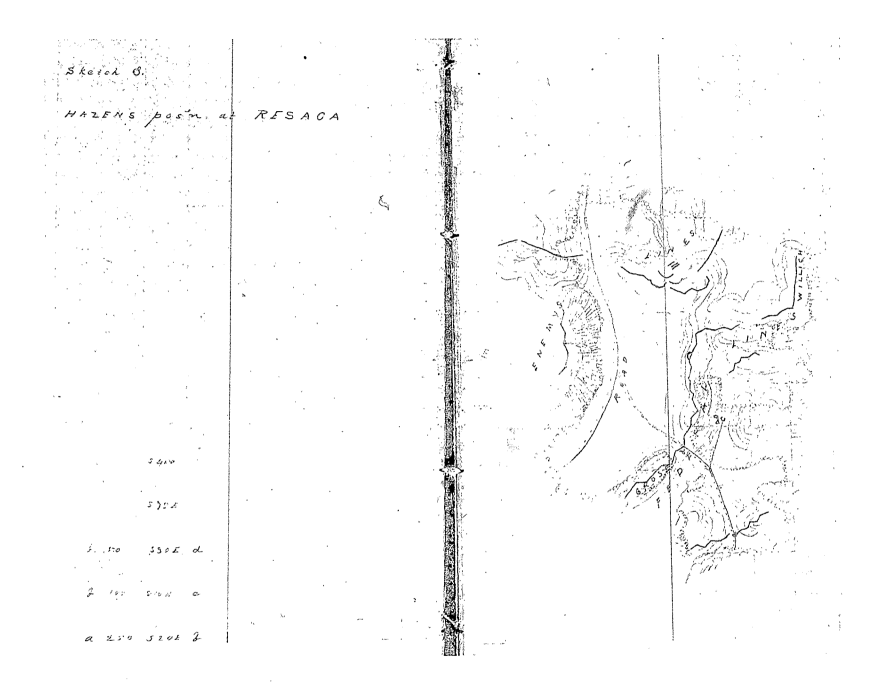

Ambrose Bierce's Sketch of Union General William B. Hazen's "Pos'n" at Resaca, Georgia, May 1864

Ambrose Bierce's field sketchbook was designed so that field notes could be made on the left page while sketches appeared on the right. This is very much a working sketch. The tiny letters on the map's roads correspond with the graph on the left page. Thus, from point "a" to point "b" is fifty yards, heading south thirty degrees east. General Hazen's *A Narrative of Military Service* features a more finished version of this map. It also appears, uncredited, in the *Official Records* (volume 38, part 1, page 426).

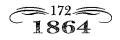

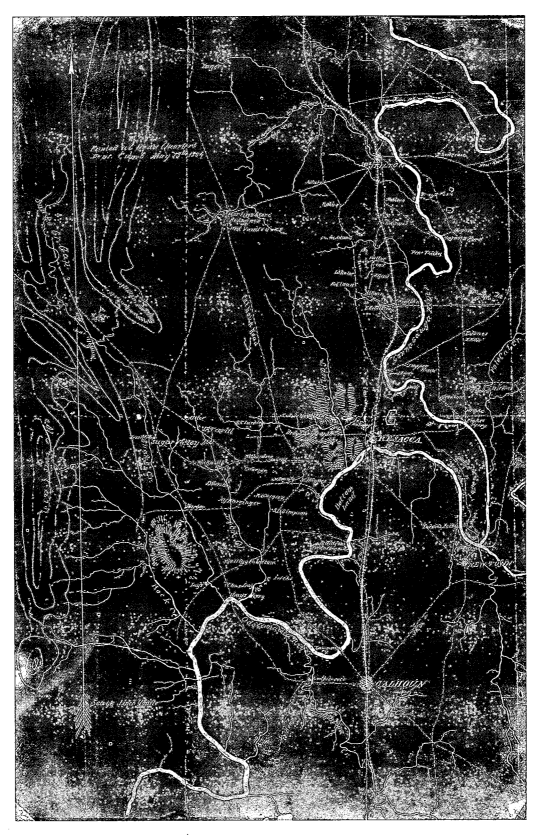

A Black Map of Northern Georgia Using William C. Margedant's Reproduction Process

This map is a photoreproduction, copied by placing chemically bathed paper beneath an original map done in heavy black ink on tracing paper. In sunlight, the blank areas of the treated paper became darkened. The ink on the original map blocked the sun, leaving the treated paper white beneath it. In this negative or black map, as it was called, the main road is hand-colored red, the Oostanaula River hand-colored blue. Union topographical engineer William C. Margedant developed the photoreproduction process.

The feature marked "Snake Creek Gap" was nearly the undoing of Confederate general Joseph E. Johnston and his Army of Tennessee. Having prepared for General William Tecumseh Sherman's expected advance against Atlanta for months, Johnston overlooked or ignored a gap through Mill Mountain. A too-cautious Federal approach wasted the opening, and a disappointed Sherman settled into his slow but ultimately successful strategy of maneuver.

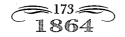

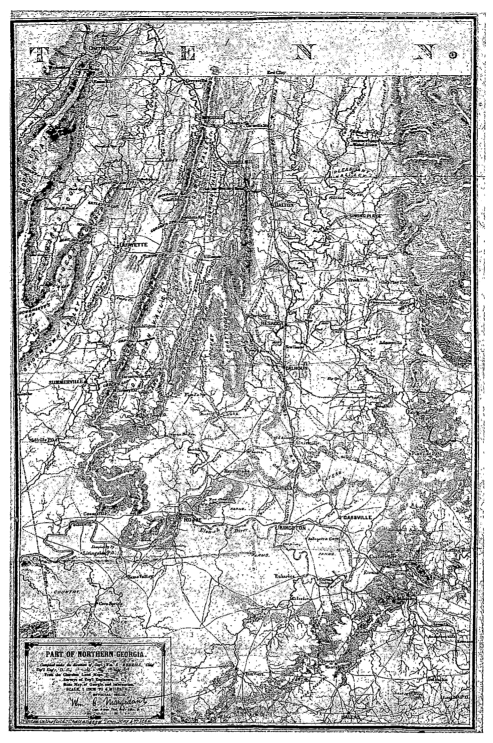

William E. Merrill's Map of Northern Georgia, May 1864

This is one of the Civil War's most famous maps, a lithographed version on muslin for the use of the cavalry. It was easily folded up to put in a saddlebag, and it could be used in all weather and washed when necessary. Work on this map was carried out all through the winter of 1863–64 by engineers under William E. Merrill, chief topographical engineer of the Union Army of the Cumberland. They were preparing for the initial phase of one of military history's great campaigns, General William Tecumseh Sherman's campaign from Chattanooga to Atlanta. Particular emphasis was put on the work of Sergeant N. Finegan and the Information Bureau, which expertly interviewed anyone and everyone who might be expected to have useful insights, information, and local knowledge of the regions where the forthcoming campaign might lead. At the beginning of May, the United States Coast Survey in Washington forwarded small-scale maps of northern Georgia to Sherman. They became the basic maps that were filled in with the data the Information Bureau had been gathering and with the mapped data gleaned from the various maps cited on this map. It was not, in spite of all the resources applied, a terribly accurate map. To a field officer such as General Jacob D. Cox, it was exasperatingly inaccurate, but it was better than anything the Confederates had: it depicted Snake Creek Gap, a feature the Confederates were ignorant of, and it was a better map than any other Union command ever carried into a campaign. A first-person account of the preparation of this map can be found in an appendix in *History of the Army of the Cumberland* by Thomas B. Van Horne, volume 2.

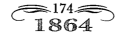

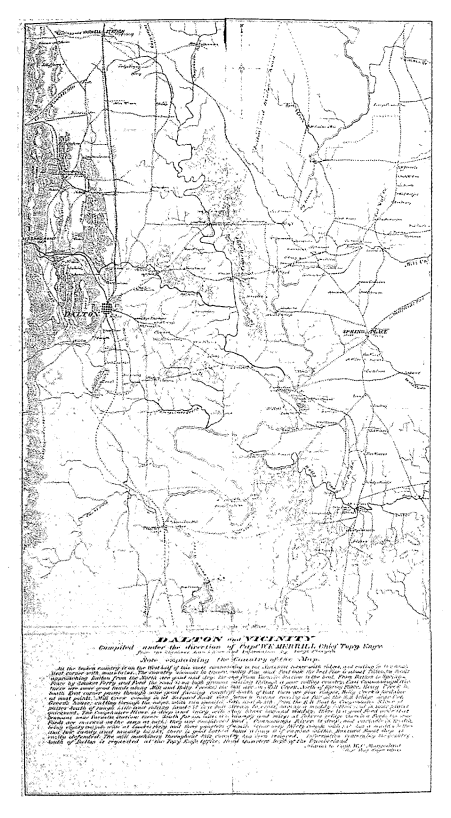

Dalton, Georgia, and Vicinity from the Cherokee Land Book and Information by Sergeant N. Finegan

This general-issue map, drawn at a standard scale of one inch to the mile, was prepared under the direction of Union captain William E. Merrill, chief of Union general William Tecumseh Sherman's topographical engineers. More detailed maps were drawn in a larger scale of two inches to the mile, another example of the standardization and uniformity in Merrill's office. The format of this map, on blue paper and with what constitutes a map memoir or detailed explanation of the militarily significant features of the region published as a sort of topographical broadside, is very unusual. Until 1832 a section of Georgia was occupied by the Cherokee Nation (they were forced to migrate to Oklahoma on what became known as "the Trail of Tears"). Dalton was held by the Confederate Army of Tennessee under General Joseph E. Johnston throughout the winter of 1863–64 and at the opening of Sherman's Atlanta campaign on May 4, 1864. The two rival generals fought one another for two and a half months, until Johnston was relieved of his command. Sherman's granddaughter later retained memories of Sherman and Johnston in their old age, "bending low over Civil War maps," mutually refighting their old battles.

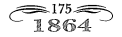

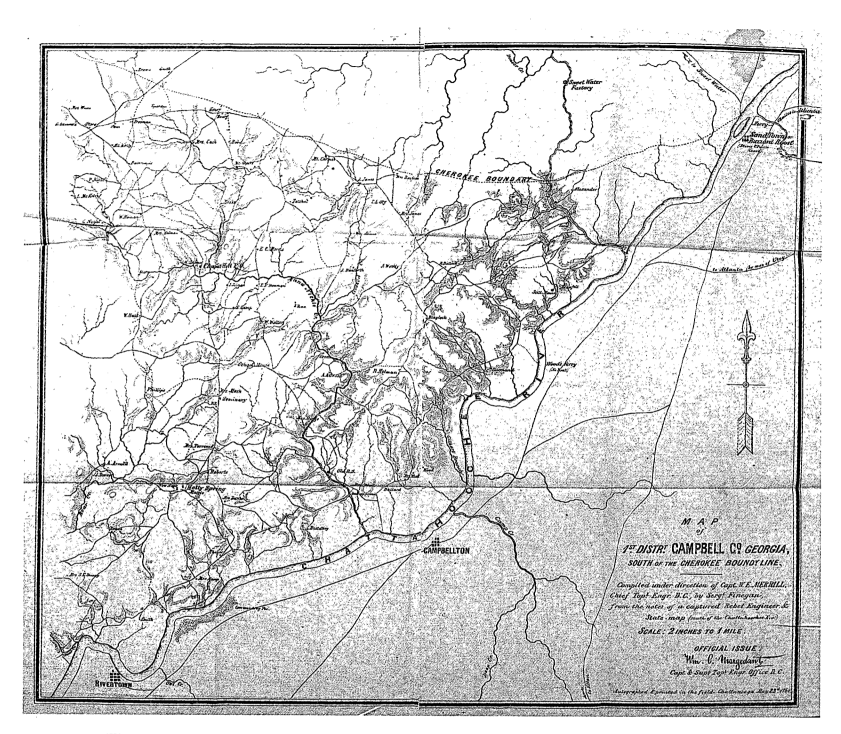

William E. Merrill and Sergeant N. Finegan's Linen, Lithographed Map of First District, Campbell County, Georgia, May 23, 1864

Union sergeant N. Finegan of the Fourth Ohio Cavalry was an expert map draftsman, but he was also an expert interrogator, eliciting crucial map information from "refugees, spies, prisoners, peddlers, and any and all persons familiar with the country. . . ." In this case, Finegan apparently made use of a captured Confederate field sketchbook to supplement the information available on extant published maps. Printed on linen or muslin for the use of cavalry, who could wad the map up and stuff it in their saddlebags and rinse it off when it was dirty, this map was prepared when Union general William Tecumseh Sherman's Atlanta campaign was well underway. "Autographed" means that the map was copied by a lithographic process known as autography. The original map was drawn in autographic ink on paper and transferred from the paper directly to the lithographic stone. This saved a lot of time and trouble because the normal lithographic process would have required that all the place-names and other writing on the map be written in reverse on the stone. By the autographic method, the "right reading" original produced a "right reading" print. One of the valuable aspects of Merrill's maps was their general distribution. Thus, even if a map were in error, all the commanders had the same wrong information and were never working at cross-purposes with contradictory information.

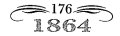

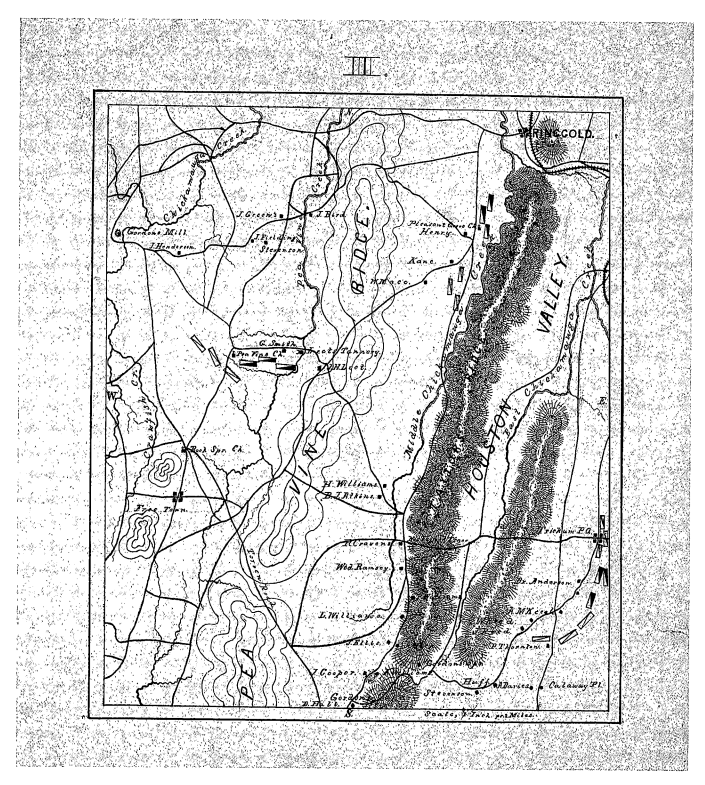

Lucius T. Stanley's Map of Ringgold, Georgia, May 6–7, 1864

Manuscript map on stiff, heavyweight drawing paper. This map was drafted by twenty-year-old Lucius T. Stanley of the 107th New York Infantry, part of Brigadier General Alpheus S. Williams's First Division, Twentieth Corps, Army of the Cumberland. At the time, William Tecumseh Sherman was just beginning his part of Ulysses S. Grant's concerted strategy to bring all the forces of the United States to bear simultaneously on the weakening and wearying Confederacy. As Stanley said in his May 4, 1864, diary entry, "The Division struck tents at 7:00 A.M. and started on this campaign." By May 6, the division was camped along Middle Chickamauga Creek at the foot of Taylor's Ridge.

Chief topographical engineer William E. Merrill had supplied the brigade, division, and corps commanders with a uniform, lithographed, rather detailed map of northern Georgia based largely on a United States Coast Survey map. A lithographed version of this map is found in the *Official Records Atlas* (plate 101, no. 3).

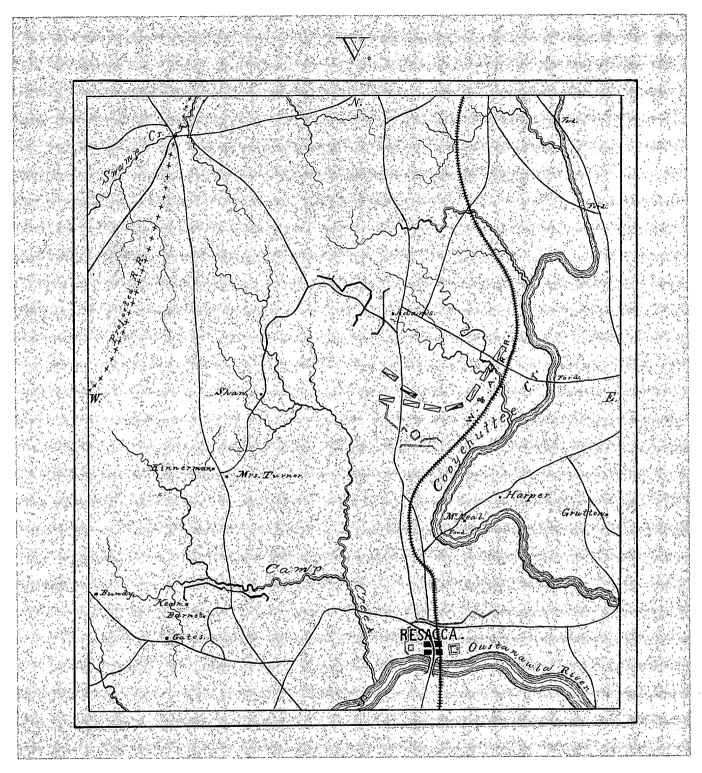

Lucius T. Stanley's Map of Resaca, Georgia, May 15, 1864

Manuscript map on stiff, heavyweight drawing paper. The armies of Union general William Tecumseh Sherman had missed and muffed a chance to end the campaign almost as it began by coming in on the flank of Confederate commander Joseph E. Johnston through Snake Creek Gap. The gap was apparently overlooked by Johnston, but Union topographical engineers discovered it during their expert interrogations of refugees, prisoners, peddlers, preachers, and others knowledgeable, or considered likely to be knowledgeable, of the country in front of the armies. Sergeant N. Finegan of the Fourth Ohio Cavalry was so adept at these interrogations that William E. Merrill held up lithographing his northern Georgia map so that Finegan could supplement the map with his invaluable data. This map, while nicely rendered, is spare on details and has several spelling errors, including "Resacca." The "Cooyehuttee Creek" is actually the Connasauga River. The Battle of Resaca on May 15 was a Union victory. A lithographed version of this map is found in the *Official Records Atlas* (plate 101, no. 9).

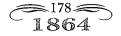

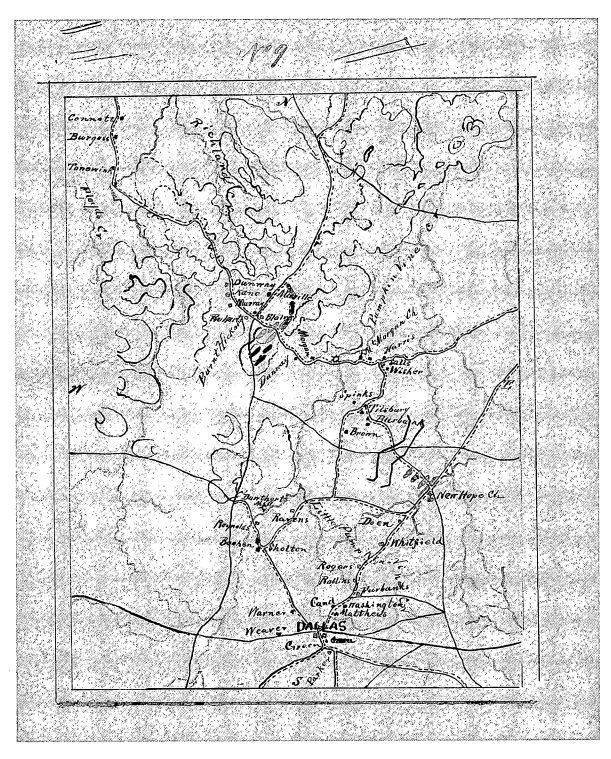

Lucius T. Stanley's Map of Dallas, Georgia (New Hope Church), May 24 through June 1, 1864

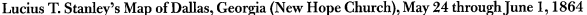

Manuscript map on tracing paper. The accuracy of the northern Georgia map issued to Union general William Tecumseh Sherman at the beginning of May grew less and less detailed, less and less accurate, as the armies moved south. Officers such as Lieutenant Lucius T. Stanley were detailed to gather topographical information, to sketch more detailed maps showing corrected cultural and physical data, and to copy and distribute the new sketch maps. In his diary, Stanley mentions a bridge near "Dawthert's Mill" that a woman living nearby saved as soon as the Rebels who had set fire to it were out of sight. This is one of the few maps of the campaign that does not include the Western and Atlantic Railroad; Sherman had swung west, away from the railroad, because of the strength of the Confederate position at Allatoona Pass. Confederate general Joseph E. Johnston moved west to counter Sherman's attempt to flank him, resulting in the Battle of New Hope Church. General Joseph Hooker, a witness to the fighting, thought it was magnificent. He encouraged Theodore R. Davis, artist-correspondent for *Harper's Weekly*, to draw an illustration of General Alpheus S. Williams's division in that action, which appeared in the magazine July 2, 1864, on page 928. A lithographed version of this map is found in the *Official Records Atlas* (plate 101, no. 13).

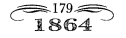

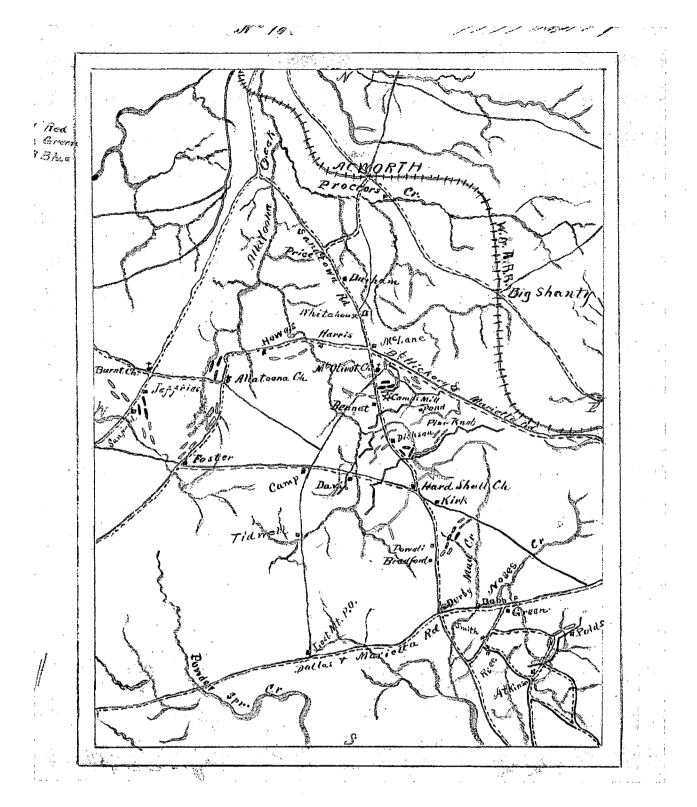

Lucius T. Stanley's Map of Lost Mountain/Big Shanty, Georgia, June 1-20, 1864

This vividly colored, tracing-paper manuscript map shows Pine Knob or Mountain, where Confederate general and bishop Leonidas Polk was killed by a Federal cannon on June 14, 1864. The Battle of New Hope Church was fought just west, off the map. Kennesaw Mountain is due east, also off the map. It was speculated that Lost Mountain received its name from the fact that it was the only topographical feature that was visible to someone otherwise lost in the vicinity. General Alpheus S. Williams described the country as "nothing but woods, woods, woods!" He also was amazed to find Yankee and Rebel pickets sitting together and talking things over on the banks of a branch of Allatoona Creek, a bit of fraternity he felt obliged to break up. These sessions were sometimes a boon to Union topographical engineers, who discreetly acquired topographical information and vital intelligence on fording sites from the talkative Rebels. General Williams's headquarters was at the Harris Jackson house immediately north of Mt. Olivet Church on the Sandtown Road, which is noted on the map. A lithographed version of this map appears in the *Official Records Atlas* (plate 101, no. 14).

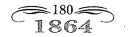

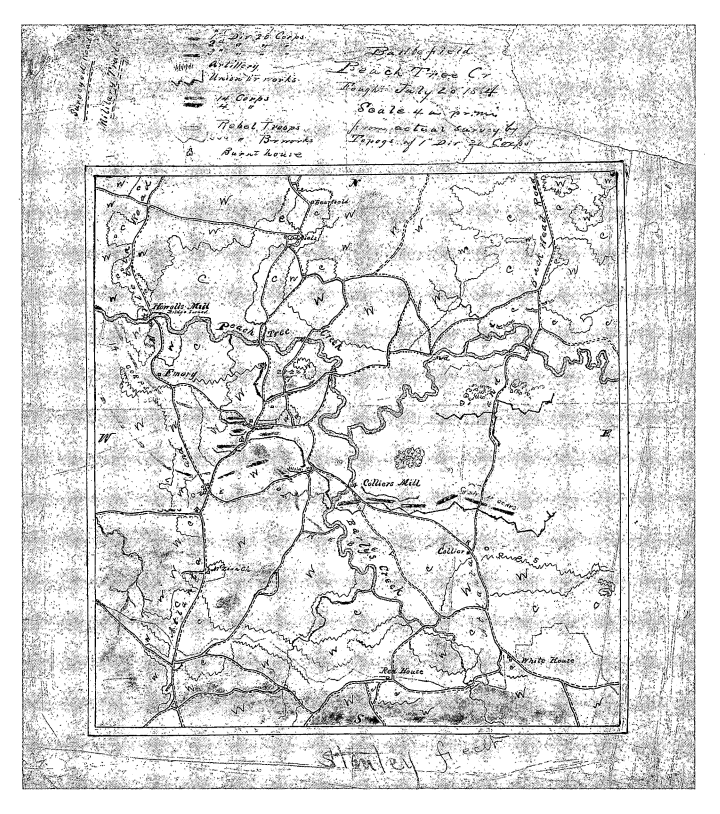

Lucius T. Stanley's Map of the Battlefield of Peachtree Creek, Georgia, July 20, 1864

This expertly drafted watercolor manuscript map carefully follows the standard symbols specified by chief topographical engineer William E. Merrill, including the vivid image for a burned house, a chimney with a black fireplace hole. Actual photographs of the campaign depict exactly the same picture. Peachtree Creek flows west to the Chattahoochee River. Confederate general John B. Hood hoped to attack the isolated Army of the Cumberland as it was in the process of crossing Peachtree Creek and separated from reinforcements by the creek's marshy headwaters. This detailed Union map indicates that Hood failed and that Union forces held the ground. A lithographed version of the map, attributed to "Capt. Samuel A. Bennett, Topographical Engineer, 20th Army Corps," appears in the *Official Records Atlas* (plate 101, no. 7), but the manuscript is clearly the work of Lieutenant Lucius T. Stanley.

Robert M. McDowell's Campaign Maps Showing Positions of Twentieth Corps on March from Chattanooga, Tennessee, to Atlanta, Georgia, 1864

This hand-printed title page contains an index or legend and the signature of Union captain Robert M. McDowell. The atlas is divided in several sections and, in fact, contains manuscript maps illustrating aspects of the marches of the Twentieth Corps with Sherman from Chattanooga to Atlanta, from Atlanta to Savannah, Georgia, and from Savannah into the Carolinas. Lithographed versions of these maps are in the *Official Records Atlas* (plates 71, 79, 80, and 101).

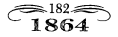

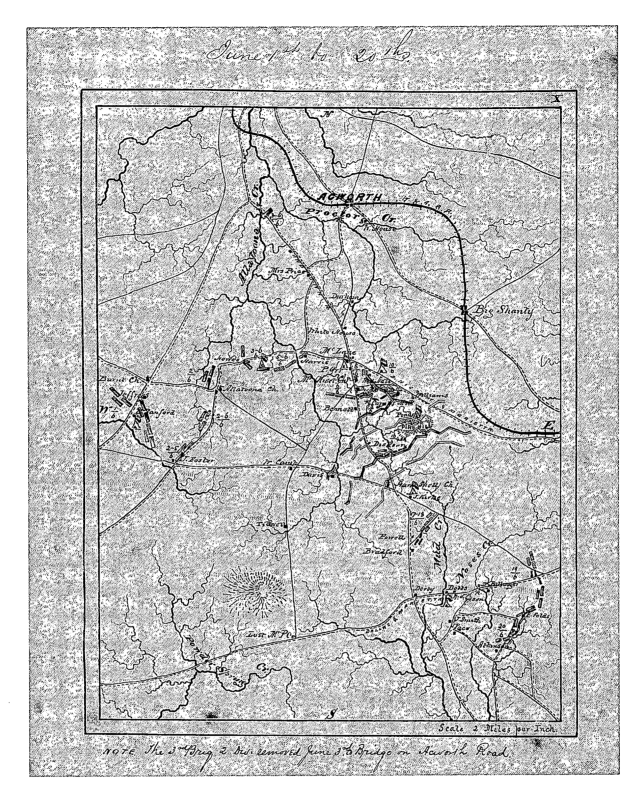

Robert M. McDowell's Map of Pine Hill and Big Shanty, Georgia, June 1–20, 1864

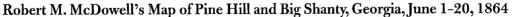

The Western and Atlantic Railroad was the spine of General William Tecumseh Sherman's campaign from Chattanooga toward Atlanta. He maneuvered back and forth, causing his opponent, General Joseph E. Johnston, to be turned out of his fortified lines, but always the Union troops shifted back to the railroad on which they depended for supplies. This map presents several locations of interest. On Pine Hill (also styled knob or mountain), Rebel general and bishop Leonidas Polk was killed by a Federal shell. At Big Shanty, earlier in the war, the locomotive General was stolen by Union scouts, resulting in the Great Locomotive Chase.

The Twentieth Corps was part of Union general George H. Thomas's Army of the Cumberland. This army was the most map-conscious army on either side during the Civil War. Its corps, division, and brigade commanders got better maps, sooner and more often, than other commands.

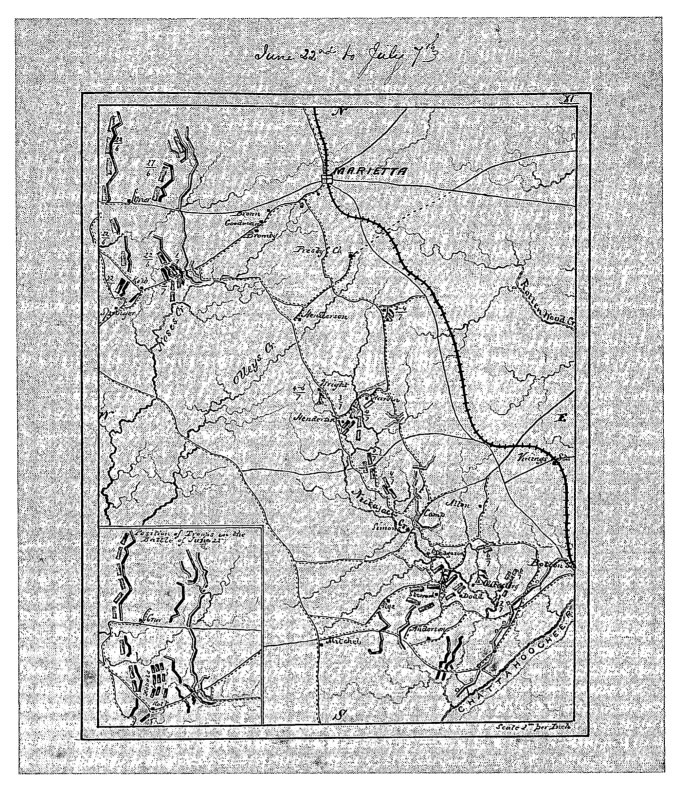

Robert M. McDowell's Map: Marietta, Georgia, to the Chattahoochee River, June 22 to July 7, 1864

From the Chattahoochee River, William Tecumseh Sherman was able to see the rooftops of Atlanta. From faraway Richmond, Virginia, the relentless progress of Sherman's command was watched with increasing alarm and dissatisfaction by the Confederate government, which blamed General Joseph E. Johnston for the way he conducted the campaign. Sherman called the fighting here "a big Indian war." By chance Sherman knew his way around the countryside. As a young lieutenant stationed in the area, he had spent his free time "exploring creeks, valleys, hills." He wrote that he "knew more of Georgia than the rebels did." When Sherman got across the Chattahoochee without Johnston delivering battle, the Rebel general's fate was sealed. He was relieved of command on July 17, replaced by his unruly subordinate, General John B. Hood. The fighting mapped in the insert was the Battle of Culp's or Kolb's Farm, as Robert M. McDowell has it. One Union participant, General Alpheus S. Williams, wrote home, "I never had an engagement in which success was won so completely and with so little sacrifice of life."

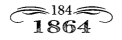

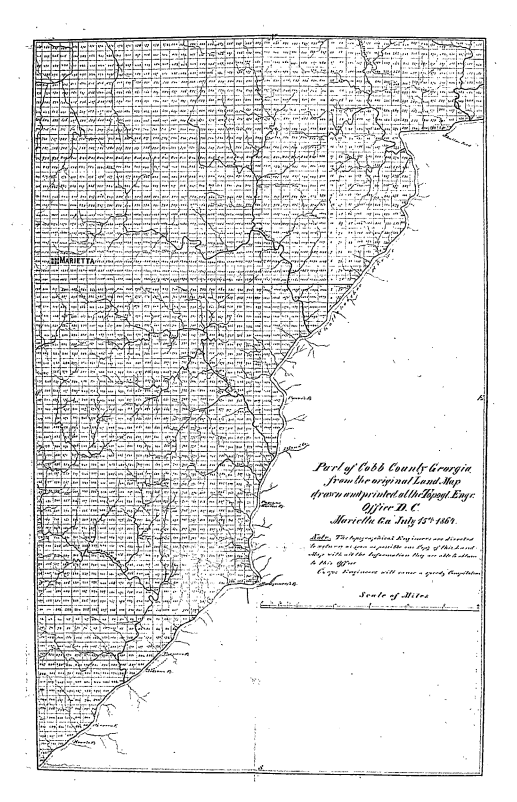

Part of Cobb County, Georgia, from the Original Land Map

This bland, unappealing, inscrutable-appearing map was a godsend for Union general William Tecumseh Sherman and his Topographical Engineer Office. In the early nineteenth century, most of Georgia was laid out in a succession of systematic, rectangular surveys. Most of its land was granted by a lottery, and the numbered squares on this map represent the property lines of Georgia's new counties surveyed in the 1840s. The square lots are six-tenths of a mile on a side, and each lot was numbered. Union topographical engineers grabbed these original maps wherever they found them and incorporated the information into their own maps, of which this is an example. These lot numbers became handy reference points. Sherman told his Army commander James Birdseye McPherson that his headquarters was in a house "not far from the northwest corner of lot 273." In the absence of other obvious cultural or physical landmarks, these were eminently satisfactory directions. Orders were given to topographical engineers to embellish, fill out, supplement, and annotate these maps and return the results to the Topographical Engineer Office. These reflect the operational basis of the most successful mapping organization of the war. (Special acknowledgment is made to Dr. Philip L. Shiman for his explication of grid lines and their significance in a draft article entitled "Groping Through Georgia" in the author's possession.)

James Birdseye McPherson's Field Map, 1864

This field map, stained with his lifeblood, was found folded in the breast pocket of James Birdseye McPherson when his body was recovered in the vicinity of Atlanta, Georgia, on July 22, 1864. The thirty-five-year-old commander of the Union Army of the Tennessee, one of Sherman's armies on the Chattanooga-to-Atlanta campaign, McPherson was the only army commander killed in battle during the Civil War. McPherson began the war as an engineer, drawing maps and planning fortifications under General Ulysses S. Grant. This uncredited and undated map on tracing linen does not show the area of McPherson's death site, which is just south of the bottom edge of the map, at about the centerfold. McPherson's death may have been the result of a map error. Movements were supposed to pivot on Howell's Mill, also named Powell's Mill on some maps. And different maps located Howell's Mill differently. McPherson may have ridden into the midst of the Rebels who had penetrated their way through a gap in the Union lines resulting from this confusion. It was the opinion of both Grant and Sherman that had he survived and had the war lasted longer, McPherson would have emerged as the greatest general of them all.

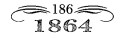

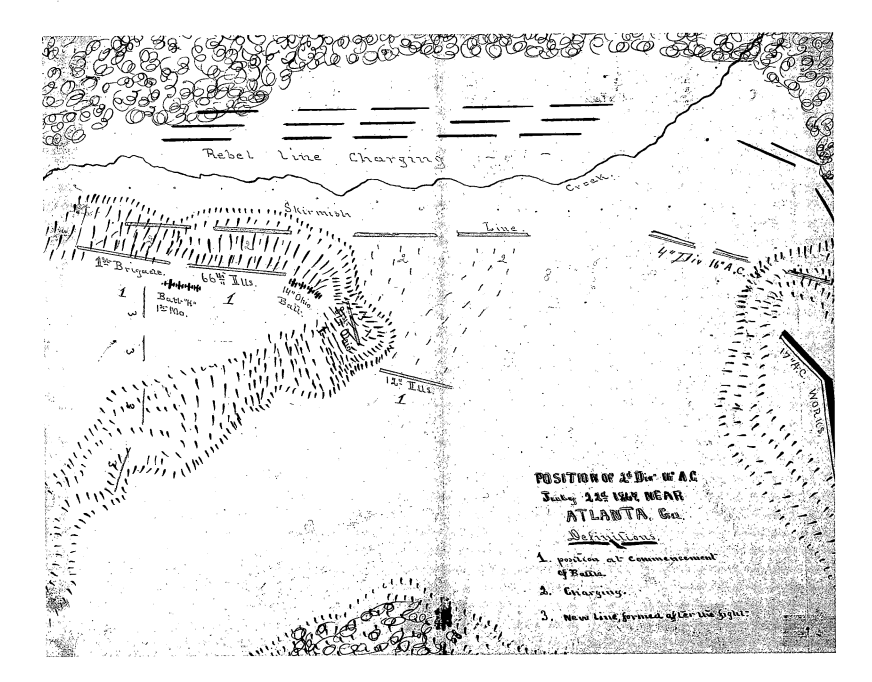

Map of Position of Second Division, Sixteenth Army Corps, Near Atlanta, Georgia, July 22, 1864

The Sixteenth Corps was part of the Union Army of the Tennessee, which lost its beloved commander, General James Birdseye McPherson, on this day in what is known as the Battle of Atlanta. The corps was commanded by General Grenville M. Dodge, a railroad surveyor and civil engineer by profession and another of the map connoisseurs in Sherman's armies. While mapping in the Army of the Tennessee never reached the levels attained in the Army of the Cumberland, considerable attention was paid to mapping, and, under the influence of William E. Merrill, specific mapping duties were delegated to each engineer. One would be designated to record the topography in the rear sectors of their division, one was assigned to handle the topography in the immediate area of field operations, and another was responsible for recording what happened and where on the battlefield. This small, rapidly executed sketch on paper is likely the work of an engineer assigned the latter duty. Elevations are quickly drawn with hachure marks, woods hastily scribbled in, and the positions of the Union and Confederate lines noted.

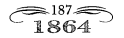

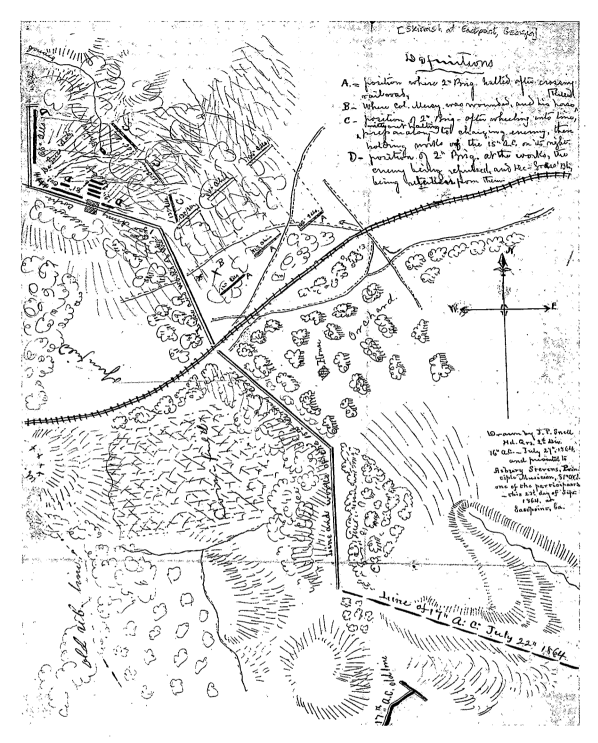

J. P. Snell's Map of Skirmish at East Point, Georgia, Drawn July 27, 1864—In Fact, a Depiction of a Segment of the Battle of Atlanta, July 22, 1864

Despite the title of the map and the date of its execution, it is not a sketch of the action that occurred at East Point, Georgia, south of Atlanta, on July 27, 1864. Rather, it depicts, in more detail than any other extant map, a phase of the Battle of Atlanta, also known as Bald Hill, fought on July 22, 1864. Through the center of the map runs the Georgia Railroad. A railroad cut separates the Union lines. De Gress's Battery is spiked and abandoned by the Union artillerists, but reinforcements, including Colonel August Mersy of the Second Division's Second Brigade, arrive to turn the tide of battle. Bald Hill is sketched, but not labeled, at the bottom of the sketch. Hand-to-hand fighting and bayonet wounds were extremely rare in Civil War combat, but both occurred here on the south slope of Bald Hill as the Confederates desperately but unsuccessfully sought to regain the victory they had within their grasp earlier in the day. Union general James Birdseye McPherson was killed several hundred yards southeast of Bald Hill. Snell's map is a very accurate, detailed rendering of this section of the battle. The East Point designation is a mystery. Of particular interest is the rail line through the cut. The Atlanta Cyclorama depicts the ground covered by this map in great detail.

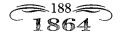

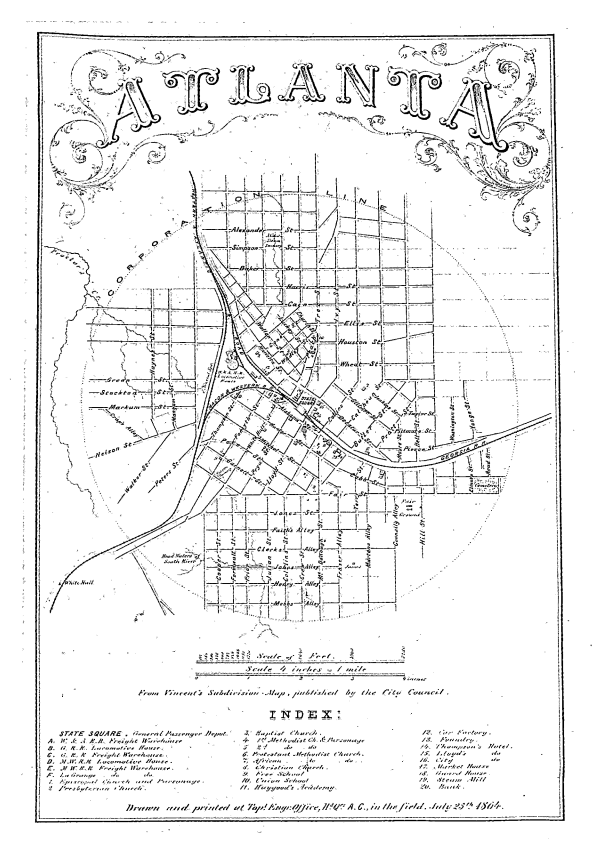

Ambrose Bierce Papers: Subdivision Map of Atlanta, Georgia, July 25, 1864

The securing, by capture or other means, of maps such as Vincent's Subdivision Map of Atlanta, from which this map was copied, was part of the responsibility of topographical engineers in Union general William Tecumseh Sherman's armies. It was printed and distributed as the Union armies began to reach to the outer limits of Atlanta, providing invaluable information, especially to the artillery, as lines were laid out and troops and guns were deployed to besiege and bombard the defended town. Maps like this were confiscated from the residences of county surveyors, from county courthouses, and from local civil engineers. When Atlanta fell, two good maps of Georgia were located in the town hall.

above:
George N. Barnard's Photograph of Buzzard-Roost Gap, Georgia, 1864

George N. Barnard traveled with Union general William Tecumseh Sherman's army as the official photographer of the Military Division of the Mississippi and later as part of the small topographical staff of Sherman's chief engineer Orlando M. Poe. Barnard arrived in Atlanta in mid-September 1864 from Nashville and was kept busy photographing maps of the army's completed marches and, secretly, photographing maps of the routes to be used en route to Savannah and the sea.

It is useful in examining Civil War maps to recall that the roads drawn on the maps could look like this on the ground.

opposite above:
Robert M. McDowell's Map: Leaving Atlanta, November 15–16, 1864

Union captain Orlando M. Poe, General William Tecumseh Sherman's chief engineer, said, "Atlanta has ceased to exist in military terms" as the Union army cut loose from its base there and began to move east, in three separate columns, toward the sea. The city of Atlanta was destroyed, as Poe said, but only in military terms. The railroads, the trains, the depots, the warehouses—any of the things that would be useful to an army—were torn up, burned down, or otherwise destroyed.

opposite below:
Robert M. McDowell's Map: Approaching Milledgeville, Capital of Georgia, November 21–23, 1864

Even as Southern newspapers were reporting that Union general William Tecumseh Sherman's march was stalled and that he was in "a woeful state," his columns appeared at Milledgeville perfectly equipped, perfectly organized, and in overwhelming force. Sherman's march probably involved more sophisticated long-term planning than any other campaign of the war. He had studied the tax maps. He had studied the 1860 census reports carefully, fully aware that the more people who lived in a region, the more easily soldiers could live off the land. The Confederate government helped too. Plantations had been encouraged to shift from cotton production to growing corn, to help the Southern war effort. The corn instead fed Sherman's columns as they moved in a sixty-mile-wide swath through Georgia. Robert M. McDowell helped map the northern edge of the campaign.

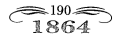

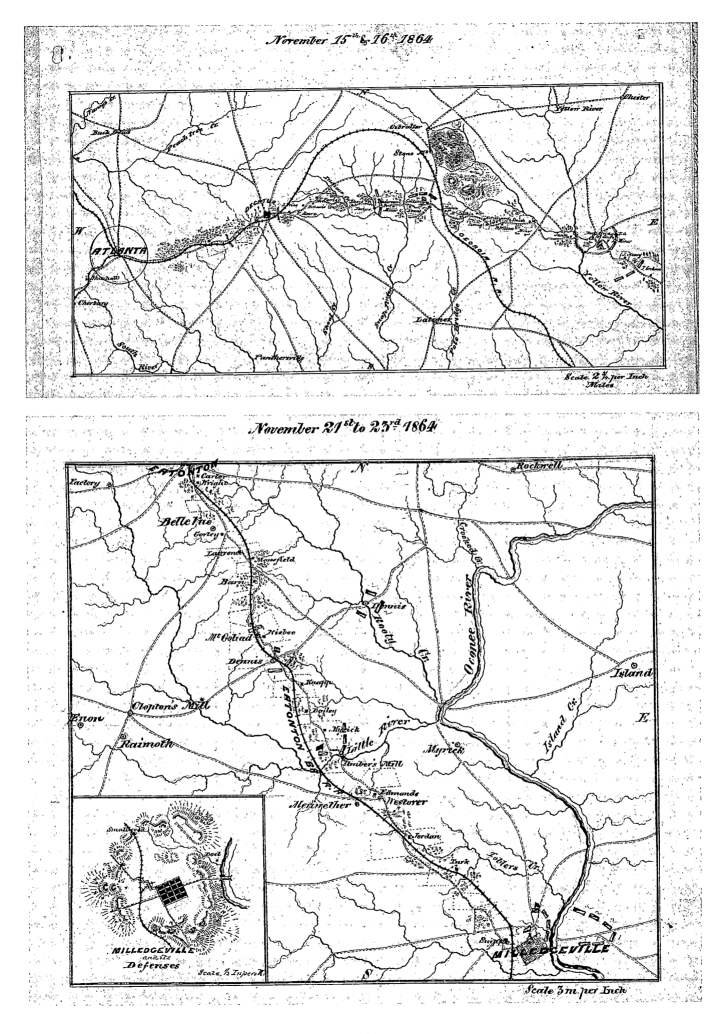

November 15th & 16th 1864

November 21st to 23rd 1864

MILLEDGEVILLE
and its
Defenses

Scale 1/2 In. per M.

Scale 3 m. per Inch

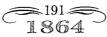

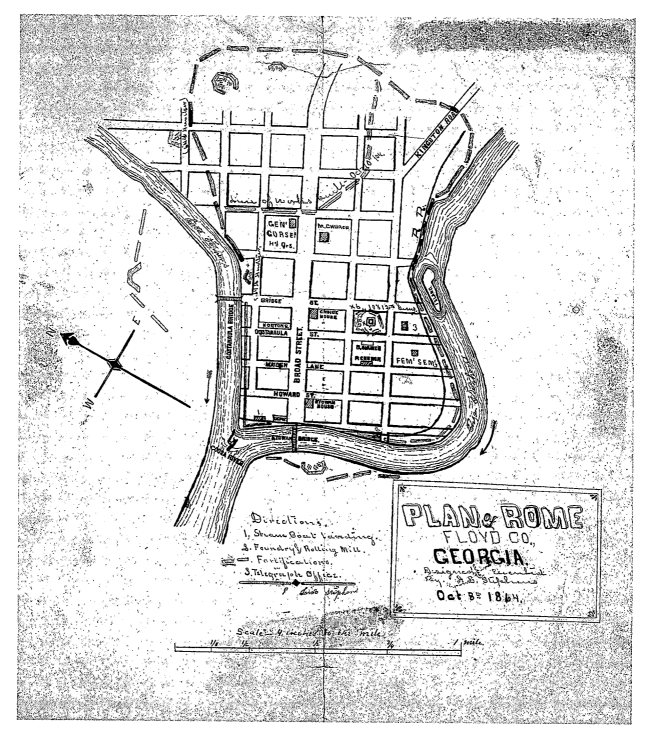

J. P. Snell's Plan of Rome, Floyd County, Georgia, 1864

Union general John M. Corse held Rome, Georgia, northeast of Atlanta at the confluence of the Coosa and Etowah rivers. A feeder rail line, the Rome Railroad, ran from Rome due east to the main Western and Atlantic Railroad. After the capture of Atlanta by Union forces, Confederate general John B. Hood attempted to prize the Federals out of Atlanta and transfer the seat of war to the valley of the Tennessee River. Essential to this plan was the cutting, or so Hood supposed, of William Tecumseh Sherman's lines of communication and supply along the Western and Atlantic Railroad. To this end, Hood detached a force to capture Union supplies at Allatoona and destroy the railroad bridge over the Etowah. General Corse was sent from Rome, via the railroad, to reinforce the small Allatoona garrison. Within hours, one of the Civil War's most dramatic military vignettes unfolded. A Confederate surrender demand was replied to with a curt, jaunty refusal. Sherman watched the proceedings from Kennesaw Mountain, fourteen miles away. Signal flags wigwagged messages across the long, smoking distance. One of these, from Sherman to the beleaguered Corse, said, "Hold fast; we are coming." It became "Hold the fort; I am coming," and in that form entered the American lexicon. This map, or plan, of Rome shows Corse's headquarters three days after the fight at Allatoona. Corse was wounded in the face but survived handily and drove the Confederates away.

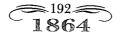

**Rodney E. Harris's Atlas of Maps Showing the Campaign of the Union Twentieth Corps
from Atlanta to Savannah, Georgia, 1864**

This is the cover to an apparently self-produced, self-published atlas of a series of maps Private Rodney E. Harris, of the 107th New York Regiment, executed en route with the left wing of Union general William Tecumseh Sherman's army on one of the most famous marches in military history, Sherman's March to the Sea. Nothing in Harris's military papers indicates his topographical engineering and/or cartographic activities. The maps are on wallpaper-weight paper or heavy linen. All are manuscript maps done in pencil, watercolor, and ink. There are several other similar compilations of maps of Sherman's marches and campaigns, but this is a singularly impressive one. Harris was from the small Finger Lakes town of Tyrone, Steuben County, New York. Lithographed versions of all the maps in this atlas appear in the *Official Records Atlas* (plate 71).

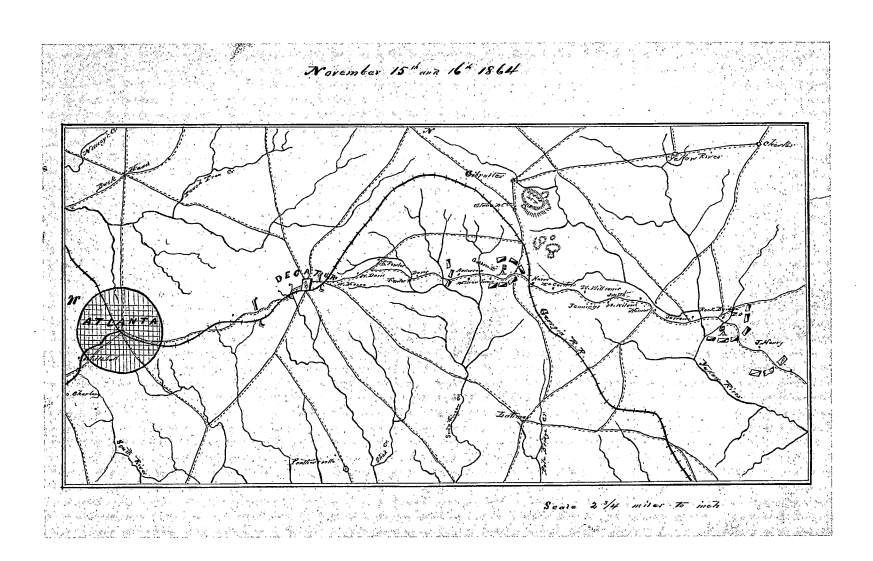

Rodney E. Harris's Campaign Map Number 1, Union Twentieth Corps, Atlanta to Savannah, Georgia, November 15 and 16, 1864

This is a small-scale map, befitting the speed of General William Tecumseh Sherman's March to the Sea. Sherman divided his army of approximately 60,000 men and 2,700 wagons into four or more columns, sometimes operating twenty miles apart, so that the army could move more quickly, forage more widely, destroy more thoroughly, and completely confuse the enemy as to its ultimate destination. The destruction of the Georgia Railroad was one of the objectives of the march. Confederate military planners, trying to evaluate what Sherman would do by determining what they supposed he could do, thought Sherman's planned march a military impossibility. Sherman himself, on November 16, 1864, "rode out of Atlanta by the Decatur Road" and turned back to see Atlanta smoldering, the woods where his favorite lieutenant, General James Birdseye McPherson, had been killed four months earlier, and his army columns spreading out behind him. "Atlanta," as Sherman wrote in his *Memoir,* "was soon lost behind the screen of trees, and became a thing of the past." A lithographed version of this map appears in the *Official Records Atlas* (plate 71, no. 1).

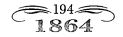

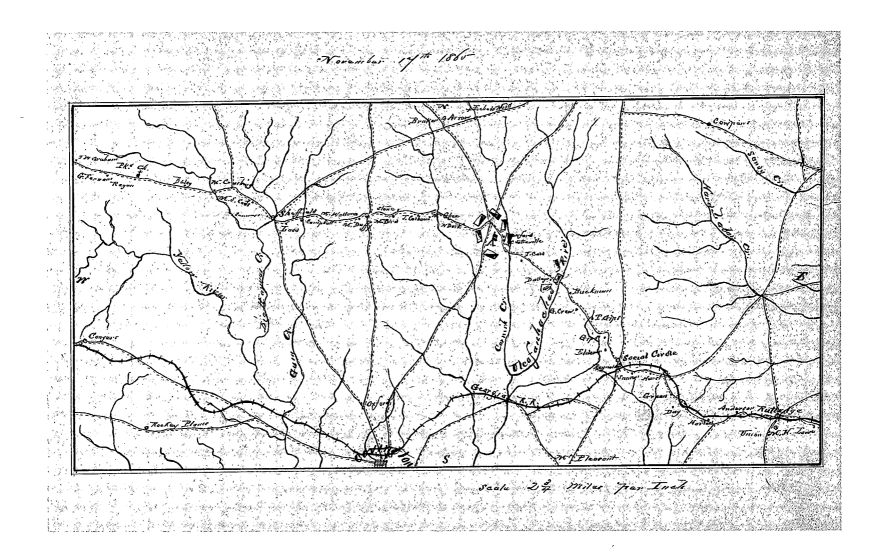

Rodney E. Harris's Campaign Map Number 2, Union Twentieth Corps, Atlanta to Savannah, Georgia, November 17, 1864

Union general William Tecumseh Sherman's armies moved at a rate of from twelve to fifteen miles per day. It would take two days for the army to cross the area indicated on this map. Sherman's men met with minimal Confederate opposition, except for occasional brushes with enemy cavalry, who could do very little but annoy veteran infantry. The wagons moved in the midst of the infantry, partly for protection and partly for convenience—the soldiers were close to their supplies and didn't have to wait for them to come up. At Covington, a soldier with a ham on his bayonet, a jug of molasses under his arm, and a piece of honeycomb in his hand caught Sherman's eye and remarked, "Forage liberally on the country," quoting general orders. A lithographed version of this map appears in the *Official Records Atlas* (plate 71, no. 2).

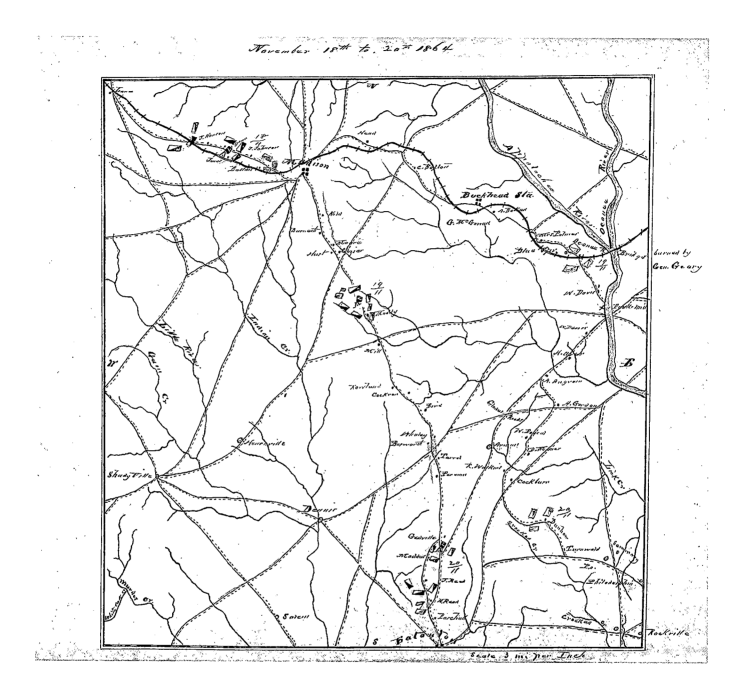

Rodney E. Harris's Campaign Map Number 3, Union Twentieth Corps, Atlanta to Savannah, Georgia, November 18–20, 1864

Drawn on a slightly smaller scale (three miles to an inch) than the two preceding maps, this map shows the Twentieth Corps moving deep into the interior of Georgia. Union general William Tecumseh Sherman deliberately cut telegraphic communications with the North, assuring Washington that the Southern papers would vividly record his progress. This they did, and their hopeful predictions as to the condition of Sherman's armies and their fate became increasingly dire. Operating on the left wing, the Twentieth Corps was working with the topographical organization of the Army of the Cumberland, the best and most efficient of its kind during the war. The "State Map of Georgia" referred to by Rodney E. Harris on the cover of the atlas would have been invaluable up to this point in the campaign. The maps of the Union Engineer Bureau had not reached all the way west to Atlanta. There were two Georgia state maps found in Atlanta town hall. The one Harris refers to may have been one of these. Harris, along the actual route of march, has heavily supplemented with his own observations and findings the meager information a very small scale state map would have presented. A lithographed version of this map appears in the *Official Records Atlas* (plate 71, no. 3).

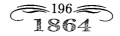

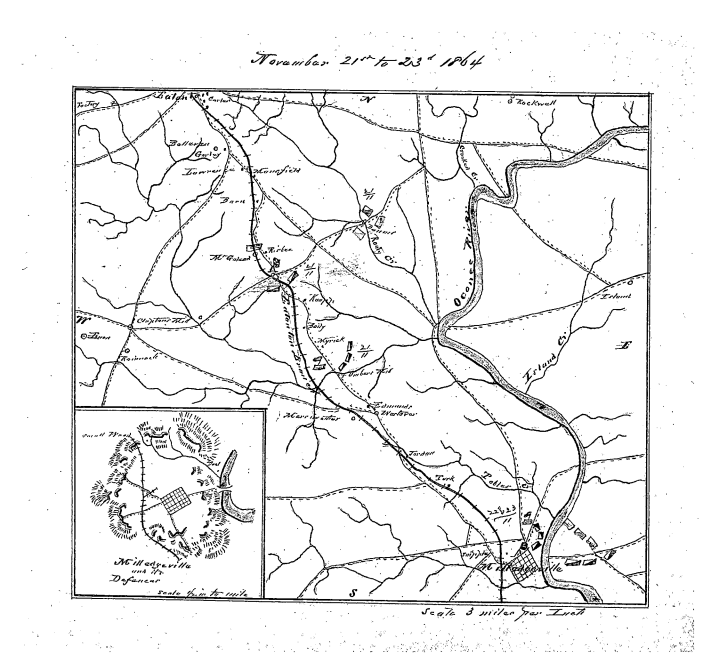

Rodney E. Harris's Campaign Map Number 4, Union Twentieth Corps, Atlanta to Savannah, Georgia, November 21–23, 1864, with Inset of Milledgeville

The Twentieth Corps detailed its topographical engineers to work in groups of three: one man with a compass estimating distances; one sketching maps; and the third gathering information. Rodney E. Harris's maps were partly based on such surveying operations. Milledgeville, Georgia's capital, was not defended, though, as the inset map makes clear, rather elaborate defenses had been prepared by the Confederates. The Union occupiers had only a brief chance to read their first newspaper in more than a week before they resumed their march. They were amused and annoyed by turns at Southern reports of the army's impending doom. A lithographed version of this map appears in the *Official Records Atlas* (plate 71, no. 4).

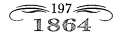

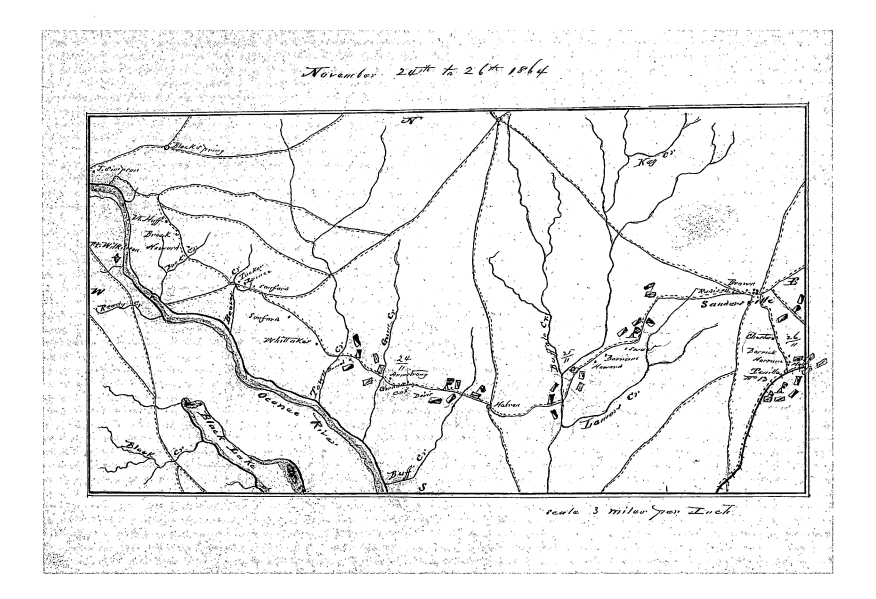

Rodney E. Harris's Campaign Map Number 5, Union Twentieth Corps, Atlanta to Savannah, Georgia, November 24–26, 1864

A small Confederate force of several hundred men had intended to contest the Federal crossing of the Oconee River, but the hopelessness of the gesture was obvious and a few desultory volleys were followed by a quick retreat. At this point in the march, Union general William Tecumseh Sherman's troops left the agriculturally rich region of Georgia, where they had traversed immense cornfields and had their fill of fowls, vegetables, and meat. On Thanksgiving Day, November 24, turkey was the preferred meal.

One of Sherman's staff officers commented that the cattle trains, a bovine commissary that accompanied the march of the armies, were becoming so large it was becoming difficult to drive them. Crossing a river or stream was a routine procedure for Sherman's fast-moving army. Each wing of the army had a pontoon train with a trained crew. The Rebels remarked that it appeared as if Sherman's men simply laid their tents under the wagons and made a bridge faster than the Confederates could burn one. A lithographed version of this map appears in the *Official Records Atlas* (plate 71, no. 5).

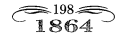

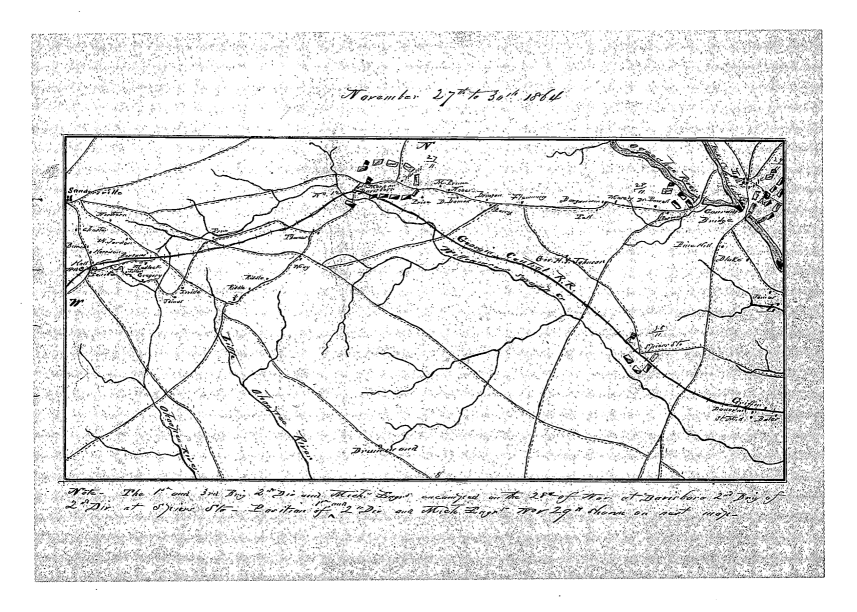

Rodney E. Harris's Campaign Map Number 6, Union Twentieth Corps, Atlanta to Savannah, Georgia, November 27–30, 1864

The Georgia roads over which Union general William Tecumseh Sherman's march proceeded were never very good. One corps commander said that they moved on roads whose condition "beggars all description." However, one Illinois soldier noted that Sherman's army made "good roads bad and bad roads good." In other words, the passage of thousands of men, wagons, and animals wrecked a practicable road, whereas an impassable road, of necessity, was covered with fence posts, fence rails, and felled trees—corduroyed—to keep the army from stalling in its tracks. A lithographed version of this map appears in the *Official Records Atlas* (plate 71, no. 6).

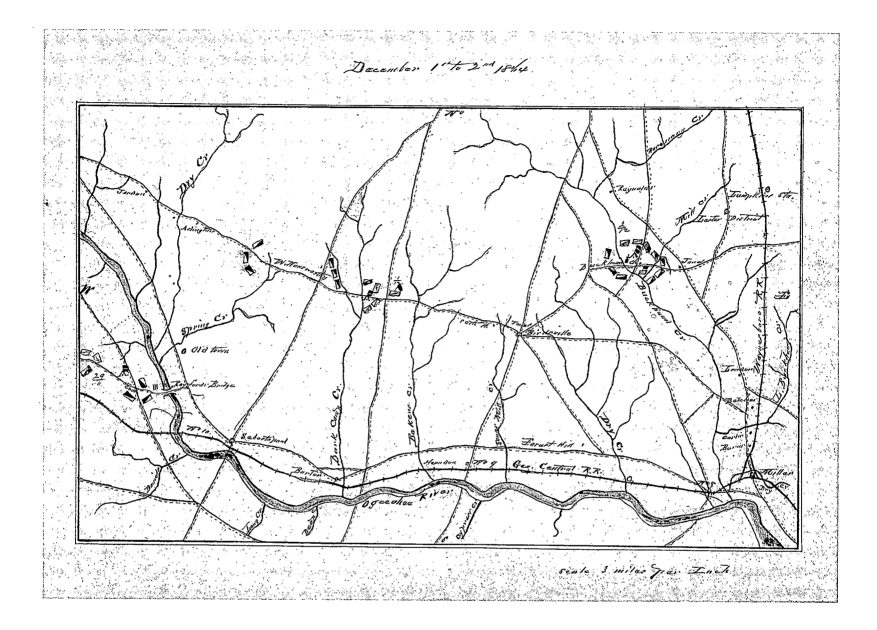

Rodney E. Harris's Campaign Map Number 7, Union Twentieth Corps, Atlanta to Savannah, Georgia, December 1–2, 1864

Union general William Tecumseh Sherman mentioned in his memoirs that on the March to the Sea, his saddlebags contained the four essential items he required on the campaign: a change of underclothing, a flask of whiskey, a bunch of cigars, and "my maps." As the Twentieth Corps moved this far east, the country became more barren and sandy and foraging became more difficult. The land was marshlike and full of pine barrens. The soldiers' life got a bit harder, but these were men who, as one of them put it, "are ready for a meal or a fight and don't seem to care which it is." And they were two-thirds of the way to the Atlantic Coast. A lithographed version of this map appears in the *Official Records Atlas* (plate 71, no. 7).

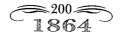

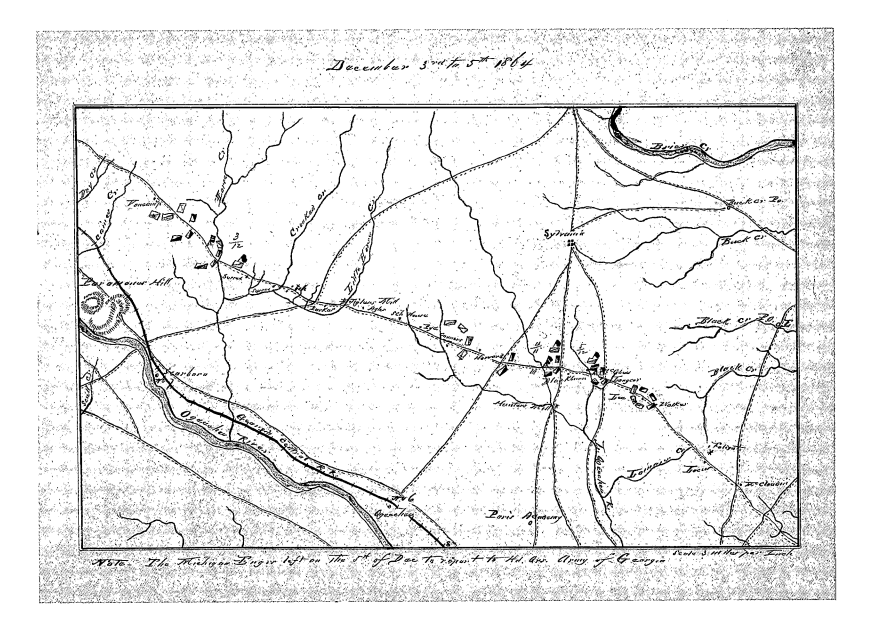

Rodney E. Harris's Campaign Map Number 8, Union Twentieth Corps, Atlanta to Savannah, Georgia, December 3–5, 1864

Unlike the Atlanta campaign, no maps were published in the field during the March to the Sea. Progress was too rapid, and, as the armies moved closer to salt water, as General William Tecumseh Sherman described their objective, the maps on hand were adequate to the needs of the columns. When maneuvering against an entrenched enemy, much more detailed, larger-scale maps were necessary than were called for here, where there was little opposition and no tactical movements. On December 3, Sherman was at Millen, Georgia, with the Union Seventeenth Corps. He took stock of the situation and, as he expressed it, "concluded to push on for Savannah." At about this time, five hundred miles to the north, President Abraham Lincoln turned around to one of his friends and said, "McClure, wouldn't you like to hear something from Sherman?" McClure was "electrified." Lincoln continued with a laugh, "Well, I'll be hanged if I wouldn't myself." A lithographed version of this map appears in the *Official Records Atlas* (plate 71, no. 8).

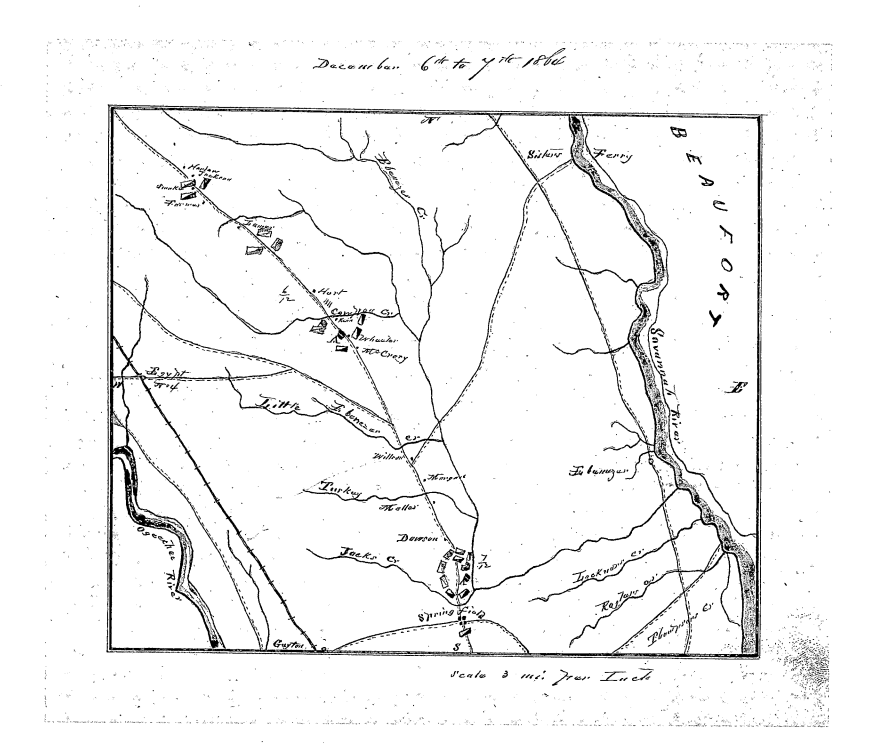

December 6th to 7th 1866

BEAUFORT

Scale 3 mi: per Inch

Rodney E. Harris's Campaign Map Number 9, Union Twentieth Corps, Atlanta to Savannah, Georgia, December 6–7, 1864

The corn and forage began to give out when the march had come to this point, but as Union general William Tecumseh Sherman was to note with satisfaction in his memoirs, rice fields began to appear and the rice "proved a good substitute" for both: "The weather was fine, the roads good, and everything seemed to favor us." Union general O. O. Howard, who commanded the Army of the Tennessee on the March to the Sea, said, "Sherman had remarkable topographical ability. A country that he once saw he could not forget. The cities, the villages, the streams, the mountains, hills, and divide—these were as easily seen by him as human faces, and the features were always on hand for use. It made him ever playing at draughts [checkers] with his adversary. Let the enemy move and Sherman's move was instant and well chosen." The stations, as they appeared along the railroad lines, were numbered by the topographical engineers and were referred to by these numbers in messages and dispatches: "We are at Station 9½—marked Barton on our map." A lithographed version of this map appears in the *Official Records Atlas* (plate 71, no. 9).

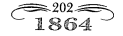

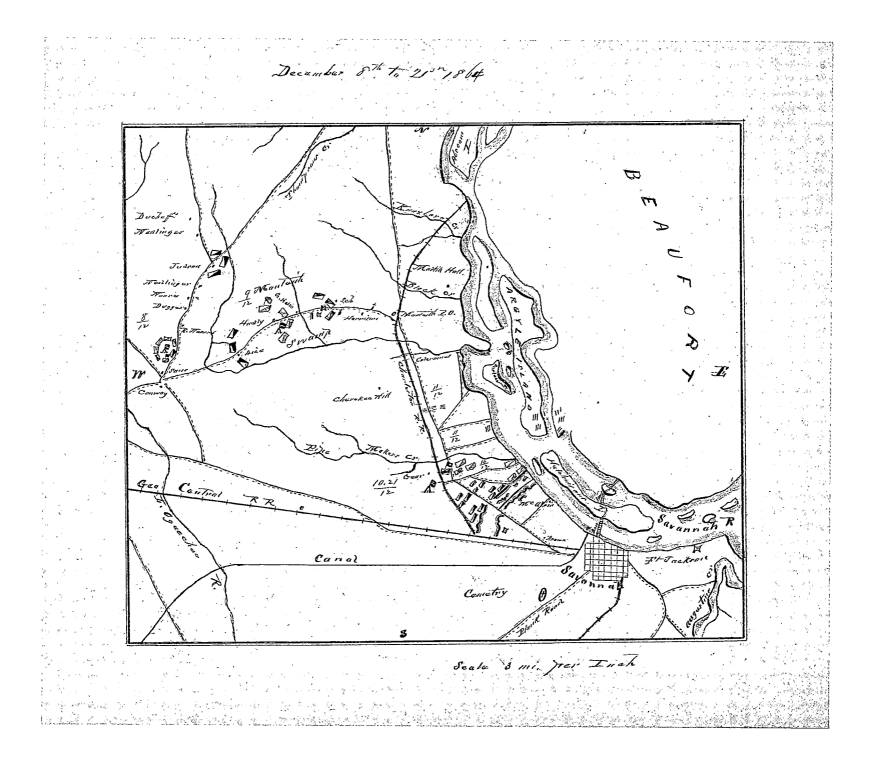

Rodney E. Harris's Campaign Map Number 10, Union Twentieth Corps, Atlanta to Savannah, Georgia, December 8–21, 1864

The city of Savannah was lightly garrisoned by Confederate forces under General William J. Hardee, but Union general O. O. Howard observed that "Savannah almost defended itself by its bays, bogs, and swamps all around, leaving only causeways to be defended." After the Twentieth Corps's Colonel Ezra A. Carman moved his Second Brigade to Argyle Island and by December 19 had his men on the South Carolina shore, threatening the Charleston and Savannah Railroad—the only line of communication Savannah's Rebel defenders had—it was confidently expected that the Confederate defense would be abandoned. Fort McAllister had fallen nearly a week earlier. On December 21, 1864, at five o'clock in the morning, Savannah surrendered. General William Tecumseh Sherman presented the city as a Christmas present to a relieved and grateful President Abraham Lincoln. A lithographed version of this map appears in the *Official Records Atlas* (plate 71, no. 10).

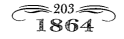

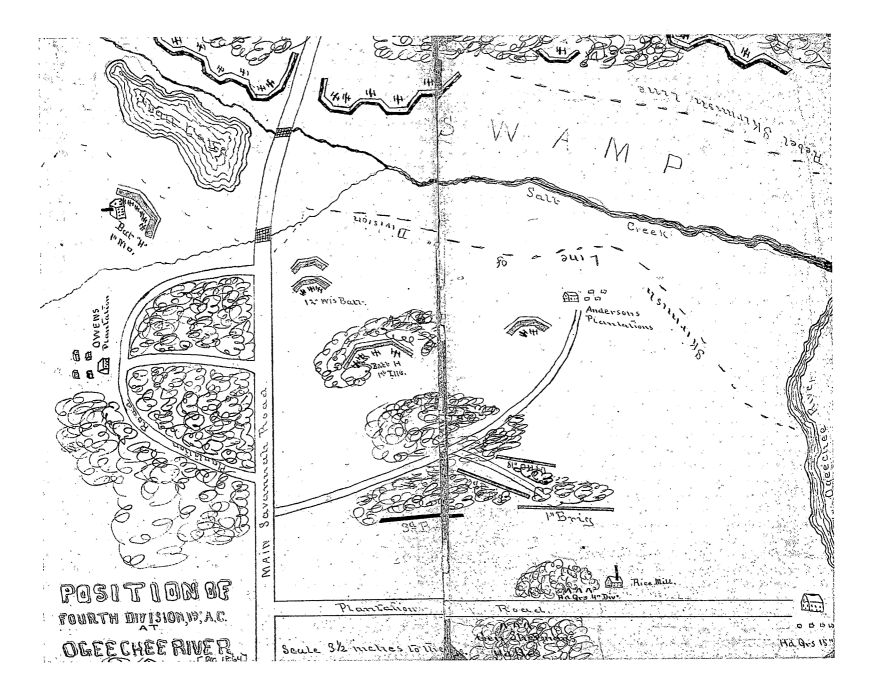

Map of Position of Fourth Division, Fifteenth Army Corps, at Ogeechee River, Georgia

Some thirty miles northwest of Savannah, Georgia, this position was taken briefly by the Fourth Division on December 10–11, 1864. By December 13, Fort McAllister had fallen to Union forces, and General William Tecumseh Sherman's March to the Sea had, in essence, been completed successfully when he made contact with a Union fleet and reestablished communications with Washington. Savannah surrendered on December 21, 1864. This very detailed map is more pictorial than a standard military map. It is oriented with north in the lower left corner, south in the upper right corner. A smaller-scale, overall representation of this position is found in the *Official Records Atlas* (plate 69, no. 4). The March to the Sea was so unstinting and rapid that there was little time for more formal mapping. The skilled members of the United States Coast Survey who accompanied the march did not attempt any maps. Sherman had learned from Ulysses S. Grant that an army could live off the land. Sherman substituted a careful analysis of the proposed routes of march for Grant's more intuitive approach. Sherman's March to the Sea was possibly the most scientifically planned operation of the war.

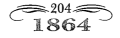

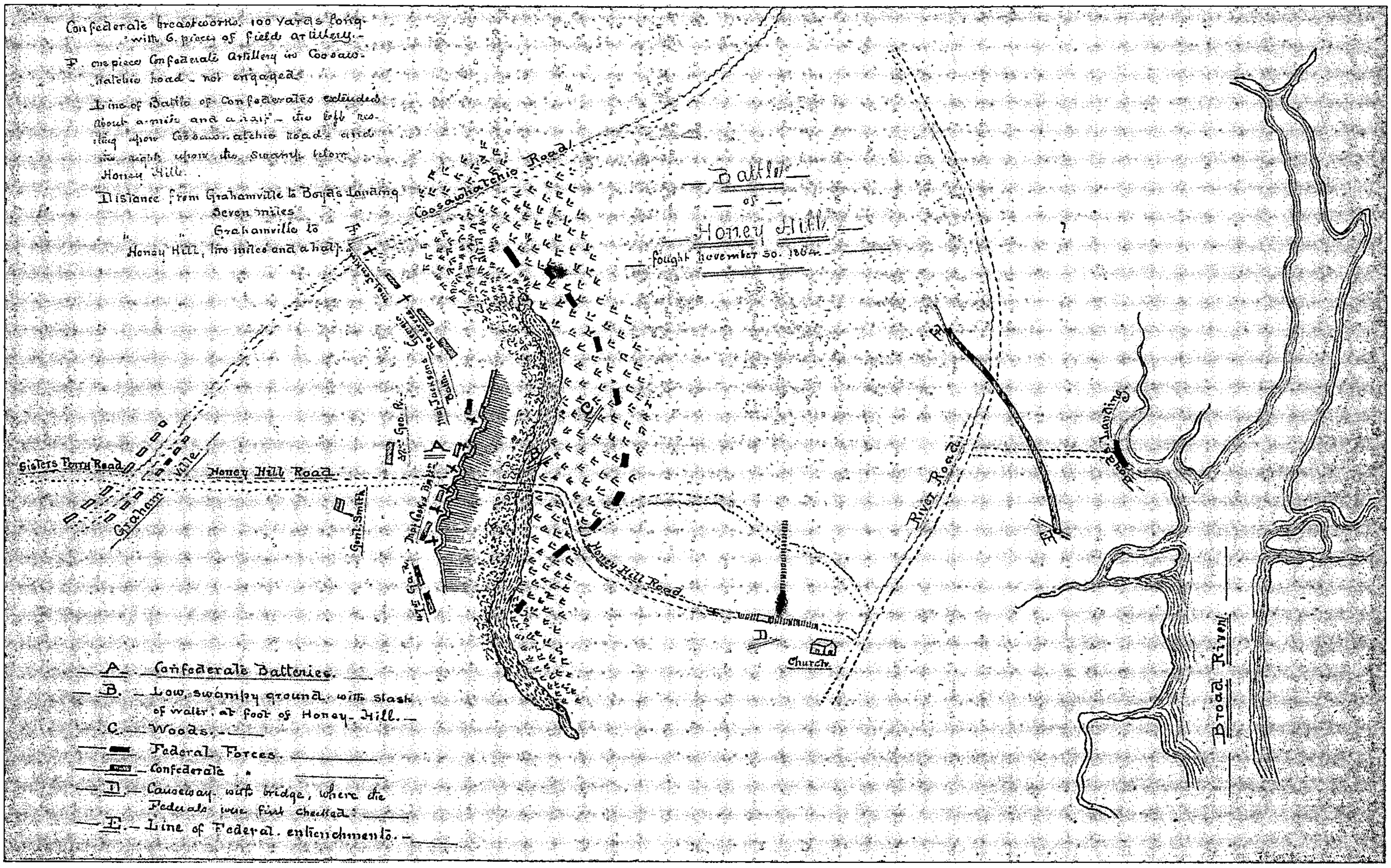

Confederate breastworks, 100 yards long with 6 pieces of field artillery.
F. one piece Confederate Artillery in Coosawhatchie road - not engaged.
Line of Battle of Confederates extended about a mile and a half - the left resting upon Coosawhatchie road - and the right upon the swamp below Honey Hill.
Distance from Grahamville to Boyd's Landing Seven miles
Grahamville to Honey Hill, two miles and a half.
Battle of Honey Hill
fought November 30. 1864.
Coosawhatchie Road
Sisters Ferry Road
Honey Hill Road
Grahamville
River Road
Church
Broad River
A. Confederate Batteries.
B. Low, swampy ground with slash of water, at foot of Honey Hill.
C. Woods.
Federal Forces.
Confederate
D. Causeway with bridge, where the Federals were first checked.
E. Line of Federal entrenchments.

The Battle of Honey Hill, South Carolina, November 30, 1864

As Union general William Tecumseh Sherman's March to the Sea was nearing Savannah, Georgia, an attempt was made by Union forces at Hilton Head, North Carolina, to establish a base to support him. Fourteen hundred Confederates in fortifications managed to withstand attacks by 5,500 Union troops, who were forced to withdraw. The informative legend explains what the map does not. This very presentable and expressive map uses only black and red ink on appropriately honey-colored paper. No cartographer is credited. Honey Hill is approximately thirty miles northeast of Savannah, Georgia.

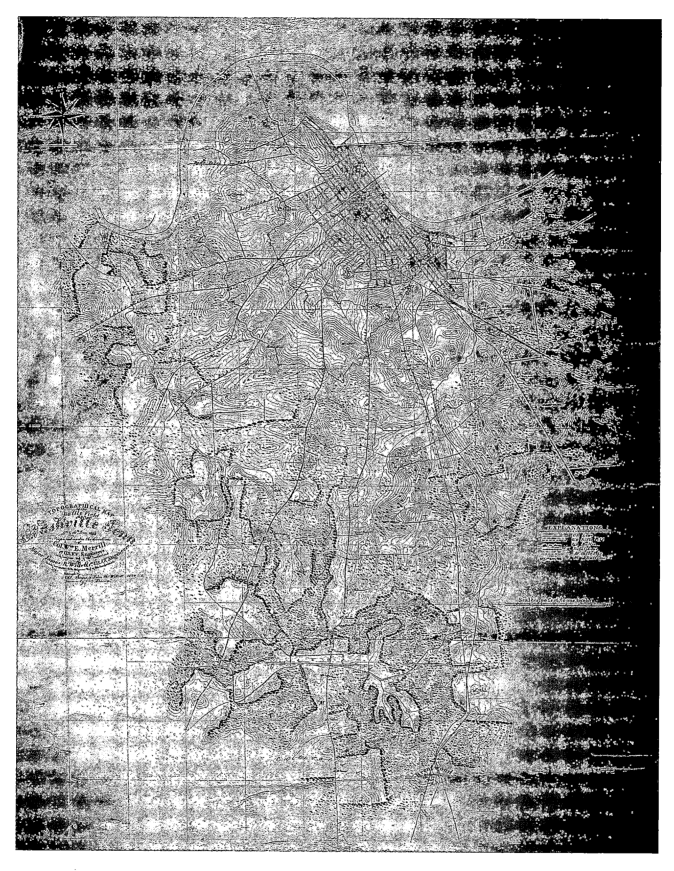

William E. Merrill's Topographical Map of the Battlefield of Nashville, Tennessee, 1864

An impressively detailed map of the Nashville battlefield. As was often the case when the field was secured after a battle, a very thorough map was possible and was prepared. That was certainly the case here. Confederate general John B. Hood's Army of the Tennessee ceased to exist after its defeat and rout at the Battle of Nashville (December 15–16) by forces under the command of Union General George H. Thomas. When William Tecumseh Sherman turned his back on Hood and marched to Savannah, Thomas had been left behind to deal with the Confederate's premier Western army. From the compass rose to the contoured relief with intervals shown in red numerals, this map represents a culmination of sorts in Civil War mapping. A lithographed version of the map appears in the *Official Records Atlas* (plate 73, no. 1).

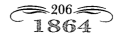

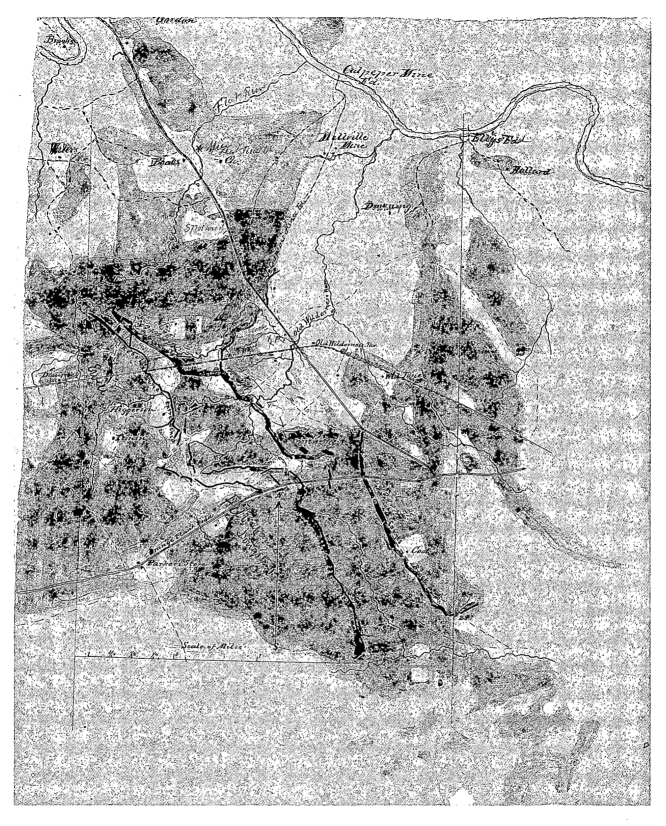

William Henry Paine's Sketch of the Battle of the Wilderness, May 5 and 6, 1864, Number 1

A work in progress, this sketch or plan of the Battle of the Wilderness is the first of three sketches documenting the first contact Ulysses S. Grant had with Robert E. Lee and the Confederate Army of Northern Virginia. The positions of different Confederate corps are written in. The broken, arching line in the lower left section of the map, running east and west, is the unfinished railroad. Lee wanted to fight Grant here because the tangled woods neutralized the Union's numerical advantage, especially in artillery. Some of the data included in this sketch may have been obtained by Paine from notes taken on the region in the late 1820s by R. C. Taylor, an employee of the State Geological Survey of Virginia.

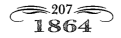

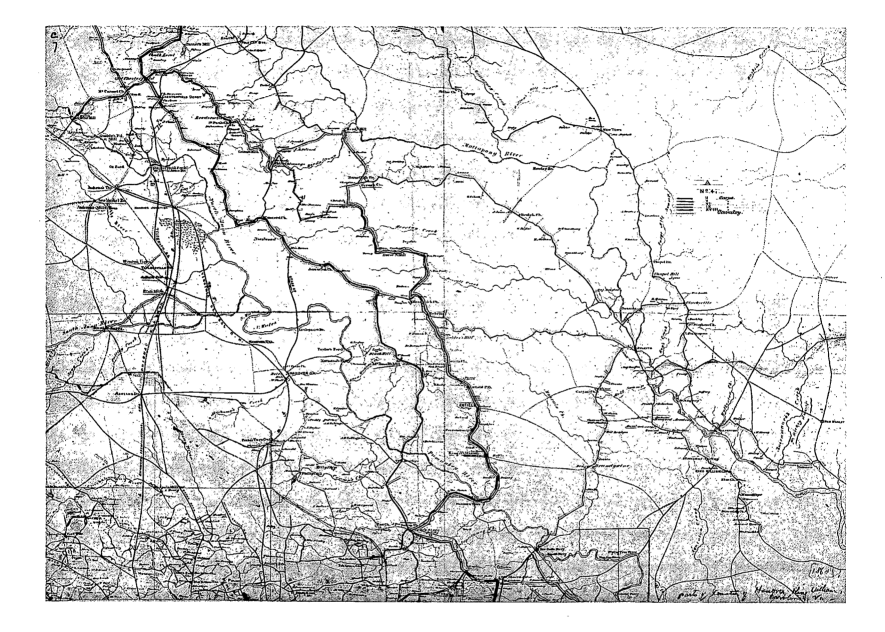

William Henry Paine's Map of Caroline, King William, and Hanover Counties, Virginia

Based on printed county maps, available state maps, captured maps, and maps and reconnaissance in the field, this large-scale theater map shows the movement of Union troops and cavalry during Ulysses S. Grant and George Gordon Meade's grueling overland campaign in May and June 1864.

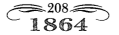

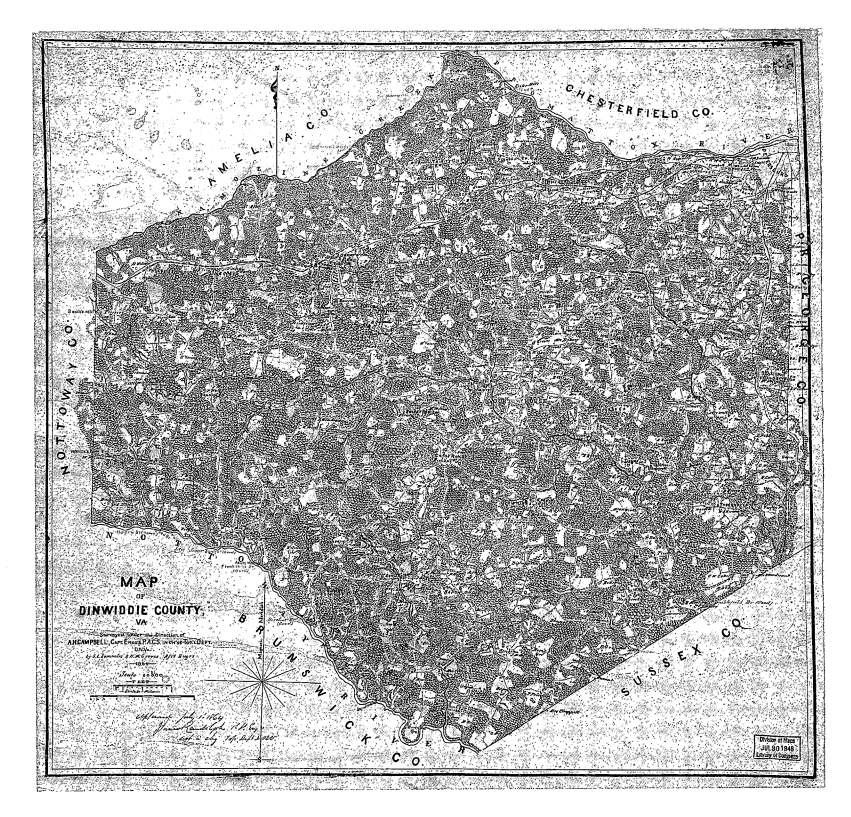

A. H. Campbell's Map of Dinwiddie County, Virginia, 1864

A colored manuscript map drawn on tracing paper in red and black ink, this Confederate map makes it vividly clear what Union general and topographical engineer G. K. Warren meant when he wrote that "the forests are the most important military feature in the battles of Virginia" and then underlined the statement.

Ironically, the fortunes of the Confederacy and of Warren came unraveled in Dinwiddie County. The Battle at Five Forks, an engagement fought on April 1, 1865, turned Robert E. Lee out of his entrenchments at Petersburg, but Warren was relieved of command of the Fifth Corps, Army of the Potomac, by General Philip Sheridan, who felt that Warren had failed to inspire his troops.

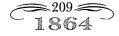

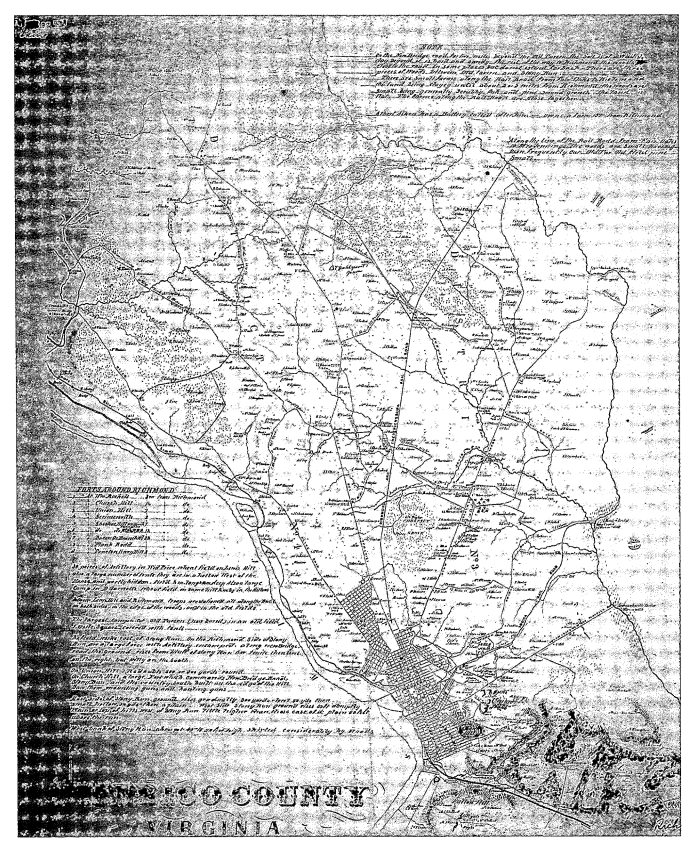

William Henry Paine's Map of Henrico County/Richmond Area, Virginia, c. 1864

This heavily annotated, printed map with some original corrections indicates that Captain Paine was actively reconnoitering in the field, as evidenced by the note "Saw them [i.e., the Confederates], mounting guns and hauling guns." Paine has sketched in pine woods, drawn more detailed terrain features, shown new Confederate redoubts, added a few cultural features, and carefully describes the woods, river-banks, and road surfaces.

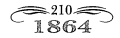

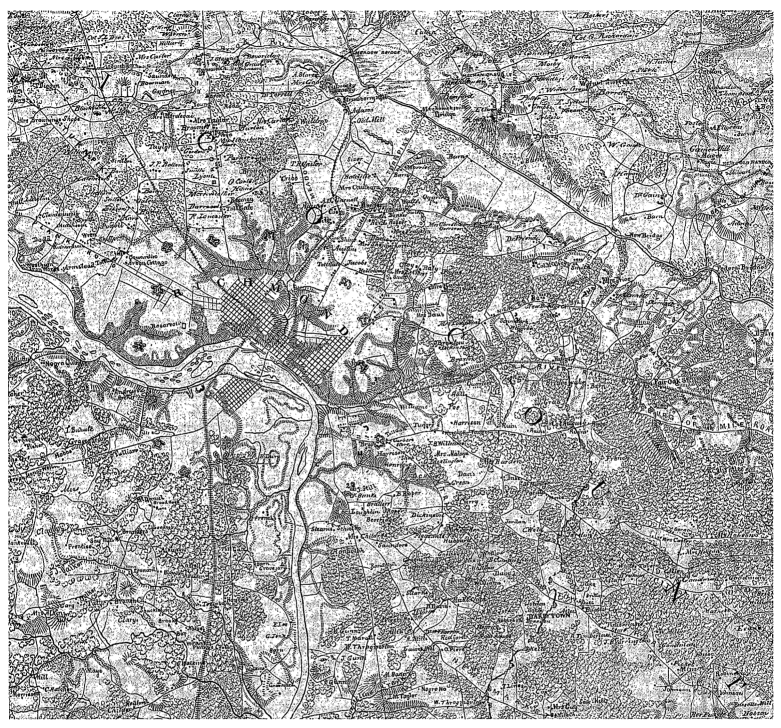

Detail: Richmond, Virginia, Henrico County, and Hanover County, 1864

The overall map is immensely detailed, very large (approximately sixteen square feet), made of linen or muslin, and drawn in red, black, and blue ink with a very weak, nearly invisible blue watercolor wash in the rivers. This is apparently a work-in-progress map from which printed versions were prepared and distributed, either photographically or by the process known as sunprinting. While it is not the most militarily valuable map, it does at least address General Richard Taylor's observation that Confederate commanders know no more about the immediate vicinity of their own capital than they did about central Africa. An early version of this map was sent by the Confederate Engineer Bureau to Robert E. Lee in February 1863 to aid in the preparation of that busy general's tardily presented report of the Peninsula or Seven Days Campaign of June/July 1862. Union cavalrymen discovered a version of this map on the body of Confederate general John R. Chambliss. He was killed in an engagement on August 16, 1864, near White's Tavern at White Oak Branch or Deep Bottom. White's Tavern is on this map, on the Charles City Road. A printed version of this map indicates that Confederate topographical engineer Captain A. H. Campbell directed its preparation and that the "Draughtsman" was S. B. Linton. A lithographed version of this map is in the *Official Records Atlas* (plate 92, no.1).

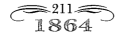

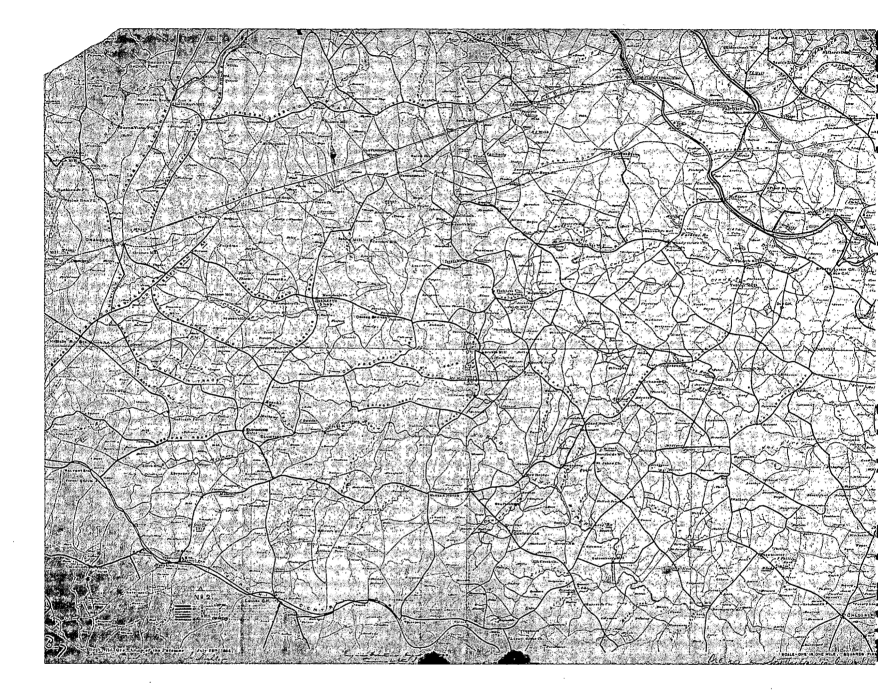

Nathaniel Michler's Map of the Union Army of the Potomac's Overland Campaign, May–June 1864

By 1864, after three years of fighting over much of the same terrain, the Union army still had no adequate maps with which to plan and/or conduct a campaign. Years after the war, Army of the Potomac commander George Gordon Meade's aide-de-camp, Theodore Lyman, prepared an exaggerated but heartfelt paper entitled "Uselessness of the Maps Furnished to the Staff of the Army of the Potomac Previous to the Campaign of May 1864." Major Nathaniel Michler, acting chief engineer of the army, made strenuous efforts to finally

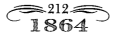

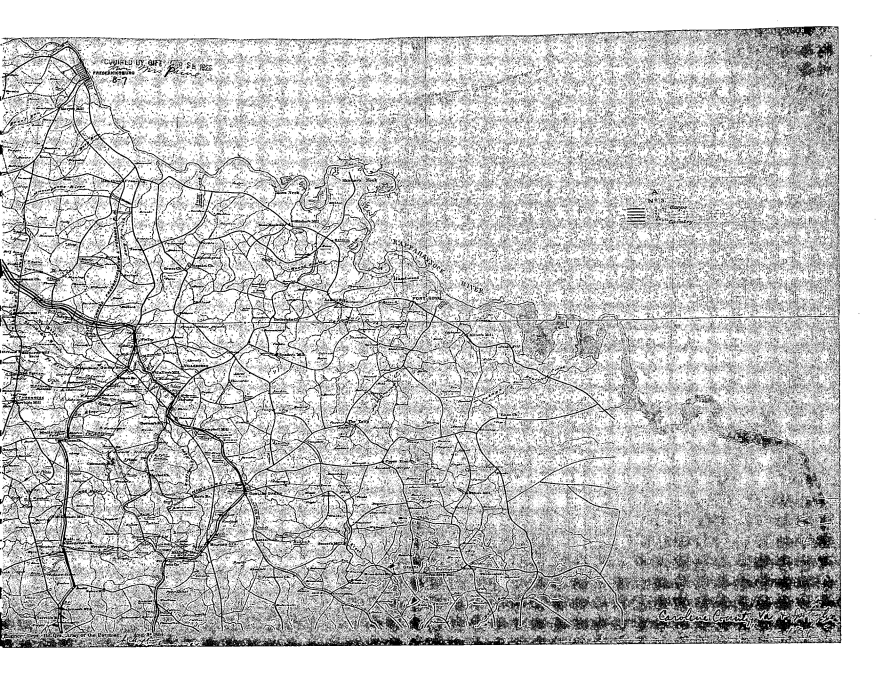

produce definitive maps as the Union army pushed its laborious way toward Richmond. He had his mapmakers, including Captain William Henry Paine, traversing and mapping every road within Federal lines. According to Michler's report, between May and August 1864, his men surveyed 1,300 miles of roads, issued 1,200 maps, and made over 1,600 photographically reproduced sketches along the routes of the campaign. These pasted-together maps, with the progress of the various army corps and cavalry watercolored by hand on the printed map, are some of the results of Michler's labors. Drawn in too small a scale to be used for tactical purposes, they are nonetheless detailed and accurate and were quickly and widely distributed.

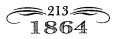

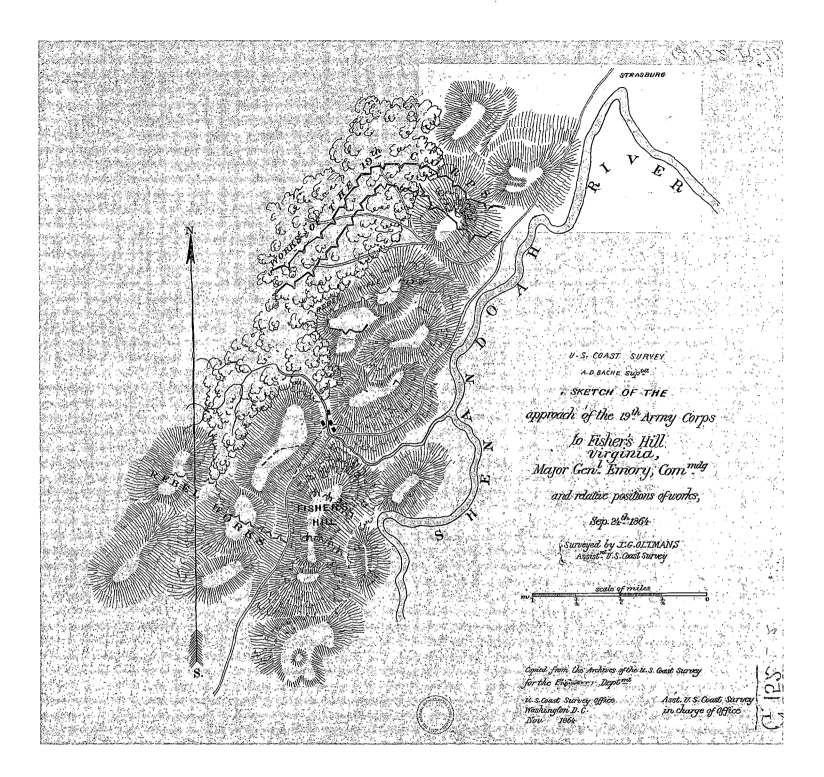

J. G. Oltmanns's Sketch of Fisher's Hill, Virginia, 1864

There were two Oltmanns (the name has been miscopied and misspelled on the map) working on maps in the immediate vicinity of Fisher's Hill on September 12, 1864. One was the surveyor of this sketch, J. G. Oltmanns of the United States Coast Survey. The other was C. W. Oltmanns, an assistant to Jed Hotchkiss, the famed Confederate mapmaker. J. G. Oltmanns was attached to the staff of General William H. Emory during the Shenandoah Valley campaign in the fall of 1864. He surveyed, executed reconnaissance, and often superintended the construction of breastworks. He had seen a great deal of the war, having been with Commander David Dixon Porter at the mouth of the Mississippi and being badly wounded while aboard the United States Coast Survey vessel *Sachem* in May 1862 in Louisiana. For a product of the United States Coast Survey, this sketch is comparatively rustic.

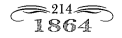

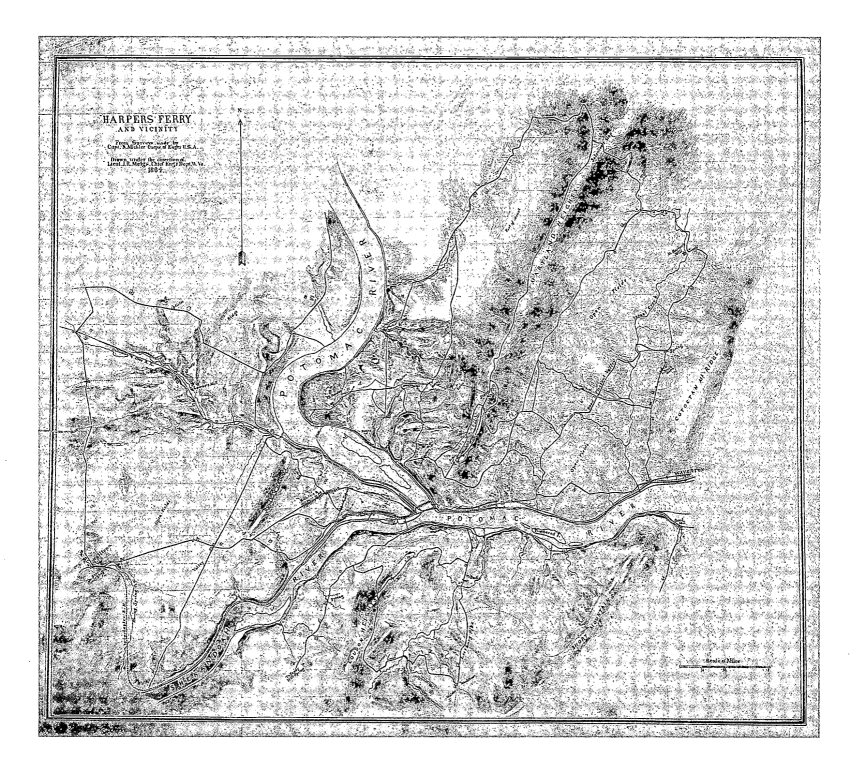

Nathaniel Michler and J. R. Meigs's Map of Harpers Ferry, West Virginia, 1864

A meticulously surveyed and beautifully drawn map of what had been Harpers Ferry, Virginia, when the war began. Some thought the Civil War began at the site of a United States arsenal and armory here when abolitionist John Brown seized the town in October 1859 and sought to start a slave insurrection. The town changed hands almost any time an opposing force made a serious effort to capture it. Note such details as the pontoon bridge across the Potomac River and the pencil shading to render the dominant terrain features. The pencil grid used to prepare this manuscript ink, pencil, and watercolor map is faintly visible. Lieutenant J. R. Meigs was the son of Union quartermaster general Montgomery C. Meigs. Lieutenant Meigs was killed under never completely determined circumstances on October 3, 1864, when he was topographical engineer on the staff of Union general Philip Sheridan. Rebels claimed Meigs was killed in a fair fight. Yankees say he was bushwacked. A variation of this map appears in the *Official Records Atlas* (plate 42, no. 1).

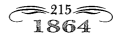

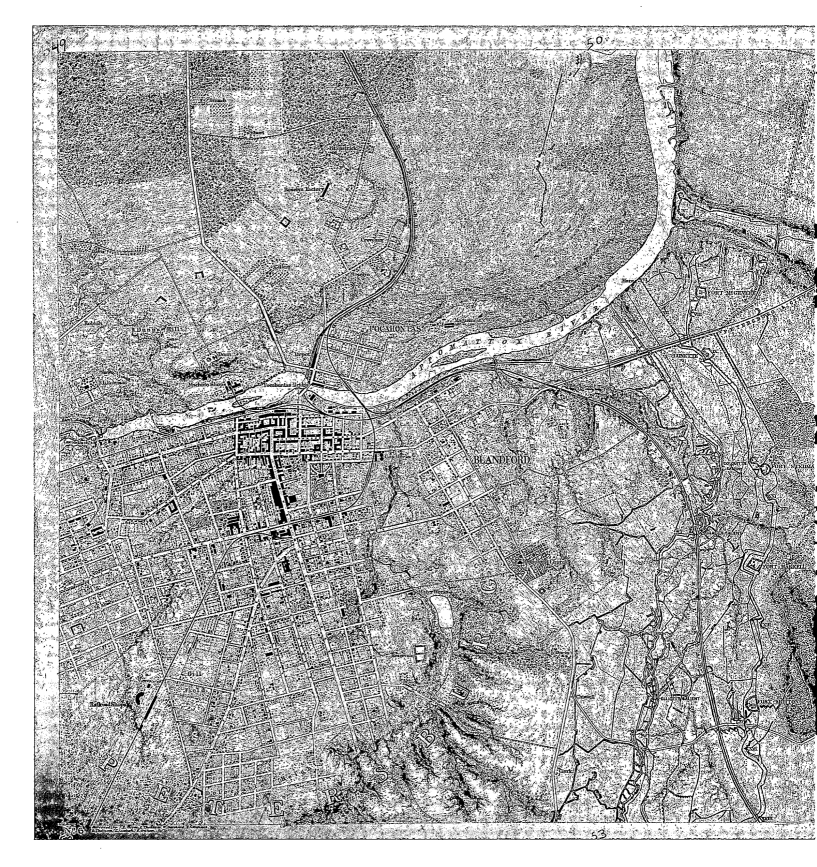

Nathaniel Michler and John E. Weyss's Map of Petersburg, Virginia, July 1864

This map is one section of what National Park Service map historian David W. Lowe called the magnum opus of Union topographical engineer Nathaniel Michler's wartime staff. Work was begun July 9, 1864, on a series of maps that expanded in scope as the Confederate and Federal lines around Petersburg and Richmond grew longer and more elaborate. John E. Weyss, a non–West Pointer, had worked as Michler's principal assistant at least as early as April 1862. According to Union general G. K. Warren, Gilbert Thompson was the best mapmaker of the war. Thompson also wrote a two-volume history, *The Engineer Battalion in the Civil War*. F. Theikuhl had been acting

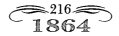

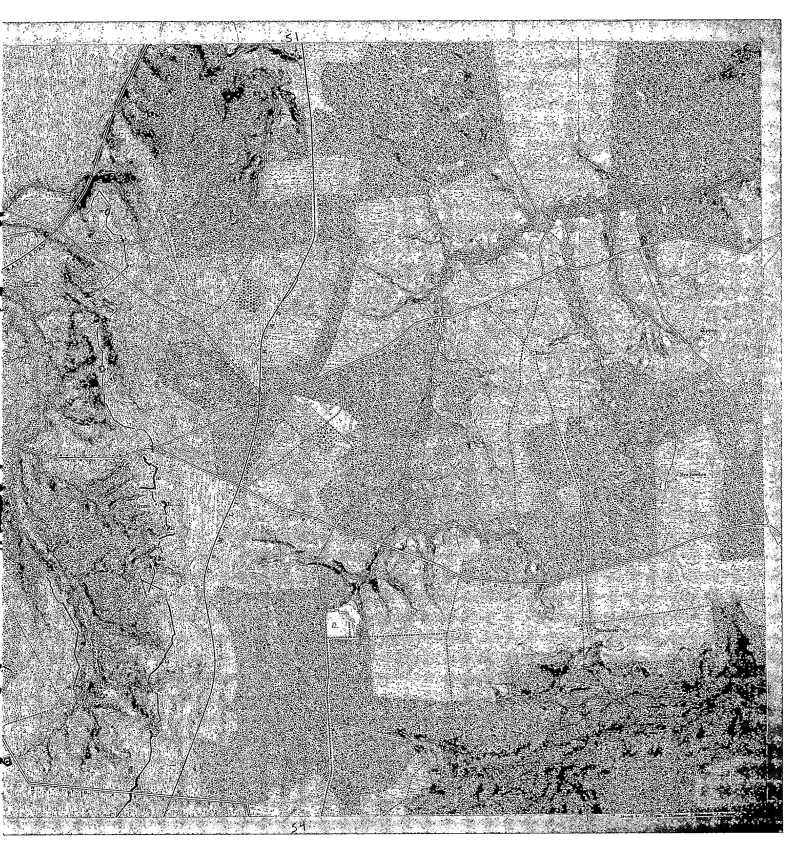

topographical engineer in the First Division, Eleventh Corp, under Warren in June 1863. The known fixed sites of Petersburg (church spires and gasworks) were used to establish the siege works using triangulation. These were juxtaposed with an existing United States Coast Survey map of the Appomattox River. "Features were drawn on a grid of one minute longitude and one minute latitude as noted in pencil in the map's margins," according to Lowe. The immense detail combined with the accurate scale of eight inches to the mile made this an invaluable military document for an army conducting a siege. When wartime mapping expertise is discussed, this map series, eventually including twenty-eight sheets and covering some 290 square miles, is among the paramount achievements of Civil War topographical engineers.

1864

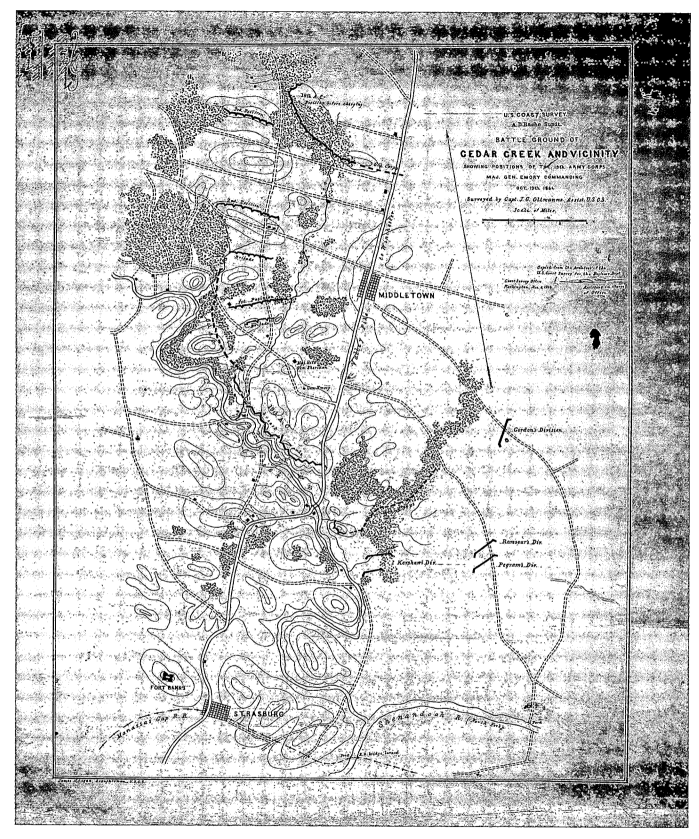

J. G. Oltmanns's Map of Cedar Creek, Virginia, October, 19, 1864

Oltmanns was a professional civilian surveyor with the United States Coast Survey. He had as eventful a Civil War career as any soldier, serving from the Mississippi River to Cedar Creek. He was attached to the staff of Union general William H. Emory during the Shenandoah Valley campaign. This manuscript map is a hand-copy of Oltmanns's original map. An initial success by Confederates under General Jubal A. Early (in an attack partly planned by Rebel mapmaker Jed Hotchkiss) was not exploited. Union forces rallied and turned a Rebel retreat into a rout, ending the Confederate threat to the Shenandoah Valley for the remainder of the war.

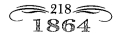

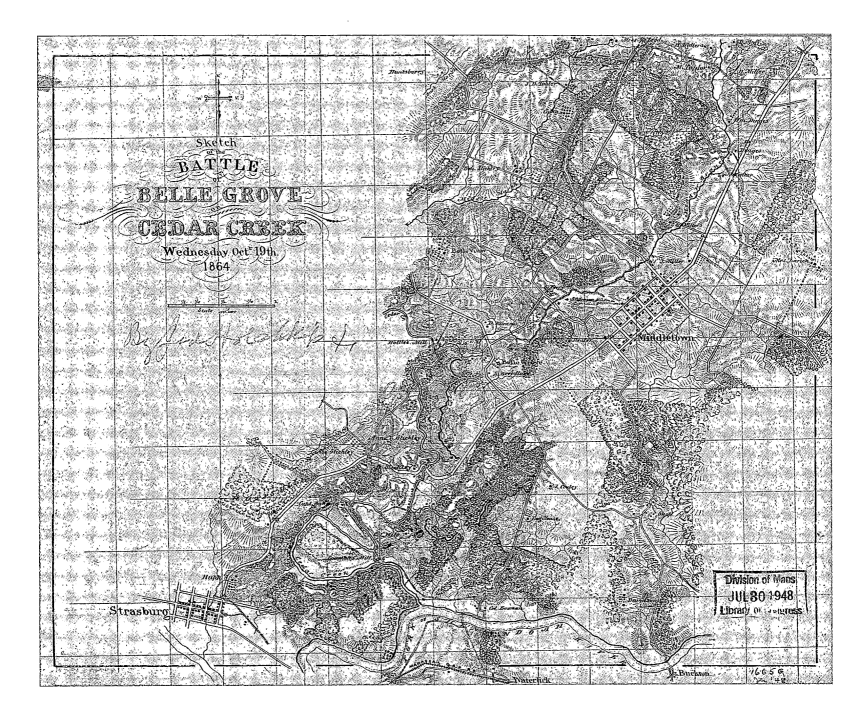

Division of Maps
JUL 30 1948
Library of ...gress

Jed Hotchkiss's Sketch of the Battle of Cedar Creek, October 19, 1864

The plan of the Battle of Cedar Creek was devised by Jed Hotchkiss and Confederate general John B. Gordon and presented to General Jubal A. Early, commander of the Second Corps, Army of Northern Virginia. The left flank of the Union army rested on what was presumed to be an unassailable position, always a dangerous assumption when in the vicinity of an enterprising foe. The Confederates completely surprised the Union forces and drove them back through Middletown. But their attack stalled, and the Union line stabilized, was reinforced, and then counterattacked, driving Confederates, Hotchkiss included, from the field. Union general Philip Sheridan provided the impetus for a decisive Union victory. Hotchkiss doubtless drew this map using battle reports and other Shenandoah Valley maps in his possession. He was working on it in late February 1865, some of the time at his home near Staunton, Virginia. The highly detailed, meticulously executed map, full of cultural and physical information—orchards, mills, shops, elevations, woods, and streams—gives no indication of the Confederacy's dire fortunes or dubious future. A lithographed version of the map is in the *Official Records Atlas* (plate 82, no. 9).

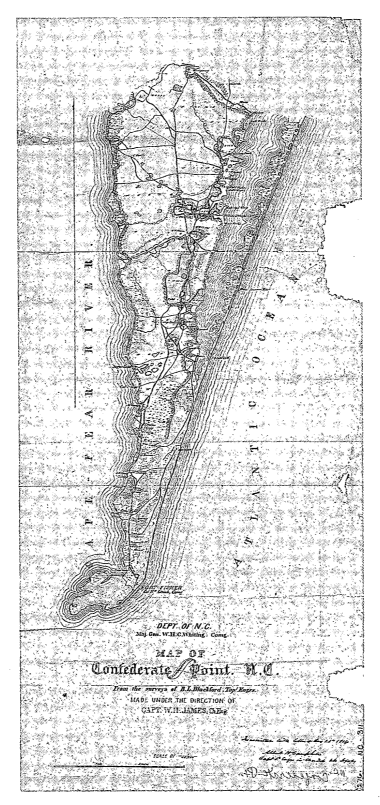

B. Lewis Blackford's Map of Confederate Point, North Carolina, c. November–December 1863

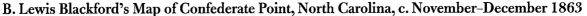

Federal Point was renamed Confederate Point for the duration of the war. It was the site of Fort Fisher, the military installation protecting the important port of Wilmington, North Carolina, a vital Southern harbor for foreign supplies. Fort Fisher was known as the Gibraltar of the South. Drawn on tracing paper in pencil and black and red ink, this map is another of Blackford's minor cartographic masterpieces. Note the creative use of the banner scroll to incorporate "New Hanover County" into the title of the map. Commanding Confederate general W. H. C. Whiting was a West Point graduate and engineer, which may explain the evident care taken on the maps drawn for him. Blackford's Smith's Island map (opposite) and this one are clearly incorporated into the maps in the *Official Records Atlas* (plate 132, no. 1) showing the approaches to Wilmington. Albert H. Campbell, chief of the Confederate Bureau of Engineers, wrote after the war that these maps of North Carolina had all been lost. Fortunately and happily, they were not. An attempt by Union general Benjamin Butler to capture Fort Fisher in December 1864 failed. A more competently managed operation under the command of General A. H. Terry in January 1865 succeeded, closing the last significant port connecting the Confederacy with the outside world.

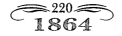

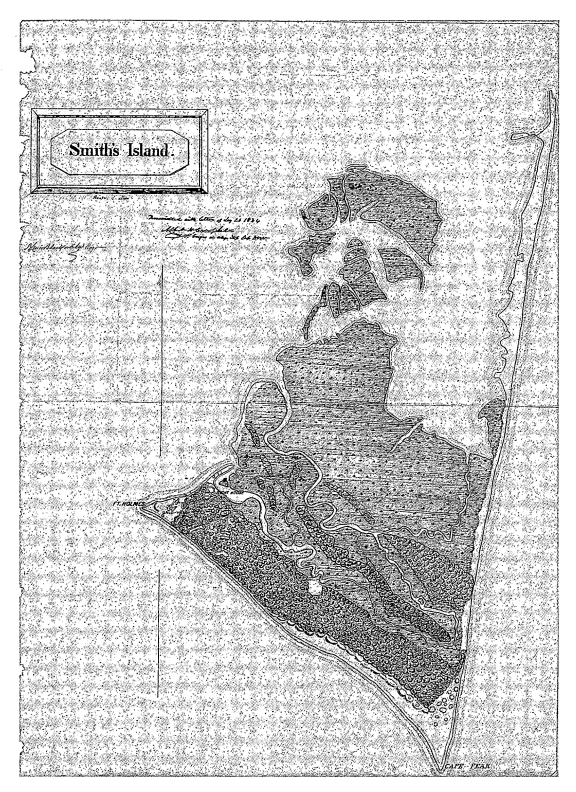

Smith's Island.

FT. HOLMES

CAPE FEAR.

B. Lewis Blackford's Map of Smith's Island, North Carolina, November 1863

An elegantly distinctive map on tracing paper in watercolor and ink. Federal Point and Fort Fisher are just north of the long, northward-extending spit of Smith's Island. The forts protected the mouth of the Cape Fear River and the important port of Wilmington, North Carolina, the destination and refuge of Confederate blockade runners. Wilmington was one of the last practicable, meaningful ports the Confederacy had by late 1863. Fort Fisher fell to the Union in January 1865. Smith's Island and its fortifications were never captured. They were abandoned in the days after the fall of Fort Fisher. Smith's Island is now named Bald Head Island. The script handwriting indicates that the map was transmitted to Albert H. Campbell of the Confederate Bureau of Engineers in Richmond, Virginia, on January 26, 1864. A lithographed version of this map appears in the *Official Records Atlas* (plate 51, no. 4). The aesthetic or cartographic differences between the lithograph and Blackford's original of the map are especially evident and noteworthy.

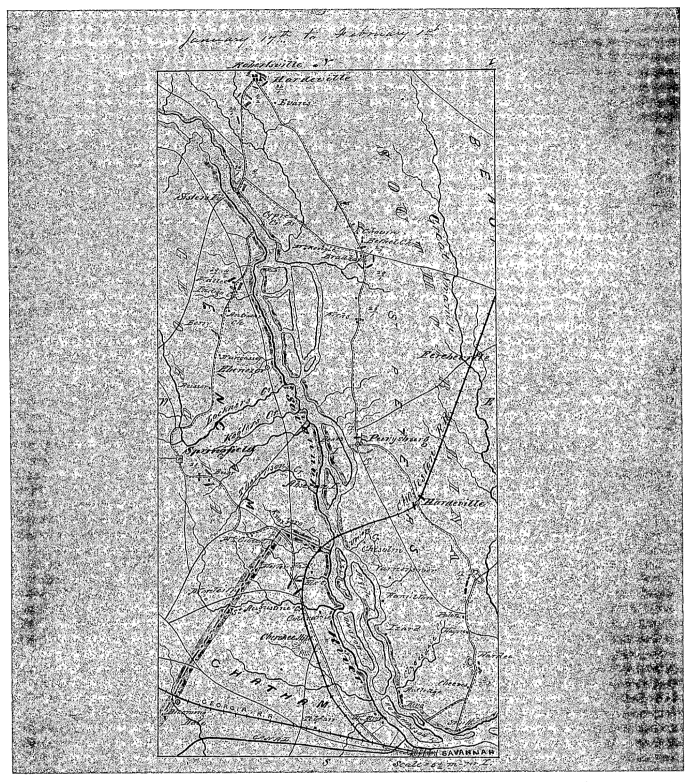

Robert M. McDowell's Map: Leaving Savannah, Georgia, for the Carolinas, January 14 to February 1, 1865

This placid map fails to give a proper idea of the difficulties Union general William Tecumseh Sherman's troops encountered as they left their comfortable bivouacs in Savannah, Georgia, and set out for the rear of General Robert E. Lee's Army of Northern Virginia, presently locked in a stalemate with Ulysses S. Grant in the fortifications around far-off Petersburg, Virginia. The mapped area consisted largely of abandoned rice plantations. The only way the troops and wagons could move was to corduroy, that is, pave with logs, the dikes. The Cheves residence, immediately north of Savannah and across the river, was a heap of smoldering ruins. The farm paths were covered with water that spread wider and grew deeper until the supposed land turned into an almost uninterrupted sea of water. It was the time of the annual "freshet" of the Savannah River. General Alpheus S. Williams, who ventured into the region, wrote, "We have a terrible job before us."

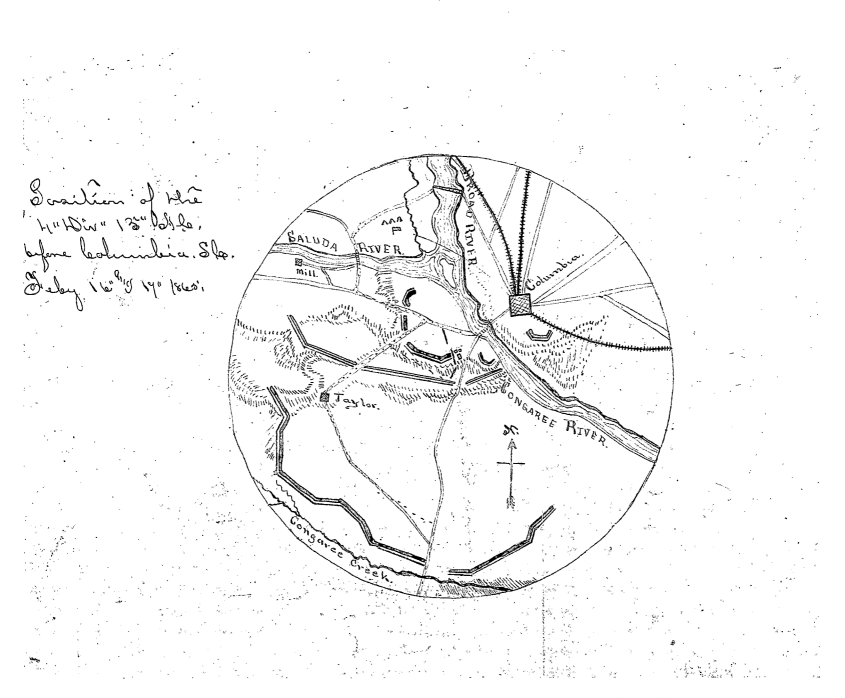

Map of the Fourth Division, Fifteenth Army Corps, Before Columbia, South Carolina, February 16 and 17, 1865

General William Tecumseh Sherman regarded the march of his Union armies from Savannah to Columbia, South Carolina, as a much more ambitious and difficult undertaking than the March to the Sea, for which he became immortalized. As the Union forces stood poised on the outskirts of Columbia, they were well fed and in fine spirits. A topographical engineer of the Twentieth Corps ended his diary account of Thursday, February 16, 1865, with "This has been a jolly day." Again, as during the March to the Sea, the United States Coast Survey officers did very little mapping and a lot of marching. Their capture of a Rebel flag at Columbia was the apex of their service during the campaign in the Carolinas. It was the duty of topographical engineers of the Army of the Tennessee to submit a map of their position at the end of each day's march. This is probably one of those required maps. The Fifteenth Army Corps, part of the right wing of Sherman's advance, was under the overall command of Major General O. O. Howard, Army of the Tennessee. It was under the direct command of Major General John A. Logan. The Fourth Division was commanded by Brigadier General John M. Corse.

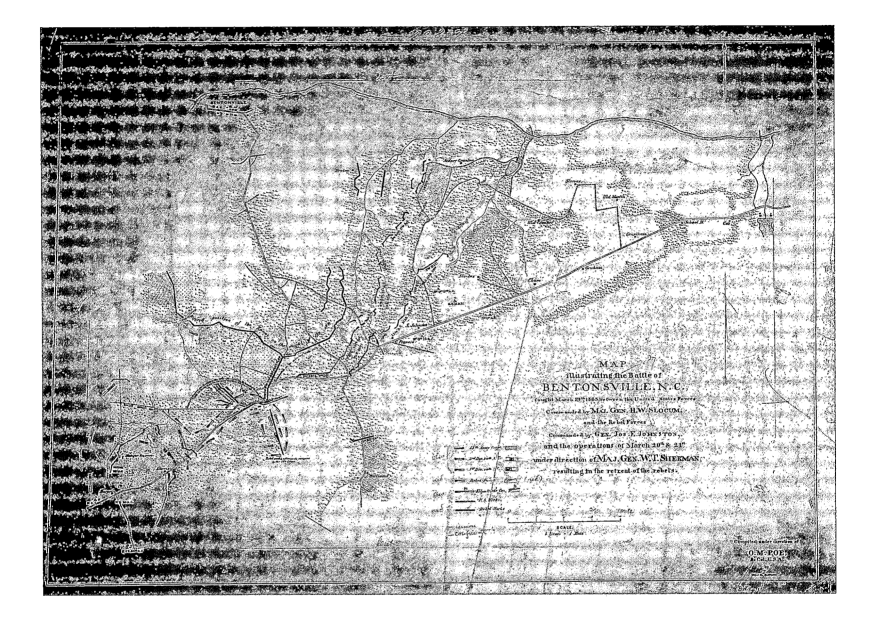

Orlando M. Poe's Map of the Battle of Bentonville, North Carolina, March 19, 1865

Union engineer Orlando M. Poe was General William Tecumseh Sherman's chief engineer throughout his campaign from Chattanooga to Atlanta to Savannah and on to the surrender at Bennitt's Farm, North Carolina. Poe was in charge of making Atlanta cease to exist in military terms. He decided which roads were practicable for the armies and then made certain they could move on them, corduroying hundreds of miles of roads; he also oversaw the mapping of the campaign. Poe had old North Carolina maps, and they misled the Union forces on the eve of the Bentonville battle. They were unaware of a second road from Bentonville, and, advancing south on this second road, the road unknown to Union forces, Confederate forces landed on the unsuspecting left flank of Sherman's army. This map is based (or vice versa) on a map of the battle prepared by Captain Robert M. McDowell. The compiler of this map, Major E. F. Hoffman, was an aide-de-camp to Army of the Tennessee commander O. O. Howard. A lithographed version appears in the *Official Records Atlas* (plate 133, no. 2).

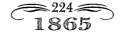

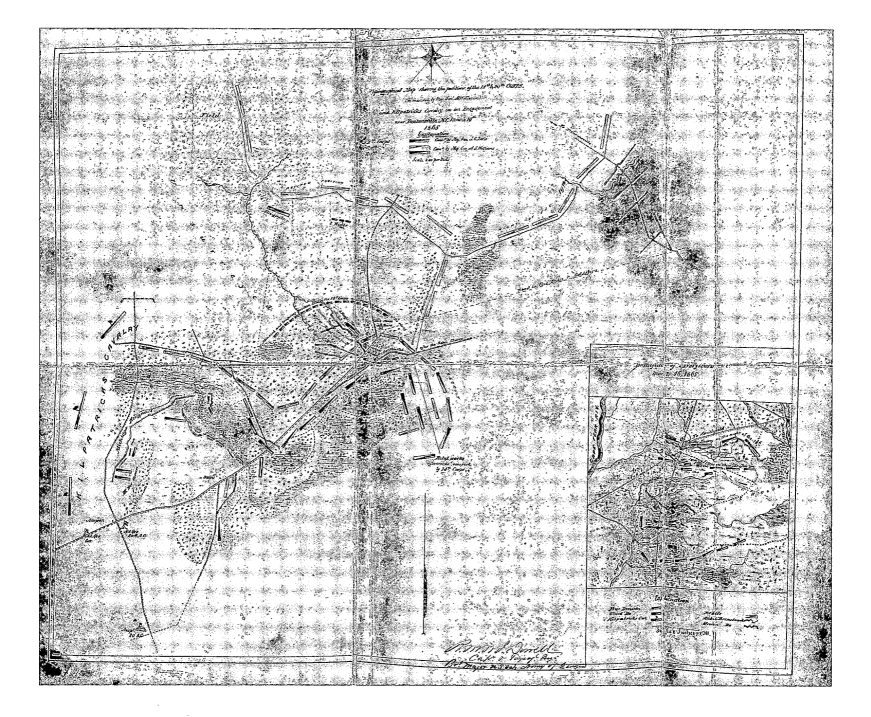

McDowell's Map of Averasborough, March 16, 1865, and Bentonville, North Carolina, March 19, 1865

Robert M. McDowell was one of several Civil War topographical engineers from New York State's southern tier (others were Lucius T. Stanley, Rodney E. Harris, and Jed Hotchkiss, all of whose maps are represented in this atlas). As McDowell entered North Carolina in March 1865, with Union general William Tecumseh Sherman's troops, he was fortunate that one of his assistants, Corporal Hueston, had captured an 1833 map of North Carolina. On the night of March 13, McDowell noted in his diary that he "worked nearly all night on maps . . . in adjoining room was a dying Confederate soldier whose moans were truly painful to hear." The battle that McDowell called "Black River" is minutely documented in the inset map. The Confederates assumed that their right flank, resting along swampy Mill Creek, was impenetrable, but Union forces pushed their way through and the Rebel forces slowly fell back. Three days later, Union forces again clashed with the Confederates near Bentonville. McDowell's attractive map reflects his description of the fight: "It was the prettiest battle I ever beheld and seemed more like a panoramic view than a current fearful reality. The nature of the country, woods, fields, and roads and spirit of the corps all added to complete a perfect picture of war." Lithographed versions of this map set are in the *Official Records Atlas* (plate 79, no. 4, and plate 80, nos. 10 and 11).

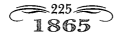

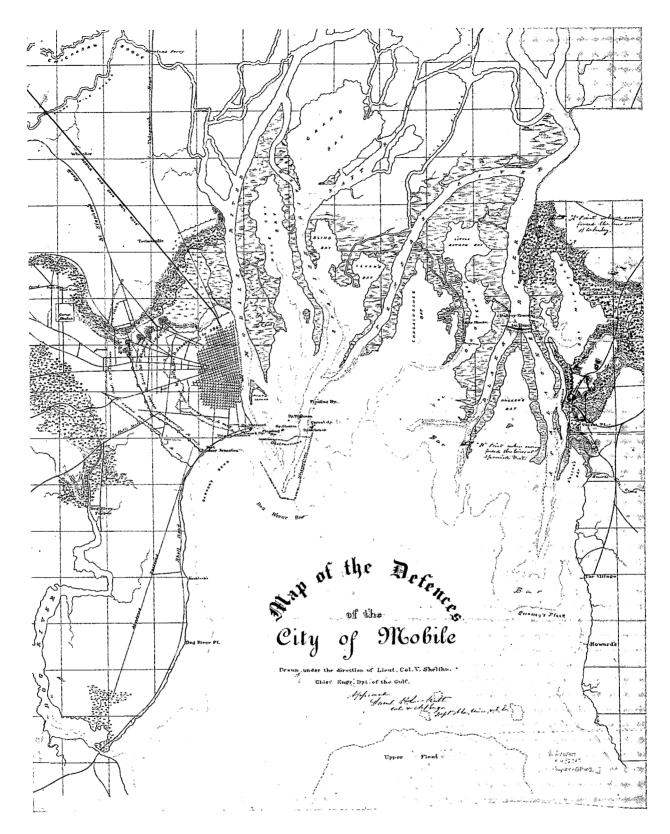

V. Sheliha's Map of the Defenses of the City of Mobile, Alabama, c. April 1865

This extremely detailed, highly accurate, beautifully drafted map of Mobile and its defenses was apparently completed after the surrender of the town. The excision, in the upper-right-hand corner of the map, of Blakely is a mystery. Blakely fell to Union forces on April 9, 1865, the same day Robert E. Lee was surrendering the Army of the Northern Virginia at Appomattox. Union forces had controlled the mouth of Mobile Bay since August 1864, when Admiral David Farragut, "damning the torpedoes," ran past Forts Morgan, Powell, and Gaines and captured the Confederate ironclad ram *Tennessee*. An extremely detailed rendering of Confederate engineer Lieutenant Colonel V. Sheliha's "Line of Works" at Mobile can be found in the *Official Records Atlas* (plates 107–9).

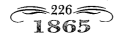

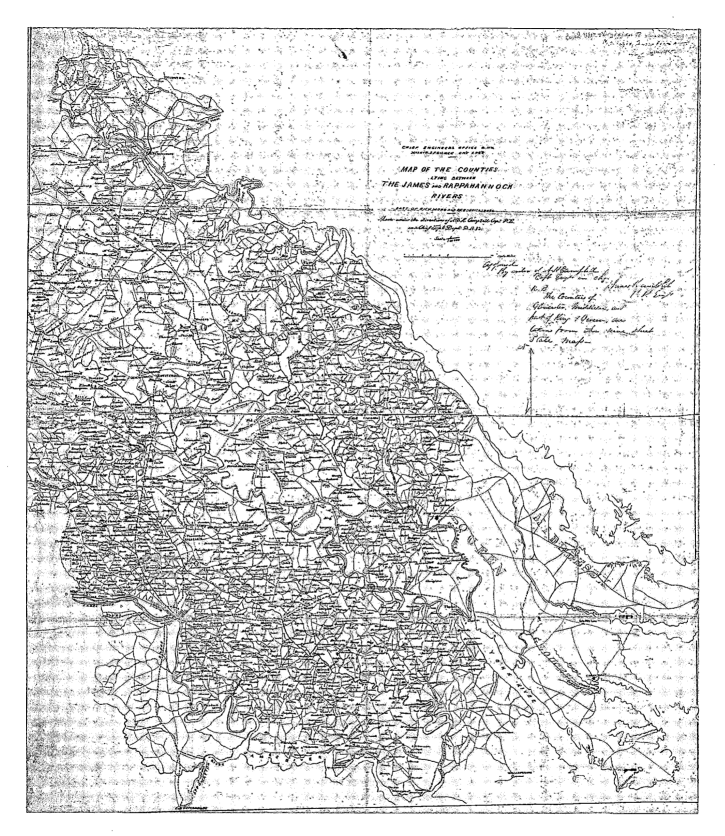

Counties Lying between the James and Rappahannock Rivers, Virginia

Highly detailed manuscript map on glazed tracing linen, partly based on the preeminent prewar map of Virginia known as the "nine sheet map" of Virginia. Approved for the Confederate Engineer Department by J. Innes Randolph, it was probably created in 1865. Jed Hotchkiss didn't hold Captain Albert Campbell, who directed the production of this map, in high regard. He said Campbell did a large amount of bad work and took a long time to do it. In some respects, this elaborate effort confirms Hotchkiss's judgment since the military value of this map is questionable.

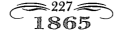

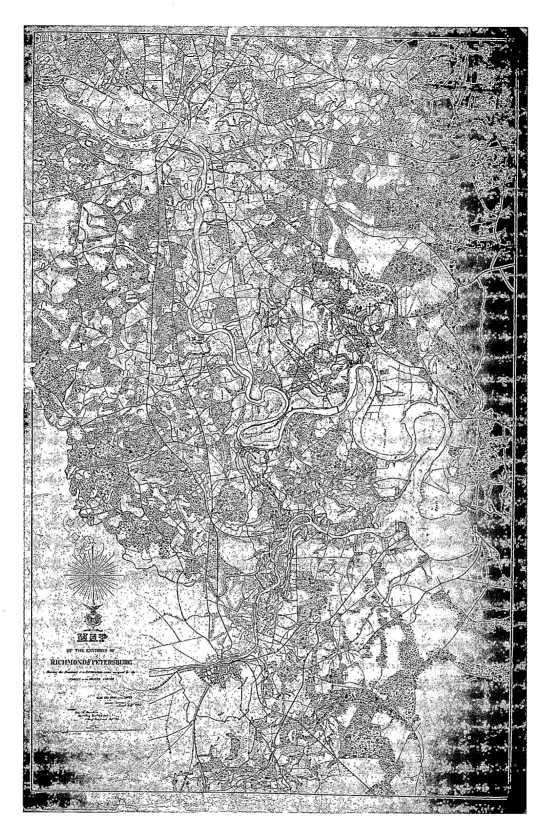

Peter S. Michie's Map of the Environs of Richmond and Petersburg, Virginia, March 1865

Peter S. Michie was described by Union general Ulysses S. Grant "as one of the most deserving young officers in the service." Michie graduated from West Point in June 1863. Less than two years later, he was a brevetted brigadier general and chief engineer of the Union Department of Virginia. This manuscript map, with Confederate defense lines in red and Union lines in blue, indicates the precarious situation of the Rebel forces. Confederate commander Robert E. Lee himself said that once his army was in this position, ultimate defeat was simply a question of time. In fact, from the date of this map to the surrender at Appomattox was slightly more than two weeks. Michie, operating in a stationary military situation, was able to prepare very thorough, highly detailed maps. Also, Michie had been trained as a military engineer, so the numerous fortifications in the Petersburg area were, in a sense, his specialty. The resourceful Michie also used photographs to help prepare his maps and to accompany his reports.

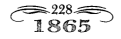

1865

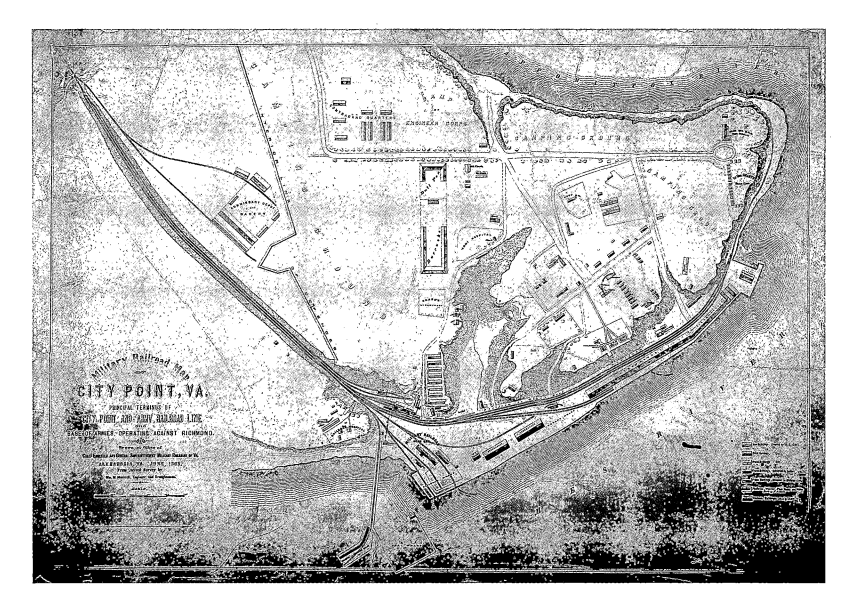

William M. Merrick's Military Railroad Map of City Point, Virginia, June 1865

City Point, northeast of Petersburg at the confluence of the Appomattox and James rivers, was the headquarters of Union general Ulysses S. Grant during most of the final year of the war. Here Grant planned the destruction of the Army of Northern Virginia, counseled with President Abraham Lincoln, and waited anxiously for the reports of his other field armies: William Tecumseh Sherman's, Philip Sheridan's, Ben Butler's, James H. Wilson's, George H. Thomas's, et. al. The subject of the map is the City Point Railroad, and a military branch of it which was added to the areas south of Petersburg immediately behind the lines of the Army of the Potomac. According to Horace Porter, Grant's aide-de-camp, "the new portion of the road was built, like most of our military railroads, upon the natural surface of the ground, with but little attempt at grading. It ran up hill and down dale, and its undulations were so marked that a train moving along it looked in the distance like a fly crawling over a corrugated wash board." On the open plains, where trains were visible to the enemy, high earthen embankments were built to shelter the engines and cars as they rolled along. The stations along the line were named for Union generals.

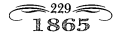

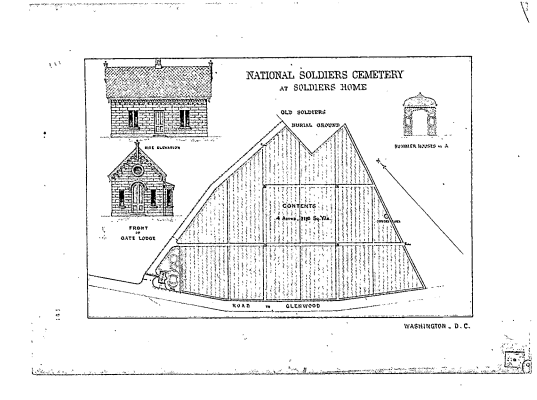

National Soldiers Cemetery at Soldiers Home, Quartermaster Corps

This manuscript watercolor map or plat is part of an overlooked series of wartime cartographic records produced for the U.S. Army's Quartermaster Corps. Soldiers Home was the refuge of the Lincolns from the pressures and heat of the White House and downtown Washington. Soldiers Home was off the Rock Creek Church Road, within the District of Columbia.

Plan of the Headquarters of Major General Christopher Columbus Augur, 1865

Union general Augur commanded the Twenty-second Corps and the Department of Washington. The Twenty-second Corps was designated to defend Washington. The department's territory changed periodically, but under Augur it included the District of Columbia, the surrounding Maryland counties, and Fairfax City, Virginia. This manuscript plan maps the ground plan of the headquarters in nearly every detail, while the accompanying note describes the construction of the buildings.

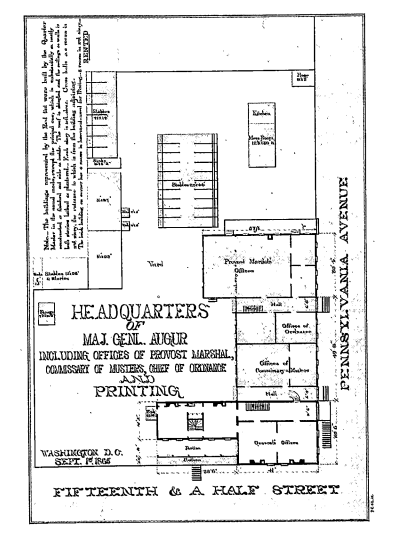

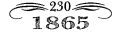

College and Trinity Church Hospitals, Quartermaster Corps

Manuscript watercolor map or plat prepared for the U.S. Army's Quartermaster Corps. The map is typical in that it not only presents a beautifully rendered record of this wartime facility but also contains a statement of repair costs. Note the "Dead house" connected with both hospital facilities. Also note the use of *do* for ditto in the itemization of repair costs. The Union Quartermaster Corps, under the very able direction of General Montgomery C. Meigs—a confidant of Lincoln—was in part responsible for the success of Union armies in the Civil War. This plan reflects the efficiency of the corps.

Buildings at Washington Arsenal

The second-oldest military post in the United States, after West Point, the arsenal was located at the confluence of the Potomac and Anacostia rivers in Washington. Mary Surratt and others convicted of conspiring to assassinate President Abraham Lincoln were imprisoned and hanged here. The site of the Washington Arsenal is the present-day Fort Leslie J. McNair.

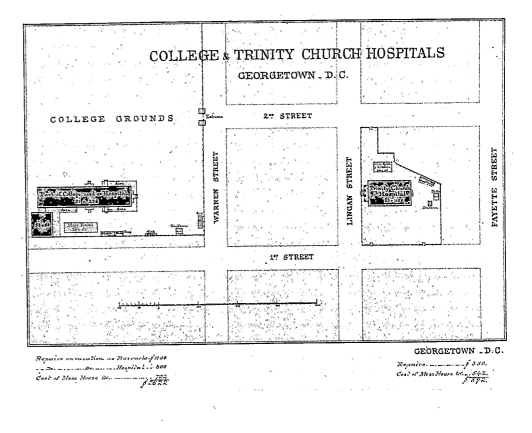

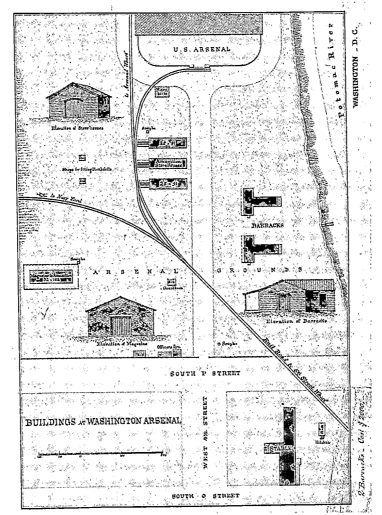

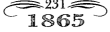

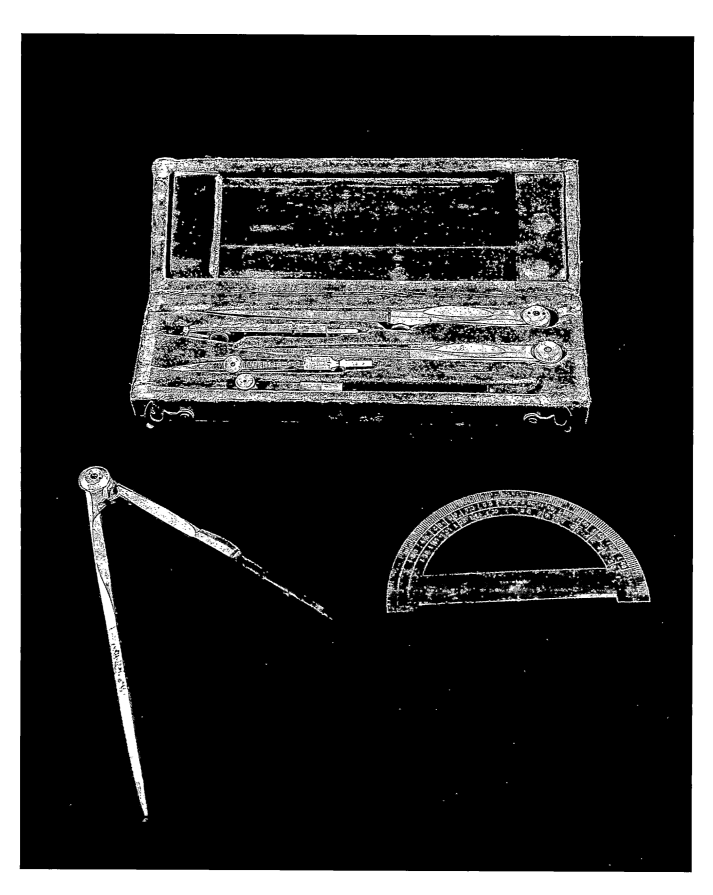

Instruments of Ambrose Bierce—Topographical Engineer

These are some of the instruments Ambrose Bierce used for preparing maps in the field: his protractor, a pair of compasses, and drafting instruments in a velvet-lined case. Bierce kept a printed catalogue of instruments and quite possibly ordered them from the Chicago Scale Company on South Jefferson Street in Chicago.

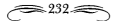

THE
MAPMAKERS

John C. Babcock
Bureau of Military Information

BORN IN Rhode Island in 1836, John C. Babcock was working as an architect in Chicago, Illinois, when the Civil War began. When President Abraham Lincoln issued one of the first calls for volunteers, Babcock joined a unit known as the Sturges Rifles.

Possibly because of some prior Chicago connection, Babcock was attached to Allan Pinkerton's intelligence operation in March 1862. Private Babcock was expected to sketch enemy fortifications from verbal descriptions provided by Confederate prisoners and deserters and anyone else who had had occasion to observe Rebel defensive sites. Babcock soon resorted to making personal reconnaissance and preparing drawings from what he could see and sketch firsthand. Pinkerton's intelligence headquarters were with General George B. McClellan's headquarters in the field on the Virginia Peninsula. There was a paucity of

John Babcock is bareheaded, second from left. As in all extant wartime photographs of him, Babcock looks away from the camera.

accurate topographical information partly because the initial map on which McClellan had based the operations of the Army of the Potomac had been incorrect in almost every detail. Thereafter, McClellan was skeptical of the information that the Corps of Topographical Engineers brought him. When it was discovered that Babcock was quite capable of producing reliable maps derived from his own personal observations, he was tapped to do just that for the Army of the Potomac.

Babcock's technique was fairly standard. He went where he could, sketched what he was able, creating a systematic set of small maps. Then he took existing maps and reconciled his details with more general published information to produce a tier or line of mapped terrain that could be relied on. He repeated the process continually, thereby expanding the area of mapped territory. This success of the civilian topographical engineer apparently created some hard feelings in the official Corps of Topographical Engineers, headed by General A. A. Humphreys, whose efforts were sometimes found to be less accurate or less detailed than Babcock's. Babcock also made regular ascents in observation balloons to supplement the intelligence he gleaned on the ground.

Babcock served in this capacity and as a scout as well, until McClellan was replaced by General Ambrose E. Burnside in November 1862. Burnside knew Babcock from both his native Rhode Island and his adopted Chicago. McClellan had mustered out the Sturges Rifles, and Babcock, at twenty-six, an ex-private, was asked by Burnside to head up the Intelligence Department that had been vacated by Allan Pinkerton. Though he took the position as a civilian, he was unofficially Captain Babcock. Little came of Babcock's service with Burnside, but under General Joseph Hooker, Babcock's further talent, mirroring the facility of Sergeant N. Finegan in the Army of the Cumberland of eliciting valuable intelligence by interrogating enemy prisoners, came to the fore. Babcock became assistant chief in the Intelligence Bureau when Hooker took command of the Army of the Potomac. Babcock's superior had the appropriate name, for the chief of the Intelligence Bureau, of Colonel G. H. Sharpe. Together they established an excellent intelligence network and a fine working relationship as well. The two, with their operatives, provided Hooker with knowledge of the enemy far superior to anything his predecessors had had. But the general lacked the resolution to act on his knowledge. Hooker's failure at Chancellorsville brought on the Gettysburg campaign.

Babcock monitored the second Confederate invasion from a base in Frederick, Maryland. A telegram from Babcock to headquarters on June 24, 1863, announcing the crossing of the Potomac by "the last of Lee's entire army," put an end to intense Federal uncertainty about Rebel intentions and soon led to the crossing of the Potomac by the Army of the Potomac—a shift that had dire consequences for Robert E. Lee. Babcock continued his services under General George Gordon Meade. Slightly wounded at Gettysburg, Babcock nonetheless managed, from interrogation of hundreds of Confederate prisoners, to ascertain that Lee had by day's end on July 2, 1863, committed his entire force to battle except for George Pickett's division. This bit of information might have been decisive in the Federal decision the night of July 2, 1863, to continue the battle at Gettysburg. Both Sharpe and Babcock believed this to have been the case.

After Gettysburg, Babcock worked sometimes out of the War Department in Washington, other times with the army in the field. The Military Intelligence Bureau flourished with the arrival in the East of General Ulysses S. Grant. The close proximity of Federal and Confederate forces in the last year of the war yielded plenty of intelligence dividends.

Babcock ended the war in Richmond, Virginia, and for a time constituted a police department there in the absence of an organized civilian government. When his military connection ended, he moved to New York City and set up practice as an architect. He helped found the New York Athletic Club, designed several prominent buildings, and was active in a rowing club and in the Grand Army of the Republic, a powerful organization of Union war veterans. Although his official rank was never higher than a private, Babcock became known in civilian life as Colonel Babcock. He was living in Mount Vernon, New York, outside of New York City, with his daughter when he died of the effects of a stroke on November 19, 1908. He was seventy-two years old.

(Edwin C. Fishel's 1996 ground-breaking study of Union Civil War intelligence operations, *Secret War for the Union,* was referred to extensively in preparing this biographical profile of John C. Babcock.)

Ambrose Bierce
Observations Entered in Red

AMBROSE GWINNETT BIERCE, born near Horse Cave Creek in Meigs County, Ohio, on June 24, 1842, moved with his family to Warsaw, Indiana, when he was four. Influenced by his illustrious uncle, General Lucius Bierce, and perhaps by a bit of personal trouble in Warsaw, Bierce was sent to the Kentucky Military Institute in 1858 and stayed for at least one year. Part of the curriculum was civil engineering, surveying, and topography. Bierce clearly had the fundamentals of mapmaking well in hand when the Civil War began.

He enlisted at Elkhart, Indiana, on April 19, 1861, in the Ninth Regiment of Indiana Volunteers and promptly found himself in western Virginia. The three-month regiment disbanded in July, but Bierce rejoined when the regiment reorganized in August. He had already distinguished himself in his three months of service and was soon sergeant major of the regiment that came to be known, with good reason, as the Bloody Ninth. The regiment was briefly stationed in western Virginia, but in February 1862 it was assigned to the Union Army of the Ohio and transferred to Nashville, Tennessee. Its brigade commander was Colonel William B. Hazen, a regular army officer, a veteran Indian fighter, and an irascible disciplinarian. Strangely enough, Bierce almost immediately developed an immense respect for and loyalty to Hazen beyond all the other Civil War luminaries with whom he had contact. Hazen somehow discovered that Bierce had a knack for topographical engineering, and by April 4, 1863, Bierce was a lieutenant and acting topographical officer on Hazen's staff. By that time, the Ninth Indiana had fought its way through Shiloh, Corinth, and Stones River, some of the most severe military actions of the war. As part of the Army of the Cumberland under General William S. Rosecrans and later George H. Thomas, Bierce did his mapping in the most map-conscious army in the war. Rosecrans and Thomas both emphasized mapping and had chiefs of topographical engineers (Nathaniel Michler and William E. Merrill) who ran highly efficient and effective mapping operations, of which Bierce was a small but efficient part.

Central to the interest in Bierce as a topographical engineer is the fact that he is the only major American author to have served in any meaningful way in the Civil War. Walt Whitman's participation in the war was ancillary and Mark Twain's was a mere snippet of experience.

Bierce's service was lengthy and eventful, a definitive period of his life. A number of Bierce's short stories clearly make use of his experiences as a topographical engineer. His dramatic accounts probably give a rather realistic portrayal of the mapmaker's activities in the field. Bierce's writing, both fiction and nonfiction, was cold, precise, and fatalistic—in short, quite grim, and sometimes sardonically amusing. At Chickamauga, in a true account, Bierce offered to guide General James S. Negley back to the battlefield after the initial Union rout there, but, as Bierce phrased it, "I am sorry to say my good offices were rejected a little uncivilly." In his story "George Thurston," the narrator is a topographical engineer who describes his vivid mental pictures fixed, in lieu of field notes, by the intense excitement of the moment: these were "observations entered in red." "Killed at Resaca" is another engineer-based story.

Bierce suffered a bullet to the head, fired by a Rebel sharpshooter, on June 23, 1864, at Kennesaw Mountain, Georgia. Left for dead, he was saved by his brother Albert. Nursed back to health, he rejoined the army at Atlanta in

September 1864. He continued his engineering duties but on a more subdued basis. He stayed in the West, missing the March to the Sea. He saw further action at Franklin and Nashville, Tennessee. Bierce obtained his discharge from the army near Huntsville, Alabama, in January 1865. After the war ended, Hazen invited him along on an inspection tour of army posts, but it was an experience that convinced him to return to civilian pursuits.

Soon after, he joined his brother Albert in San Francisco and began a career in journalism and writing that led to a column in the *Examiner* and short stories such as "An Occurrence at Owl Creek Bridge." Bierce spent time in England, married and separated from his wife, Mollie Day, had two sons, was known as "the wickedest man in San Francisco," and perhaps died in January 1914 when he disappeared in revolutionary-war-torn Mexico, at age seventy-one. The circumstances and date of his death remain a mystery.

W. W. Blackford
Bold, Fearless, Cool, and Correct

WILLIAM WILLIS BLACKFORD was born in Fredericksburg, Virginia, on March 23, 1831. He was a first-generation Virginian on his father's side, though his mother was from an established Virginia family, the Minors. In 1861, the family moved to Bogotá, in what is now Colombia, where his father, William

M., was chargé d'affaires. They returned to Virginia after two years. The family had five sons, of which William was the eldest.

William chose civil engineering as a profession and worked to put himself through the University of Virginia, later becoming resident engineer for the Virginia and Tennessee Railroad.

With a wife, four children, and a prosperous business in 1861, Blackford reflected that he was "settled for life." However, John Brown's assault on the United States Armory and Arsenal at Harpers Ferry, Virginia, in October 1859 greatly disturbed both Blackford's and the nation's political equanimity. Anticipating trouble, Blackford set about raising a cavalry company in Washington County and was elected its first lieutenant. The young family man was against Virginia's secessionist movement but he ended up, as he wrote, "going with the South in her mad scheme, right or wrong." When the war commenced Blackford's company came under the command of Lieutenant Colonel Jeb Stuart.

Stuart appointed Blackford his adjutant—the military equivalent of an executive secretary—after personally observing Blackford's courage under fire. He served with aplomb in the First Manassas battle, but in the organizational shake-up that followed the battle he found himself without a position. After a few unsettled and unpleasant months, the trained civil engineer secured a captaincy in the Confederate Engineers. Jeb Stuart, by then a brigadier general, needed a staff engineer and applied for Blackford. The captain rejoined his old commander in time for the Seven Days campaign on the Virginia Peninsula.

Blackford's memoir of his service, *War Years with Jeb*

Stuart, describes his Civil War career in a low-key, self-effacing style that can't disguise the fact that its author was an arm's length away from a thrilling cavalry commander through three action-packed years of war.

But Blackford's memoir is more than merely exciting. His service as an engineer provides a view of wartime Virginia, Maryland, and Pennsylvania from the practical perspective of an engineer. Blackford surveys the battlefield in the wake of the action not to philosophize on the horrors of war but to figure out what lessons the terrain can teach in order to fight across it more effectively. His men push him up a tall tree with poles and he sketches a map of Union general George B. McClellan's position at Harrison's Landing on the peninsula. He makes an exhilarating reconnaissance in force, sketching positions within fifty yards of a swarming field full of "blue jackets." He passes along such intelligence insights as the fact that mounted cavalry "show" much larger than infantry and are generally accounted to be three times their actual number by untrained observers. In true topographical engineer spirit, Blackford enthuses about his powerful field glasses as much as his horses. Then, in typically flamboyant cavalry fashion, he strides into a sitting room in Mercersburg, Pennsylvania, and slices a county map off the wall, possibly overdramatizing this fairly routine requisitional proceeding. In another interesting anecdote, Blackford recalls an exhausting and futile attempt to make headway on his horse along a road full of infantry going in the same direction he was. It seems possible, given Blackford's experience, that Jeb Stuart's controversial decision to ride around the maneuvering Union army in order to reach the head of the advancing Confederate columns during the Gettysburg campaign may have reflected this logistical difficulty.

The maps Blackford prepared showed a relatively large area with few details. His maps concentrated on the road systems rather than on ground-cover details—a nod to the greater mobility of Civil War cavalry—though cavalry's speed, in fact, averaged only about twice that of the infantry.

Captain Blackford and civilian Jed Hotchkiss were friendly, Hotchkiss taking the occasion to mention Blackford rather frequently. They shared maps regularly, both their own productions and their map resources, and they compared notes, and they also shared soup, dinner, and conversation. Blackford was unlike Hotchkiss in that he thoroughly enjoyed the life of a soldier and was involved in far more military action than was his civilian counterpart.

Like Hotchkiss, Blackford survived the war, only slightly the worse for wear. He left Stuart's staff in February 1864 to join a newly formed engineer regiment. He made the retreat to Appomattox and surrendered there on April 9, 1865.

Blackford's wife died soon after the war, and the engineer had his share of the vicissitudes of postwar life in the defeated South. He worked as chief engineer on a railroad, tried his hand at running a sugarcane plantation in Louisiana, taught at Virginia Polytechnic Institute, again went back to railroading, and died April 30, 1905, near Norfolk, Virginia, at the age of seventy-four.

James Keith Boswell
Peace to His Memory

THE OVERRIDING image of young Confederate engineer James Keith Boswell is of a small, lined notebook containing a careful topographical sketch of "Bank's Ford, four miles above Fredericksburg, April 1863." A jagged hole through the middle of the sketch records the fate of Boswell on May 2, 1863, because he carried the memorandum book in his side pocket. The captain of Confederate engineers died instantly in the same misdirected Rebel fusillade that mortally wounded General Stonewall Jackson at Chancellorsville, Virginia. Boswell was shot twice in the heart and once in the leg. He was twenty-four years old.

Boswell was born in Fauquier County, Virginia, on November 18, 1838, the second son of Dr. Thomas Hamlin Boswell and Lucy Ann Steptoe Skinker, and was one of five living children. He attended private schools and trained to be a civil engineer. He worked in Missouri for the Pacific Railroad and in Alabama for the Montgomery and Florida Railroad. When Virginia seceded, Boswell volunteered his services. He was made civil assistant engineer in charge of the defenses of the Rappahannock. In short order, he was appointed first lieutenant of engineers in the Provisional Army of Virginia. He served under the command of General John B. Magruder and was engineer of the

defenses around Williamsburg on the Virginia Peninsula. In February 1862, he was appointed first lieutenant of engineers, Provisional Army of the Confederate States. At his own request, he was appointed to the staff of General Stonewall Jackson and became chief engineer of the Second Army Corps. In September 1862, Boswell was promoted to captain of engineers.

Boswell was a close friend of Jed Hotchkiss, as well as the civilian topographical engineer's nominal superior. His duties were broader than Hotchkiss's, as they included, routinely, work on fortifications, roads, and batteries, and he was frequently consulted about the meaning and consequences of enemy engineering activities. When Union gunboats were reported steaming up the Rappahannock River toward Fredericksburg, Virginia, in April 1863, General Robert E. Lee requested Jackson "to send some reliable officer to reconnoiter their position and attack them." Jackson sent Boswell, though the gunboats had already turned back. One of Boswell's great days of the war came on August 25, 1862, when he led Stonewall Jackson's command through Fauquier County, using every little known ford, bypath, and shortcut to get the Rebels onto Union general John Pope's flank as quickly, as secretly, and as soundly as possible during the Second Manassas campaign.

The young engineer captain approached his final battle with premonitions of death. Boswell wrote in his journal on April 21, 1863, "Strange as it may seem, not one of Genl. J's [Jackson's] staff has ever been killed." That was soon to change.

Jackson led a brilliantly successful attack against the unsuspecting right flank of Union general Joseph Hooker's Army of the Potomac in the late afternoon of May 2, 1863, but Jackson's attack lost momentum in the gathering darkness. There was uncertainty as to the enemy's whereabouts as well as the inevitable confusion and disorganization incidental to military operations. Unaccountably, the general and his staff, including Boswell, rode into a dark tangled thicket of woods in front of his own confused lines to see what was before him. As Jackson approached the Union picket line, several warning rounds rang out. These shots were answered by the edgy Confederate lines, which fired volley after volley at the mysterious group of horsemen in the black woods. Boswell, in advance of Jackson and the rest of the staff, probably fell in the first Confederate volley.

Jed Hotchkiss, who survived the melee, returned the next afternoon and found his friend's body. He wrapped Boswell in his overcoat and covered him with a tent he had picked up on the field. He laid him to rest in the small family cemetery at the Lacy house, Ellwood, at the western edge of the Chancellorsville battlefield, next to Stonewall Jackson's amputated left arm. Boswell's body was moved after the war to the Confederate cemetery in Fredericksburg, Virginia.

John S. Clark
Delineator

JOHN S. CLARK was born in Mentz, New York, on November 2, 1823. His birth name was John Swartout Smith Duvall. His maternal grandfather despised his son-in-law, a French Huguenot, and changed John's name when the boy's father, Israel S. Duvall, died. John Clark went south for an education and attended Bethany College in Virginia. He inherited his grandfather's farm near Throop, New York, and settled into farming with his new wife, Mary Ann Crofoot. Like any number of his contemporaries, including Abraham Lincoln, abolitionist John Brown, and naturalist Henry David Thoreau, Clark learned surveying and took up the profession. He grew quite proficient at it and had just moved to nearby Auburn, New York, as city engineer when the Civil War intervened. Some freelance heroics early in the war earned Clark local renown and he was given the command of an Auburn area regiment, the 19th New York. Clark had the wrong temperment for a field commander. His tenure was contentious, unsuccessful and short. Soon thereafter, General Nathaniel P. Banks invited Clark to join his staff—the exact wartime situation for which he was best suited. Clark enthusiastically served in that capacity for the balance of the war.

Clark mapped Cedar Mountain and Second Manassas for Banks and had his moment of personal glory the morning of August 25, 1862, when, peering through his telescope, he spotted Confederates in significant numbers moving northwest. Two hours after Stonewall Jackson began his secret flank march, Clark was reporting its size and direction to Banks.

Clark was a better cartographer than he was a topographical engineer. His maps, in other words, look better than they are. The nondescript pencil maps that William Henry Paine drew, as one example, were nonetheless much more reliable than Clark's more finished productions. Clark's reputation was based on his being a fine companion, a brave scout, and a man who never lost his bearings. Clark was also a great favorite of the otherwise virulently anti-northern ladies of New Orleans when he was in the Crescent City on the staff of Banks in 1863. Clark was a friend and colleague of topographical engineer David Hunter Strother, who mentions Clark frequently and favorably in his Civil War writings. Clark was severely wounded during the 1863 Port Hudson, Louisiana, campaign. He read and carefully saved the obituaries that appeared when it was supposed he had been killed.

John S. Clark survived the war and devoted himself to historical pursuits in upstate New York, including a study of the state's Indian tribes and the Revolutionary War campaign of General John Sullivan and Sir Henry Clinton, which he meticulously mapped. He died in Auburn on April 7, 1912, at the age of eighty-eight.

George Armstrong Custer
Topographical Engineer

GEORGE ARMSTRONG CUSTER, always associated with Michigan, was born in New Rumley, Ohio, December 5, 1839. He attended school in Monroe, Michigan, and it gradually became his adopted home. He briefly taught school back in Ohio and then decided to try for an appointment at West Point. He had boyish notions of military glory, so West Point made sense. Strangely enough, Custer also thought a West Point education would bring ample monetary rewards in civilian life. Custer attained an appointment partly, it is thought, because an unhappy parent of an object of Custer's lovestruck attention wanted him out of town for a few years.

Custer's West Point career was a legendary one of demerits, scrapes, near expulsions, a court-martial, and sheer exuberance. Except for drawing, where he stood in the upper half of his class, Custer was near the bottom. The onset of the Civil War did more for the troubled cadet's prospects than his studying did. His class was scheduled to graduate in 1862, but the date was moved up to 1861 after the war began. The fact that Custer graduated at the bottom of his class came about, to be fair, in part because a number of Southerners had withdrawn from the academy.

Custer purchased a lieutenant's uniform in New York and hurried on to Washington. He reported for duty on July 20, 1861, and by that evening, Custer's luck, an elemental force for him, was in evidence. He was carrying dispatches from Lieutenant General Winfield Scott to General Irvin McDowell on the virtual eve of the First Battle of Manassas, thus making himself known to the top command of the Union army on his first day at war. Illness kept Custer out of the war for four months, and when he returned, General George B. McClellan was in top command. Custer followed the Army of the Potomac to the Virginia Peninsula.

Rank in the class at West Point normally guided the selection of which branch of the army a graduate served in. Engineers were the army's elite, and Second Lieutenant George Custer was certainly not one of them. However, his West Point training included several hundred hours of study in mapping and topographical engineering. Custer's experience in this field, among volunteers with no training whatsoever, resulted in his being appointed an assistant engineer on the staff of the army's chief topographical engineer, A. A. Humphreys. Amid the dangers and privations of service in the field in a wartime army, Custer found what he described beatifically as perfect happiness. He made ascents in observation balloons to monitor the activity of the Rebels and to sketch their positions. Custer wrote an excellent, self-deprecating account of his balloon ascents: he huddled on the deck of the wicker basket as the balloon rose. The uneasy lieutenant failed to appreciate the efforts of the "civilian aeronaut" who accompanied him to reassure the novice flier. The civilian jumped up and down on the floor of the basket to demonstrate its sturdiness until Custer prevailed on him to stop. Custer got out a small notebook and sketched what he could see of the Rebel positions. His surviving sketch maps are very creditable productions, instantly communicating to the viewer the feel for the land and its essential characteristics that a naturally gifted topographical engineer always conveys in his maps. The colored pencil work is delicate but steady, deft but thorough, indicative of the work of a methodical craftsman whose whole attention is fixed on the job at hand.

Custer's raw courage and cool confidence soon brought him notice and advancement and ultimately legendary fame at the Little Bighorn in June 1876, where he died at the age of thirty-six. The map sketches help balance the glare of Custer's ultimate notoriety with their evidence of his earlier earnest competence.

Rodney E. Harris

Regimental Postmaster, Executor of Maps

RODNEY E. HARRIS was born in Tyrone, Schuyler County, New York, in 1844, the second of four sons of Monroe and Mercy Harris. He enlisted in the army at Elmira, New York, on August 7, 1862, listing his occupation as clerk. He served as a private in Company A, 107th New York Infantry. Harris spent nearly his entire enlistment on detached duty, detailed as regimental postmaster, or "post boy," until August 1863, when he was appointed headquarters clerk of the First Division Artillery Brigade of the Twelfth Corps. This transfer seemingly brought Harris to the attention of General Henry W. Slocum, who commanded the Twelfth Corps and signed the paperwork relating to Harris's transfer.

Both Slocum and Harris traveled west in October 1863, when the Eleventh and Twelfth Corps of the Army of the Potomac were sent to reinforce the besieged Union troops in Chattanooga, Tennessee, following their defeat in the Battle of Chickamauga in September 1863. Slocum continued west to Vicksburg because he refused, after the Union debacle at Chancellorsville in May, to serve under General Joseph Hooker. When Hooker resigned over a seniority issue in July 1864, Slocum returned to command the Twentieth Corps, a consolidated unit consisting of his old Twelfth Corps and the Eleventh Corps. Slocum had missed the dramatic Union victory at Chattanooga in November 1863 and the initial operations of General

William Tecumseh Sherman's Atlanta campaign. Rodney E. Harris had not missed these events, serving in the Twelfth Corps throughout.

There was apparently a hotbed of topographical talent in the south-central section of New York State in the early 1860s. Captain Robert M. McDowell of Elmira, New York, a member of the 141st New York Infantry, drew an excellent series of maps recording the operations of the Union Twentieth Corps under Sherman in the West. Lucius T. Stanley, like Rodney E. Harris, served not only in the 107th New York Infantry but in Company A. Stanley was responsible for a series of expertly rendered maps starting with Chancellorsville, Virginia, thereafter following the activities of his regiment and corps to the western theater. Not too many miles to the east of Elmira is the village of Windsor, New York. This was the boyhood home of mapmaker for the Confederate army, and the Civil War's most noted topographical engineer, Jed Hotchkiss.

To this list of cartographers and engineers can be added Private Rodney E. Harris. At Savannah, Georgia, in December 1864, Harris executed, in an impressive handmade volume, a series of ten watercolor maps. They traced the progress of the Twentieth Corps along its route of march in one of military history's most famous campaigns, Sherman's March to the Sea, from Atlanta to Savannah, Georgia. There is nothing in Harris's career to explain how he could prepare this startling "atlas" of military maps, except possibly the fact that as a clerk he would presumably write with a steady, legible hand.

Harris was mustered out of the army in Washington on June 5, 1865. He returned to south-central New York, living for a period in Bath, Steuben County, where he was active in the Union Free School and worked as bookkeeper in a local hardware store. In 1868 he moved to nearby Liberty (now Cohocton), New York, where he joined two of his brothers in their own hardware store, Harris Brothers and Co.

Rodney E. Harris married Ella Lyon and had two children. In 1875, the Harris family moved to Naples, an Ontario County, New York, town about twelve miles north of Cohocton. Two years later, in April 1877, Harris was stricken with typhoid pneumonia. He died April 21, 1877, at the age of thirty-three. He was buried in Maple View Cemetery in Cohocton.

Jed Hotchkiss

Mister Hotchkiss of the Second Corps

JED* HOTCHKISS was born in a fieldstone house on the right bank of the Susquehanna River in Windsor, New York, on November 30, 1828. He attended local schools and the Windsor Academy. In 1846 he interrupted a tour to take a teaching position near Harrisburg, Pennsylvania, for a year. Hotchkiss then continued his walking tour south and accepted a teaching position in Mossy Creek, in Virginia's Shenandoah Valley, where he soon settled permanently after marrying a Pennsylvanian, Sara Ann Comfort, whom he had met on the first leg of his walking tour. In his spare time, Hotchkiss educated himself in subjects that were taught formally only in military schools: mapmaking, surveying, and engineering.

Hotchkiss started the Loch Willow Academy in Churchville, near Staunton, and it was here that the Civil War found him. As a resident of Virginia for all of his adult life, Hotchkiss offered his services to the Confederacy as a matter of course. Strangely, Hotchkiss never received a military rank from the Confederate government. As a consequence, he once was nearly drafted into the Confederate army as a private, though he was, at the time, an invaluable aide on the staff of Stonewall Jackson. For three years of the war he was plain Mr. Hotchkiss, a hired civilian topographical engineer.

The first maps that Hotchkiss drew were painstakingly detailed productions of Camp Garnett at Rich Mountain in western Virginia. The camp was attacked and captured by invading Federals in mid-July 1861. In the process, Hotchkiss lost his maps and mapmaking equipment. General Robert E. Lee arrived for a brief, unhappy western Virginia command experience and set Hotchkiss to work on a map of Tygart's Valley. The map completed, Hotchkiss suffered a physical collapse and returned home to recuperate for the winter.

By then, there were no pupils to teach, and Hotchkiss's only employment prospects were with the army. With his brief résumé from the previous summer as a topographical engineer, Hotchkiss headed back to the war. He reported to Jackson and in a few days was detailed by him as a member of the engineering staff. It was a momentous decision, fraught with consequences for both men. Hotchkiss's maps provided the literal-minded general with the representational information he needed to visualize and grasp terrain and topography. Hotchkiss won Jackson's complete confidence, and the general's appreciation of Hotchkiss's work galvanized the mapmaker's talents. On March 26, 1862, Jackson asked Hotchkiss to prepare a map of the Shenandoah Valley. This map, the size of a door, remained a work in progress throughout the war and was still unfinished at the time of Jackson's death in May 1863. Hotchkiss performed his final duty with his revered commander when he guided the ambulance of his mortally wounded chief to safety behind the lines to Guinea's Station, Virginia, after the Chancellorsville battle.

Hotchkiss remained with the Army of Northern Virginia under various commanders, including Generals Richard S. Ewell and Jubal A. Early. His map portfolio is one of the great archival treasures of the Civil War. The fact that Hotchkiss managed to preserve his original maps as well as record his day-by-day experiences as a protégé of Jackson, Lee, Ewell, and Early established his place in the canon of the Civil War.

When the war ended, Hotchkiss restored his Loch Willow Academy but eventually moved to Staunton to teach school. Summoned, unexpectedly, to surrender his wartime maps, Hotchkiss ended up being paid by the U.S. government to supply copies of them to the United States War Department for use in the *Official Records Atlas,* then in preparation. Hotchkiss's diary, as well as his original maps, came into demand as books about the war began to be contemplated. His schoolteaching gave way to engineering, and Hotchkiss became very active—traveling twice to England—in developing Virginia's resources after obtaining the investment capital to do so. He was a primary source of information for British military historian G. F. R. Henderson's classic 1898 biography *Stonewall Jackson and the American Civil War.* Hotchkiss's own *Virginia* appeared as volume 3 of *Confederate Military History.* Hotchkiss died on January 17, 1899, in Staunton, Virginia, at the age of seventy.

*Hotchkiss used this short form of his name almost exclusively. On one known occasion, on a handmade, hand-lettered pamphlet in 1850, Hotchkiss used a very unusual spelling of his full first name, viz. Jedadiah. On his gravestone he is referred to as Jed.

William E. Merrill
Innovative, Invaluable, Unknown

WILLIAM E. MERRILL was born October 11, 1837, at Fort Howard, Wisconsin. Perhaps the most innovative and conscientious exponent of mapping during the Civil War, Merrill was not even a topographical engineer but a member of the Corps of Engineers.

Merrill was a West Pointer, graduating from the academy at the top of his class in 1859. A lieutenant in the Corps of Engineers, he served in Georgia and Florida but was back at West Point working as an instructor in the Department of Engineering when the Civil War began. As with Orlando M. Poe, Merrill's early war service came under Union general George B. McClellan in Maryland and western Virginia and on the Virginia Peninsula, where he was wounded. He was a prisoner of war in Richmond for six months. Upon his release, he was assigned as engi-

neer in the Army of the Potomac on the Virginia Peninsula and worked on the fortifications around Washington. He was present at Cedar Mountain near Culpeper, Virginia, and at Second Manassas. When invaders led by Rebel general E. K. Smith threatened Newport and Covington, Kentucky, Merrill was sent to superintend the defenses of the towns. He served under General William S. Rosecrans, one of the few senior officers on either side who gave top priority to maps and mapmaking. Rosecrans's Chickamauga campaign was a brilliant and successful campaign of maneuver—always a sign of good mapping—but it ultimately came to grief at Chickamauga, Georgia, on September 19, 1863. Thereafter, Merrill served under General George H. Thomas in the Army of the Cumberland. The two were a perfect match because Thomas believed wholeheartedly in mapping and engi-

neering. As one of the three armies under the command of General William Tecumseh Sherman in the campaign from Chattanooga to Atlanta, the Army of the Cumberland set the standard for mapping for all of Sherman's forces.

Merrill built pontoon boats whose gunnels were attached so they swung together. He built blockhouses that protected the railroad bridges on the vital railroad supply lines that maintained Sherman's army in the Atlanta campaign. And not only did Merrill see to it that Sherman's armies had the best maps of any Civil War army, he made certain that the maps were continually updated and promptly distributed. Further, Merrill insisted on maintaining control of the assignments of his topographical engineers so that their talents were not squandered locating likely campsites and carrying dispatches. Merrill's operation was well equipped—with printing presses, lithographic presses, cameras and plenty of photographers,

map mounters, clerks, lithographers, and draftsmen. Sherman's army was indisputably the best supplied with maps of any that fought in the Civil War. If the maps were often rudimentary and frequently wrong, they were nonetheless on hand and constantly being corrected. Merrill remained in the West, protecting the railroads and constructing defensive works in Tennessee, Alabama, Georgia, and Ohio, ending up at Nashville in January 1866. He ultimately reached the rank of brevet colonel.

After the war, Merrill was on Sherman's staff as chief engineer of the Division of the Missouri and later on General Philip Sheridan's staff in Chicago, Illinois. His postwar career was largely devoted to the improvement of the Ohio River for navigation and management of the uses of rivers and bridges. Merrill died while taking a train, on official business, from Cincinnati, Ohio, to Shawneetown, Illinois, on December 14, 1891. He was fifty-four.

Peter Smith Michie

A Born Soldier

A NATIVE of Scotland, Peter Smith Michie was born in 1839. His family immigrated to the United States in 1843 and settled in Cincinnati, Ohio. Michie had already embarked on a successful career as a mechanic when he seized upon a vacancy at West Point in 1859 and won an appointment there. Michie graduated second in his class on June 11, 1863, at the very height of the Civil War: the outcome of the Gettysburg and Vicksburg campaigns was still in doubt. Michie was assigned to the elite Corps of Engineers and almost immediately was put on duty at Hilton Head, North Carolina. During a brief furlough home, Michie had just enough time to get a marriage license, arrange an evening wedding to fiancée Marie Louise Roberts, and catch a train back East.

By June 24 he was on duty, in charge of constructing artillery batteries for use in operations against Charleston, South Carolina. That campaign was a failure, but Michie had more success in the capture of Fort Wagner. He was assigned the position of chief engineer in the Northern District of the Department of the South. He served briefly in Florida at the Battle of Olustee and worked on fortifications there. At the end of May 1864, he was at Fortress Monroe, as assistant engineer of the Army of the James. The Union forces, hitherto scattered and ineffective, were brought under the unified control of General Ulysses S. Grant and directed to operate in concert.

His new duties brought Michie into close personal contact with the luminaries of the eastern theater, from his commander, General Ben Butler, to George Gordon Meade, Winfield Scott Hancock, and Grant himself. For services performed while he was still a lieutenant, Michie was given brevet ranks of captain, major, lieutenant colonel, and brigadier general.

Michie was an indispensable man in the summer of 1864 on the Virginia Peninsula. He was laying pontoon bridges, constructing defensive fortifications, planning offensive operations, preparing surveys, and drawing maps. He was a source of contention because a number of generals wanted him on their staffs. He was talented, conscientious, tireless, intrepid. In the understaffed Civil War armies, he was "worth his weight in gold," as one general commented.

Michie cheerfully performed the dangerous but undramatic work of an engineer in an entrenched army—extending the lines, overseeing bombproofs, building bridges, corduroying roads, undertaking reconnaissances and surveys, making maps and sketches, observing and reporting—and as the days and months passed, he never became discontented or downcast.

Finally, in the spring of 1865, the relentless reach of the Union army prized the Confederates out of their Petersburg fortifications, and the final campaign to Appomattox got under way. Michie organized a flying column of pontoniers (temporary bridge builders) and engineer troops and led them by an alternate route to Farmville, ahead of the rest of the Army of the Potomac. His column laid a pontoon bridge over the Appomattox River, which the Union Second Corps crossed. Their subsequent attack on the Confederates fatally delayed the Rebels in their exhausted attempt to elude the relentless Yankees.

The war ended here for Michie, but immediately after the cessation of hostilities, he was posted to Richmond, the erstwhile Confederate capital, where he spent a year thoroughly and accurately mapping the nearby battlefield and area of operations. His first permanent home as a family man, ironically, was Richmond. One of his mapping collab-

orators was famed New York Confederate mapmaker Jed Hotchkiss. The war was indeed over.

He later served as an instructor at West Point and from 1871 until his death he was a professor at his alma mater. He was the author of two books about his Civil War colleagues: *The Life of Major General Emory Upton* and *General McClellan.* Michie died, at age sixty-two, at West Point.

Nathaniel Michler
Compiled under the Direction of . . .

BORN in 1827, Nathaniel Michler was a Pennsylvania native. He graduated from West Point in 1848, seventh in a class of thirty-eight. He was made a lieutenant of topographical engineers and was soon involved in making surveys and reconnaissance in Texas and in what is now New Mexico. In 1860, he was sent to the West Coast as engineer of the Twelfth Lighthouse District.

When the Civil War began, he was assigned to duty as a captain of topographical engineers. Sent east to join the Army of the Potomac, he was captured by Jeb Stuart's cavalry on the outskirts of Rockville, Maryland, in the midst of the Gettysburg campaign. By July 28, 1863, he was paroled and at work surveying at Harpers Ferry in western Virginia. In charge of the Topographical Department in the Army of the Potomac from September 1863 until the surrender at Appomattox in April 1865, Michler was connected with all the movements of the army until the war's end. By that time he had reached the rank of brevet brigadier general.

After the war, Michler was involved for two and a half years in surveying and mapping the operations and fields of battle of the Armies of the Potomac and the James.

Michler's postwar career took him on various engineering and surveying projects throughout the United States, from Oregon and Idaho to Michigan and Vermont. He served from 1878 to 1880 as military attaché to the United States Legation at Vienna, Austria.

Michler died July 17, 1881, at Saratoga Springs, New York, at the age of fifty-four.

William Henry Paine
Pathfinder

ONE of the finest topographical engineers in the Union Army of the Potomac was Captain William Henry Paine. A hardworking, unassuming staff officer, Paine stands at General George Gordon Meade's left shoulder in the headquarter staff's photographic portrait. The army commander, himself an old topographical engineer, obviously trusted and relied on Paine. When Meade made his perilous night ride from Taneytown, Maryland, to the Gettysburg battlefield on July 1, 1863, it was Paine who led the way. Arriving at midnight, Paine then led Meade on a tour of the field to acquaint the general with the Union position and to prepare a map of it.

A native of Chester, New Hampshire, Paine was born May 27, 1827. In his early twenties, he moved west to Wisconsin. Like his Rebel counterpart, Jed Hotchkiss, Paine taught himself the basics of engineering and surveying and was soon laying out the Johnson route, one of several proposed for a railroad from Utah to Sacramento, California. Paine invented a light, sleek, flat tape line to replace the unwieldy and heavy link chain then in use by surveyors.

Back in Wisconsin when the Civil War began, Paine joined the Fourth Wisconsin Regiment. His profession brought him to the attention of General Amiel Whipple, chief topographical engineer of the short-lived Union Army of Northeast Virginia. Paine made a daring reconnaissance, bringing back a report on the status of various bridges in Rebel-held Virginia. This crucial intelligence helped Union planners develop their strategies. It also impressed another surveyor and keen appreciator of maps, Abraham Lincoln, who appointed Paine captain of engineers, a rare honor for a non–West Point soldier. Paine eventually served on the staffs of Generals John C. Frémont, Irvin McDowell, John Pope, Joseph Hooker, and Meade.

Paine formed a close relationship with another topographical engineer, a non–West Pointer like himself, Washington Roebling. On June 4, 1863, the two men rode in an observation balloon to peer at the Rebels across the Rappahannock River. They noted and reported the first stirrings in the Confederate lines that signaled the beginning of the Gettysburg campaign.

Paine was a fine draftsman, a quick and accurate mapmaker, a reconnaissance expert, and also a very adept interrogator. He was recognized as one of the hardest-working and most valuable officers on the Army of the Potomac staff. It was said the entire army was often dependent upon the sole judgment of this mere captain. Paine became a colonel by brevet on April 9, 1865. On this last day of war, but before an official cessation of hostilities, he took advantage of Confederate confusion to slip behind their lines and sketch a final map on the hostile field in the midst of fully armed, uncertain, and unhappy enemy troops.

His postwar career was in construction and engineering. He was hired by fellow veteran Roebling to assist in building the Brooklyn Bridge. Paine died December 21, 1890, at age sixty-three, in Cleveland, Ohio.

Orlando M. Poe
Sherman's Engineer

ORLANDO METCALFE POE was born at Navarre, Stark County, Ohio, on March 7, 1832. He was teaching at a district public school when he learned that his district's appointee to West Point had dropped out and was returning home. Poe secured the appointment and went to West Point. After an inauspicious start, he graduated sixth in his class. Poe remained briefly as an instructor at West Point and served as a lieutenant in the Corps of Topographic Engineers. He worked under George Gordon Meade doing surveys on the Great Lakes.

After Fort Sumter was fired upon, the governor of Ohio asked Poe to recommend a commander for Ohio's troops. Poe suggested George B. McClellan, and shortly thereafter Poe became chief topographical engineer of the Department of the Ohio under McClellan, Poe married Eleanor Carroll Brent in Defort on June 17, 1861, and went off to war the same day. In September of that year, he was appointed a colonel and commanded troops during the Virginia Peninsula fighting, at Second Manassas, and after the Battle of Antietam, from October to November 1862. He was in the Battle of Fredericksburg, commanding a division of the Ninth Army Corps, moving with that corps to the Department of the Ohio in March 1863.

Nominated as a brigadier general, Poe resigned his post as colonel. His nomination was not acted upon by the Senate, and Poe went in one stroke from a nominated brigadier general in the volunteer army to a lieutenant of engineers in the regular army. He was made chief engineer on the staff of General Ambrose E. Burnside in eastern Tennessee. At Knoxville, he had the idea of adding a new item to the usual obstructions put up to disrupt an enemy assault. He strung surplus telegraph wire at knee level around the fortifications, a first in warfare.

During the last two years of the war, Poe became friends with General William Tecumseh Sherman. Poe served as an engineer in the Military Division of the Mississippi, Sherman's command. He was Sherman's chief engineer during the Atlanta campaign, the March to the

Sea, and the campaign in the Carolinas. A brevetted or wartime brigadier by war's end, Poe remained on Sherman's staff into peacetime and until Sherman's retirement in 1884. He continued as a military engineer because he believed that his usefulness was at the service of his government. Poe died October 2, 1895, at the age of sixty-three, from an infection following a leg injury he suffered while inspecting a lock at Sault Ste. Marie, Michigan.

Washington Roebling
Necessity's Soldier

WASHINGTON AUGUSTUS ROEBLING was born in 1837 in the farming community of Saxonburg, northeast of Pittsburgh, Pennsylvania. His father, John, a Prussian, had immigrated there six years earlier in an effort to find a more congenial life as a farmer in a free land than as an engineer with limited prospects in the stifling political climate of Prussia. The elder Roebling was a born engineer but a poor farmer on what turned out to be a poor farm. He therefore returned to engineering. The family moved to Trenton, New Jersey, in 1848.

"Wash" Roebling, as he was known, attended Rensselaer Polytechnic Institute in Troy, New York—the leading engineering school at the time in the United States. When the Civil War began, Roebling was working in his father's wire rope business and helping his father build the Allegheny River suspension bridge at Pittsburgh. Probably as a result of his college residence in New York, Roebling joined the Ninth Regiment, New York State Militia. He served briefly with an artillery unit and then, as Lieutenant Roebling, he became an aide to General Joseph Hooker. He worked hard during the August 1862 campaign of Second Manassas, literally riding his horse to death, surveying near Madison Court House, Virginia, and accompanying the expedition that nearly captured Rebel Jeb Stuart at Verdiersville. He was soon assigned to the Engineer Brigade of the Army of the Potomac, where he oversaw the construction of (and philosophically endured the destruc-

tion of) military suspension bridges across the Rappahannock River in Virginia and across the Shenandoah River at Harpers Ferry.

With his highly objective, straight-thinking engineer's mind, Roebling was a mystified observer of the conduct of the war and a low-key cynic when commenting on the claims and representations of the army's principal commanders. Promoted to major, he began a highly satisfactory position on the staff of Major General G. K. Warren, chief topographical engineer of the Army of the Potomac. The Union army, occupying the left bank of the Rappahannock River opposite Fredericksburg, Virginia, was caught flatfooted with respect to maps when the Confederate Army of Northern Virginia began its second invasion of the North, the Gettysburg campaign. A scramble to procure maps ensued. Roebling and his colleague William Henry Paine rode to Washington together to see what they could turn up there. Roebling continued on alone to Baltimore, Philadelphia, and then to the family home in Trenton, New Jersey. He borrowed a good general topographical map of Pennsylvania that his father had, and as he later recounted in his memoirs, with a "quantity" of maps rode warily through an "eerily deserted" Maryland to rejoin the army, now at Gettysburg. On an initial reconnaissance of the Union position, Roebling made a curious oversight. He failed even to notice Little Round Top. "It was hidden," he said, "by the woods on the side of the Taneytown Road, and so overshadowed by Big Round Top that no one expected it." Soon, however, the new commanding general George Gordon Meade heard a "little peppering going on in the direction of that little hill yonder" and sent Warren and Roebling off to investigate. Their timely action in reinforcing Little Round Top was a major factor in the Union victory at Gettysburg.

The laconic, slightly stooped Roebling was noted for always "poking about in the most dangerous places" and for his no-nonsense conversational style. Once Meade asked him, "What's that redoubt doing there?" Roebling replied, "Don't know; didn't put it there."

Roebling left the army in January 1865 and was brevetted a colonel soon after. He married Warren's sister Emily and rejoined his father in building bridges. His father's untimely death in 1869 left Roebling in charge of one of the great engineering projects of the nineteenth century, the Brooklyn Bridge. He tapped a fellow topographical engineer, William Henry Paine, as an assistant.

The eleven-year project left Roebling an invalid. He moved with his family to Troy, New York, and then back, finally, to Trenton, New Jersey. Roebling died in Trenton on July 21, 1926. It is often mistakenly supposed that Roebling died on the *Titanic*. In fact, it was a nephew and namesake, not the great engineer, who went down with that ship.

D. H. Strother
Virginia Yankee

DAVID HUNTER STROTHER was born at Martinsburg, western Virginia, in the lower Shenandoah Valley, on September 26, 1816. He wanted to go to West Point but could not get an appointment there, eventually attending Jefferson College in Canonsburg, Pennsylvania. He later studied in New York and at various art centers in Europe: Paris, Florence, and Rome. When he returned to the United States in 1843, he gained recognition for his book illustrations. His civilian career was born when *Harper's* asked him to write something to accompany his sketches. He chose the pen name "Porte Crayon"—a French pencil-holding instrument—and was soon turning out a series of very popular travel books.

Strother clearly saw a national crisis looming, but with ambivalent feelings he hoped to remain neutral and withdrew to Berkeley Springs, a western Virginia spa, where his family ran the famous Strother Hotel. Neutrality proving impossible, Strother volunteered and went to work as a civilian topographical engineer with the Union army. He was a skilled artist, knew the ground thoroughly, and understood the locals well. He served under a string of lackluster commanders—generals Robert Patterson, Nathaniel P. Banks, and John Pope—in, respectively, the First Manassas campaign, the Valley campaign, and the Second Manassas campaign, all major Union defeats. His one winning campaign was under General George B. McClellan at Antietam. Strother, by now commissioned a lieutenant colonel, rejoined Banks for his expedition against Port Hudson in 1863. Port Hudson fell to Banks after the capture of Vicksburg by U. S. Grant rendered the fortress useless, but Strother by that time had already gone back east, returning to Washington on May 10, 1863. He was on a sixty-day leave and missed the Gettysburg campaign. Strother's rather luckless army service then resumed as he served under generals Benjamin Kelley, Franz Sigel, and distant kinsman David Hunter, all marginal commanders at best.

Hardly any of Strother's maps seem to have survived the war, though he made minor contributions to several that are extant. His Civil War–era writings have fared better: his contemporaneous diary was edited by Cecil D. Eby, Jr., and published as *A Virginia Yankee in the Civil War* in 1961. Strother's voluminous "Personal Recollections of the War by a Virginian" appeared serially in *Harper's New Monthly Magazine* from June 1866 to April 1868. Both are eminently readable and provide the most insightful look at the activities and concerns of a Civil War topographical engineer available.

After the war, Strother was briefly adjutant general of Virginia. He returned, with somewhat disappointing results, to his writing. After a successful term as consul general of Mexico, he came back to the United States in 1885 and died in Charleston, West Virginia, on March 8, 1888, at the age of seventy-one.

G. K. Warren
Savior of Little Round Top

GOUVERNEUR KEMBLE WARREN was born on January 8, 1830, at Cold Spring, New York, across the Hudson from, and in sight of, West Point. He entered the academy at sixteen, graduated second in his class, and was assigned to the Corps of Topographical Engineers.

Warren was immediately involved in the grueling work of surveying the Mississippi Delta. In 1854, Lieutenant Warren was ordered to compile a general map of the country west of the Mississippi River. He did this in uncharted, vast expanses of wilderness amid the dangers of hostile Indians and while performing the other regular duties of a frontier army officer. The map he produced is one of the great cartographic milestones in nineteenth-century

American history. Warren's first taste of combat came in an encounter with the Sioux at Blue Water Creek, near the North Platte, in September 1855.

Ordered back to West Point after nine dramatic years of service, Warren was teaching mathematics there at the outbreak of the Civil War. The Corps of Topographical Engineers reacted very slowly and with great reluctance to the necessities of wartime. There was an inborn bureaucratic fear that the war would disrupt—of all things!—the normal functions of the corps. Wishing to distinguish himself and rightly judging that remaining in the topographical corps in the regular army was no way to do that, Warren received a commission in the volunteer army and was soon promoted to colonel of the Fifth New York Zouaves Regiment.

Warren and his well-drilled regiment were conspicuous in the fighting on the Virginia Peninsula in the spring of 1862. The regiment was sent in August to reinforce John Pope's Army of Virginia at Second Manassas. In the aftermath of the Union defeat, Warren had the horrific experience of standing by, nearly helpless, while his command was ripped to pieces. Through no fault of Warren's, fighting for ten minutes in what one of the regiment's survivors called "the vortex of Hell," the Fifth New York sustained the largest loss of life of any one regiment in any one battle of the Civil War. Warren commanded the Third Brigade, Second Division, of the Fifth Army Corps, Army of the Potomac, through the Antietam campaign and the Battle of Fredericksburg. In February 1863, after General Joseph Hooker took over the command of the army, Warren, now a brigadier general, was ordered to join Hooker's staff as chief of topographical engineers. He was subsequently named chief engineer of the Army of the Potomac when the Corps of Topographical Engineers was merged with the Corps of Engineers.

Warren's tenure was not entirely successful. Though there were many reasonable excuses for it, the army was ill prepared for the Confederate invasion of Pennsylvania as far as maps were concerned. Warren had his men—including Captain William Henry Paine and Major Washington Roebling—hurrying hither and yon, securing maps that might reasonably have been expected to be at hand. Warren had managed to fit his wedding in Baltimore into the time frame that saw the Rebel Army of Northern Virginia at the banks of the Potomac River, which indicates some sort of complacency at the army's map headquarters. Part of the blame also rested on the acrimony that accompanied changes of command in the Army of the Potomac. Their maps went with the old staff, leaving Warren bereft of information. Eventually, though, Warren was on hand at the Battle of Gettysburg itself to make sure that the inadvertent failure to occupy Little Round Top was quickly and effectively remedied.

Shortly after Gettysburg, Warren was back in field command—of the Second Corps. At Bristoe Station, near Manassas Junction, Warren, now a major general of volunteers, held off a superior number of Confederate attackers and inflicted twice as many casualties as he sustained. This October 1863 success was followed by a November 30, 1863, decision in the field by Warren at Mine Run, Virginia, to suspend a planned attack because daylight revealed an unexpectedly formidable entrenched Rebel position. Warren undoubtedly displayed great moral courage in putting his own reputation on the line rather than the lives of his men.

When General Ulysses S. Grant arrived in the East, Warren was given permanent command of the Fifth Corps and fulfilled his duties satisfactorily throughout the Overland campaign to Petersburg. He likely fulfilled them more than satisfactorily at Five Forks, Virginia, in the action that crumbled the nearly yearlong Rebel defense of Petersburg. Yet in an episode that probably stemmed partly from an inability to see and understand what was going on and partly from a visceral dislike of Warren, General Philip Sheridan on April 1, 1865, summarily relieved Warren of his command and ordered him from the field. Grant expressed dismay that this was done, but upheld Sheridan's action.

The last seventeen years of Warren's life were in thrall to that one eclipsing moment. Fourteen years after being relieved at Five Forks, Warren was granted a Court of Inquiry. The court concluded that Sheridan had acted unjustly, but no other action was called for in the matter. Meanwhile, Warren had reverted to the regular army immediately upon the cessation of hostilities and served in the Corps of Engineers.

He died, it was said, of a broken heart at Newport, Rhode Island, on August 8, 1882. He wanted no military trappings at his funeral and no military uniform on his body. He was fifty-two years old.

In his memoirs, Grant wrote a tempered explanation of the relief of Warren at Five Forks. He indicated that Warren had "a defect which was beyond his control. . . . He could see danger at a glance before he had encountered it. He would not only make preparations to meet the danger which might occur, but he would inform his commanding officer what others should do while he was executing his move." One senses, if Grant's assessment is just, that Warren may have relived the annihilation of the Fifth New York Zouaves at Second Manassas and thereafter took elaborate precautions to avoid a repetition of that singularly grim experience.

Epilogue

I seem to possess a knowledge of men and things, rivers, and roads . . .
—General William Tecumseh Sherman

MAPS had an immense, sometimes definitive impact on the various campaigns and battles of the American Civil War. There were never enough maps, and the maps produced were never quite as good as they needed to be. The possession of a good map was never an assurance of success nor was the lack of a good map a foregone conclusion of failure. There were very good generals (such as Robert E. Lee), who seemed to pay little heed to maps, and there were quite terrible generals (such as John Pope), who valued maps highly and studied them expertly.

George B. McClellan probably had as great an influence on the conduct of the Union's Civil War as anybody except Abraham Lincoln and Ulysses S. Grant. His petulant, self-serving persona cast a pall over the North's principal army (the Army of the Potomac) throughout the four years of conflict. McClellan rose to prominence based on his performance as a commander in western Virginia. But his minor yet timely victories there were owed almost entirely to his subordinate, General William S. Rosecrans. By chance, Rosecrans had surveyed the area before the war and was therefore the only officer on either side who was intimately familiar with the area of operations. This minor coincidence was the basis of McClellan's amazing, fateful prominence.

There were occasions when a small mapping error had major consequences. Three quarters of Stonewall Jackson's 1862 Valley campaign was over when his troops began to recede up the valley. The Union tried to catch him in a pincer movement, but General James Shields made a little noted yet cataclysmic mistake. As he set off from Front Royal to intercept Jackson, Shields inadvertently moved out on the Winchester Road. But it was the Strasburg Road that would have got him to the Valley Turnpike ahead of Jackson's vulnerable, retreating army. Shields eventually realized his error, but in the resulting confusion and delay, Jackson managed to squeak between both of his pursuers, Shields to the east and General John C. Frémont to the west.

Robert G. Tanner, eminent historian of the 1862 Valley campaign, indicated that it was unlikely that Shield's error was caused by poor maps; more likely it was caused by Shields having no maps at all. Had Shields caught Jackson on the Valley Turnpike, it is almost certain that the Valley campaign would now be viewed as the inevitable failure of a flawed eccentric rather than one of the great campaigns of military history.

Knowledge is power, as they say, and maps provided the kind of knowledge that enabled a general to bring power to bear where it was needed, when it was needed, so that it could be applied with maximum efficiency as it was needed.

War has an implacable, economical logic all its own. Lincoln certainly recognized it. On the morning of April 7, 1865, the president was in the just-fallen Confederate capital, Richmond, Virginia, awaiting word from the erstwhile Confederate Legislature of Virginia. The legislature was trying to decide whether it should act to recall Virginia's soldiers from the Rebel armies. The president remarked that Union general Philip Sheridan "seemed to be getting Rebel soldiers out of the war faster than the Legislature could think."[1]

General William Tecumseh Sherman possessed the same gritty, practical sense. One soldier noted with admiration that Sherman wore a "dirty dickey with points wilted down, black old-fashioned stack brown field officer's coat with high collar and no shoulder stripes, muddy trousers and one spur."[2] The general wore only one spur because if the spurred side of his horse got moving, the other side was ineluctably in motion too.

It was said of Sherman that he loved the earth as a sailor loves the sea. He said of himself that he had a way "of going across the country direct to the object regardless of water, roads or paths."[3] He went "across the country" in much the same way he spurred his horse—in a manner so intensely practicable that he caught and held the world's startled attention. As a young officer in Georgia twenty years before the war, Sherman spent his leisure time studying the ground—exploring the creeks, the valleys, the hills—learning the lay of the land. For a soldier, this made perfect sense: the terrain was his shop, his office, his farm; it was where he worked, and in a lot of cases it was what he loved.

The military maps of the Civil War itemized in very down-to-earth detail the American landscape over which the armies of the Blue and Gray marched and fought for four long years. The tens of thousands of soldiers, North and South, who marched millions upon millions of collective miles over strange and unfamiliar ground, saw far more of their country than they would ever otherwise have done. Military maps recorded the entire journey—from Fort Sumter to Bennitt's Farm—in pencil, ink, and watercolor. The maps leave behind a precise and intimate portrait of a nation as it was at perhaps the most significant time of its history. In some places, the Civil War maps remained the best available well into the twentieth century.

The soldiers slept on the pine-needle floors of the Carolina forests; they bathed in Virginia's Rappahannock River. They climbed into the clouds up Tennessee's Lookout Mountain; they gorged on branchfuls of Pennsylvania cherries. These very experiences had the ironic effect of accomplishing what the previous fourscore years of shared government had failed to achieve: an awakened—a realized—sense of a common country and a genuine unbreakable Union.

Notes

1. The Necessity of Military Maps

1. Cecil D. Eby, Jr., *"Porte Crayon": The Life of David Hunter Strother* (Chapel Hill: University of North Carolina Press, 1960), 35.

2. [David Hunter Strother,] "Personal Recollections of the War by a Virginian," *Harper's New Monthly Magazine* (July 1866): 154. These recollections appeared serially in the magazine from June 1866 to April 1868.

3. Thomas J. C. Williams, *A History of Washington County, Maryland* (1906; reprint, Baltimore: Regional Publishing, 1968), 359–60.

4. Glenn Tucker, *Lee and Longstreet at Gettysburg* (1968; reprint, Dayton, Ohio: Morningside, 1982), 244.

5. John G. Moore, "Mobility and Strategy in the Civil War," *Military Analysis of the Civil War* (Millwood, N.Y.: KTO Press, 1977), 106–14.

6. John W. Schildt, *Roads to Gettysburg* (Parsons, W.Va.: McClain Printing, 1978), 422.

7. Ibid., 67.

8. Ibid., 71.

9. Edward Porter Alexander, *Fighting for the Confederacy: The Personal Recollections of General Edward Porter Alexander,* ed. Gary W. Gallagher (Chapel Hill and London: University of North Carolina Press, 1989), 273.

10. John Townsend Trowbridge, *The South: A Tour of Its Battlefields and Ruined Cities* (1866; reprint, New York: Arno Press, 1969), 476.

11. John J. Hennessy, *Return to Bull Run: The Campaign and Battle of Second Manassas* (New York: Simon and Schuster, 1993), 42.

12. G. F. R. Henderson, *Stonewall Jackson and the American Civil War* (1898; reprint, New York: DaCapo, 1988), 360.

13. Moore, "Mobility and Strategy," 107.

14. Kenneth P. Williams, *Lincoln Finds a General* (New York: Macmillan, 1949–59), 3:321.

15. Alpheus S. Williams, *From the Cannon's Mouth: The Civil War Letters of General Alpheus S. Williams,* ed. Milo M. Quaife (1959; reprint, Lincoln and London: University of Nebraska Press, 1995), 240.

16. George Ward Nichols, *The Story of the Great March* (1865; reprint, Williamstown, Mass.: Corner House, 1972), 388.

17. R. Pearsall Smith, "Communication Regarding the Published County Maps of the United States," *Proceedings of the American Philosophical Society* 9 (March 1864): 350–52.

18. Jubal A. Early, *Narrative of the War Between the States* (1912; reprint, New York: DaCapo, 1991), 264.

19. United States Coast and Geodetic Survey, comp., *Military and Naval Service of the United States Coast Survey, 1861–1865* (Washington, D.C.: U.S. Government Printing Office, 1916), 68.

2. Familiar Territory

1. W. Bart Greenwood, comp., *The American Revolution, 1775–1783: An Atlas of Eighteenth Century Maps and Charts* (Washington, D.C.: Naval History Division, 1972), 4.

2. James I. Robertson, Jr., *Stonewall Jackson: The Man, the Soldier, the Legend* (New York: Macmillan, 1997), 416.

3. Ibid., 392.

4. David A. Lilley, "Mapping in North America, 1775 to 1865,

Emphasizing Union Military Topography in the Civil War" (master's thesis, George Mason University, Fairfax, Va., 1982), 61.

5. Richard Taylor, *Destruction and Reconstruction: Personal Experiences of the Late War* (1879; reprint, New York, London, and Toronto: Longmans, Green, 1955), 98.

6. United States War Department, *Atlas to Accompany the Official Records of the Union and Confederate Armies* (1891–95; reprint, New York: Fairfax, 1983), plate 19, no. 1, and plate 20, no. 1.

7. Taylor, *Destruction and Reconstruction,* 98.

8. David Hunter Strother, *A Virginia Yankee in the Civil War: The Diaries of David Hunter Strother,* ed. Cecil D. Eby, Jr. (Chapel Hill: University of North Carolina Press, 1961), 72–73.

9. W. C. King and W. P. Derby, comps., *Camp-Fire Sketches and Battle-Field Echoes of 61–5* (Springfield, Mass.: King, Richardson, 1888), 362.

10. [Strother,] "Personal Recollections," September 1866, 412.

3. The Topographical Engineer

1. W. W. Blackford, *War Years with Jeb Stuart* (1945; reprint, Baton Rouge: Louisiana State University Press, 1993), 63.

2. Paul Fatout, "Ambrose Bierce, Civil War Topographer," *American Literature* 26, no. 3 (November 1954): 391.

3. J. R. Perkins, *Trails, Rails, and War: The Life of General G. M. Dodge* (New York: Arno Press, 1981), 93.

4. Gilbert Thompson, *The Engineer Battalion in the Civil War* (Washington Barracks, D.C.: Press of the Engineer School, 1910), 1:5.

5. Strother, *A Virginia Yankee,* 205, 211.

6. James L. Nichols, *Confederate Engineers* (Tuscaloosa: Confederate Publishing, 1957), 81.

7. Robertson, *Stonewall Jackson,* 381.

8. T. M. R. Talcott, "General Lee's Strategy at the Battle of Chancellorsville," *Southern Historical Society Papers* 34 (January–December 1906): 6.

9. Jacob Hoke, *The Great Invasion of 1863, or General Lee in Pennsylvania* (Dayton, Ohio: W. J. Shuey, 1887), 481.

10. Horace Porter, *Campaigning with Grant* (1897; reprint, New York: Blue and Gray Press, 1984), 65.

11. Oliver Otis Howard, *Autobiography of Oliver Otis Howard* (New York: Baker and Taylor, 1907), 2:113.

12. Harry W. Pfanz, *Gettysburg: The Second Day* (Chapel Hill and London: University of North Carolina Press, 1987), 206.

13. Mark M. Boatner III, *The Civil War Dictionary* (New York: David McKay, 1959), 917.

14. William J. Miller, *Mapping for Stonewall* (Washington, D.C.: Elliot and Clark, 1993), 33–37.

15. United States Coast and Geodetic Survey, *Military and Naval Service,* 16.

16. Ibid., 61.

17. Jed Hotchkiss, *Make Me a Map of the Valley: The Civil War Journal of Stonewall Jackson's Topographer,* ed. Archie P. McDonald (Dallas: Southern Methodist University Press, 1973), 10.

18. Ibid., 10.

19. Ibid., 264.

20. David L. Ladd and Audrey J. Ladd, eds., *The Bachelder Papers: Gettysburg in Their Own Words* (Dayton, Ohio: Morningside, 1994–95), 1:458.

4. Making the Maps

1. Adrian G. Trass, *From the Golden Gate to Mexico City: The U.S. Army Topographical Engineers in the Mexican War, 1846–1848* (Washington, D.C.: Center of Military History, 1993), 222.

2. Hartman Bache, *Report of the Secretary of War,* 37th Cong., 2d sess., Sen. Exec. Doc. 1 (Washington, D.C.: U.S. Government Printing Office, 1861), 2:129.

3. Albert H. Campbell, "Lost War Maps," *Century* 35 (January 1888): 479.

4. James L. Nichols, "Confederate Map Supply," *Military Engineer* 46, no. 309 (January–February 1954): 28.

5. Peter W. Roper, *Jedediah Hotchkiss: Rebel Mapmaker and Virginia Businessman* (Shippensburg, Pa.: White Mane Publishing, 1992), 193.

6. [Strother,] "Personal Recollections," October 1866, 556.

7. Wendell Upchurch, interviews by author, telephone, August 7 and 18, 1995.

8. Wilbur F. Foster, "Battle Field Maps in Georgia," *Confederate Veteran* 20 (1912): 369.

9. [Strother,] "Personal Recollections," September 1866, 411.

10. Hotchkiss, *Make Me a Map,* 22.

11. Captain William Willoughby Verner, *Rapid Field Sketching and Reconnaissance* (London: W. H. Allen, 1889), 12, 14.

12. Philip L. Shiman, undated, untitled paper in the possession of the author.

13. J. E. Taylor, *With Sheridan up the Shenandoah Valley in 1864: Leaves from a Special Artist's Sketch Book and Diary* (Dayton, Ohio: Morningside, 1989), 439.

14. Ibid., 429–39.

15. [Strother,] "Personal Recollections," July 1866, 157; September 1866, 415; January 1867, 187; and March 1867, 427; Richard W. Stephenson, comp., *Civil War Maps: An Annotated List of Maps and Atlases in the Library of Congress* (Washington, D.C.: Library of Congress, 1989): 302.

16. Pfanz, *Gettysburg,* 487.

17. [Strother,] "Personal Recollections," September 1866, 413.

18. W. W. Blackford, *War Years with Jeb Stuart,* 85.

19. Ibid., 83.

20. Ambrose Bierce, *Ambrose Bierce's Civil War,* ed. William McCann (1956; reprint, Washington, D.C.: Regnery Publishing, 1996), 214.

21. Hazard Stevens, *The Life of Isaac Ingalls Stevens* (Boston and New York: Houghton Mifflin, 1900): 2:329.

22. David C. Edmonds, *The Guns of Port Hudson* (Lafayette, La.: Acadiana Press, 1983), 1:88, 91; Strother, *A Virginia Yankee,* 156.

5. Roads

1. Robertson, *Stonewall Jackson,* 713.

2. United States War Department, *The War of the Rebellion: A Compilation of the Official Records of the Union and Confederate Armies* (Washington, D.C.: U.S. Government Printing Office, 1880–1901), 32 (part 3): 163–64.

3. Howard, *Autobiography,* 2:206.

4. John S. Clark Papers, Cayuga Museum, Auburn, N.Y.

5. Susan Leigh Blackford, comp., *Letters from Lee's Army* (New York: Charles Scribner's Sons, 1947), 194.

6. George Gordon Meade, *With Meade at Gettysburg* (Philadelphia: War Library and Museum of the Military Order of the Loyal Legion of the United States, 1930), 186.

7. Schildt, *Roads to Gettysburg,* 260.

8. A. S. Williams, *From the Cannon's Mouth,* 242.

9. W. A. Neal, comp., *Illustrated History of the Missouri Engineers and the Twenty-fifth Infantry Regiments* (Elkhart, Ind.: W. A. Neal, 1888), 35, 36, 42.

6. Military Fords

1. Robertson, *Stonewall Jackson,* 545.

2. Jac Weller, "The Logistics of Nathan Bedford Forrest," *Military Analysis of the Civil War,* editors of *Military Affairs* (Millwood, N.Y.: KTO Press, 1977), 177.

3. W. W. Blackford, *War Years with Jeb Stuart,* 223.

4. Stephen W. Sears, *Chancellorsville* (Boston: Houghton Mifflin, 1996), 166.

5. W. P. Conrad and Ted Alexander, *When War Passed This Way* (Greencastle, Pa.: Besore Memorial Library, 1982), 203.

6. United States War Department, *The War of the Rebellion,* vol. 19, part 1, 277.

7. Ibid., vol. 27, part 3, 272.

8. John H. Worsham, *One of Jackson's Foot Cavalry: His Experience and What He Saw during the War, 1861–1865* (New York: Neale Publishing, 1912), 237.

9. William Todd, *The Seventy-ninth Highlanders: New York Volunteers in the War of the Rebellion, 1861–1865* (Albany, N.Y.: Press of Brandow, Barton, 1886), 33.

10. Lloyd Lewis, *Sherman: Fighting Prophet* (New York: Harcourt, Brace, 1932), 379.

11. Sam R. Watkins, *"Co. Aytch": A Sideshow of the Big Show* (1882; reprint, New York: Collier Books, 1962), 113.

12. Alexander, *Fighting for the Confederacy,* 261.

13. Strother, *A Virginia Yankee,* 277.

14. Worsham, *One of Jackson's Foot Cavalry,* 192.

15. Robert K. Krick, *Stonewall Jackson at Cedar Mountain* (Chapel Hill and London: University of North Carolina Press, 1990), 46.

7. Woods and Forests

1. Gouverneur K. Warren, *General G. K. Warren's Report of the Operations in which He Took Part Connected with the Chancellorsville Campaign, May 12, 1863,* Joint Committee on the Conduct of the War, 38th Cong., 2d sess., Serial Set 5, 1212 (Washington, D.C.: U.S. Government Printing Office, 1863), 51–61.

2. Henderson, *Stonewall Jackson,* 696.

3. Lewis, *Sherman,* 336.

4. Ernest B. Furgurson, *Chancellorsville, 1863: The Souls of the Brave* (New York: Alfred A. Knopf, 1992), 188.

5. Warren, *General G. K. Warren's Report,* 54.

6. Robert Ross Smith, "Ox Hill: The Most Neglected Battle of the Civil War, September 1862," *Fairfax County and the War between the States—Ox Hill* (1961; reprint, Fairfax County, Va.: Office of Comprehensive Planning, 1987), 60, 62.

7. Ambrose Bierce, "A Little of Chickamauga," in *Ambrose Bierce's Civil War,* 32.

8. William M. Lamers, *The Edge of Glory: A Biography of William S. Rosecrans, U.S.A.* (New York: Harcourt, Brace and World, 1961), 328.

9. Charles A. Dana, *Recollections of the Civil War* (1898; reprint, Lincoln and London: University of Nebraska Press, 1996), 112.

10. Theodore F. Dwight, ed., "Case of Fitz-John Porter," in *The Virginia Campaign of 1862 under General Pope* (1886; reprint, Wilmington, N.C.: Broadfoot Publishing, 1989), 230, 231.

11. Robert Underwood Johnson and Clarence Clough Buel, eds., *Battles and Leaders of the Civil War* (1887–88; reprint, Secaucus, N.J.: Castle, n.d.), 2:696, 697.

12. James Harrison Wilson, *Under the Old Flag* (1912; reprint,

Westport, Conn.: Greenwood Press, 1971), 1:139, 140.

13. Porter, *Campaigning with Grant*, 66.

14. Douglas Southall Freeman, *R. E. Lee: A Biography* (New York and London: Charles Scribner's Sons, 1935), 3:348.

15. Ulysses S. Grant, *Memoirs and Selected Letters: Personal Memoirs of U. S. Grant/Selected Letters, 1839–1865* (1885; reprint, New York: Library of America, 1990), 342.

16. Ibid., 345.

17. Robert G. Tanner, *Stonewall in the Valley* (Mechanicsburg, Pa.: Stackpole Books, 1996), 202.

18. M. Lefebvre, *Military Landscape Sketching*, trans. Captain W. V. Judson (Washington Barracks, D.C.: Press of the Engineer School, 1905), 10.

8. Local Knowledge

1. Thomas B. Van Horne, *History of the Army of the Cumberland* (1875; reprint, Wilmington, N.C.: Broadfoot Publishing, 1988), 2:457.

2. Ibid., 457.

3. Horace Cecil Fisher, *A Staff Officer's Story: The Personal Experiences of Colonel Horace Newton Fisher in the Civil War* (Boston: 1960), 51.

4. [Strother,] "Personal Recollections," August 1867, 274.

5. Oliver T. Reilly, *The Battlefield of Antietam* (Hagerstown, Md.: Oliver T. Reilly, 1906), 26.

6. S. L. Blackford, *Letters from Lee's Army*, 195.

7. [Strother,] "Personal Recollections," January 1867, 176.

8. [Strother,] "Personal Recollections," August 1867, 277.

9. Strother, *A Virginia Yankee*, 254.

10. Admiral David Dixon Porter, *Incidents and Anecdotes of the Civil War* (New York: Appleton, 1885), 155.

11. Frederick A. Mitchel, *Ormsby Macknight Mitchel: Astronomer and General* (Boston and New York: Houghton Mifflin, 1887), 317.

12. Grant, *Memoirs and Selected Letters*, 318.

13. Ibid., 321.

14. Todd, *The Seventy-ninth Highlanders*, 459.

15. Ibid.

16. Robertson, *Stonewall Jackson*, 477.

17. Trowbridge, *The South*, 41.

18. J. Willard Brown, *The Signal Corps, U.S.A. in the War of the Rebellion* (1896; reprint, New York: Arno Press, 1974), 641.

19. Early, *Narrative of the War*, 377.

20. A. S. Williams, *From the Cannon's Mouth*, 224.

21. John O. Casler, *Four Years in the Stonewall Brigade* (Dayton: Morningside, 1971), 70.

22. Bell Irvin Wiley, *The Life of Billy Yank: The Common Soldier of the Union* (Baton Rouge and London: Louisiana State University Press, 1978), 98.

23. Alan T. Nolan, *The Iron Brigade* (Berrien Springs, Mich.: Hardscrabble Books, 1983), 47.

24. Walter H. Taylor, *General Lee: His Campaigns in Virginia, 1861–1865* (1906; reprint, Lincoln and London: University of Nebraska Press, 1994), 66.

25. G. K. Warren Papers, New York State Library, Albany, N.Y.

26. Freeman, *R. E. Lee*, 4:74.

27. Alexander, *Fighting for the Confederacy*, 525.

28. Henry H. Humphreys, *Andrew Atkinson Humphreys: A Biography* (Philadelphia: John C. Winston, 1924), 284.

29. Shiman, undated, untitled paper, no pagination.

30. Schildt, *Roads to Gettysburg*, 171.

31. Randolph Abbott Shotwell, *The Papers of Randolph Abbott Shotwell*, ed. J. G. DeRoulhac Hamilton (Raleigh, N.C.: North Carolina History Commission, 1929), 1:498.

32. Hanover Chamber of Commerce, *Encounter at Hanover* (Hanover, Pa.: Hanover Chamber of Commerce, 1963), 35.

9. Hills, Mountains, Rolling Terrain, Gaps . . .

1. W. W. Blackford, *War Years*, 44.

2. J. F. J. Caldwell, *The History of a Brigade of South Carolinians* (1866; reprint, Dayton, Ohio: Morningside, 1984), 77, 115.

3. Peter C. Hains, "The First Gun at Bull Run," *Cosmopolitan* 51 (1911): 389.

4. Krick, *Stonewall Jackson at Cedar Mountain*, 84.

5. Early, *Narrative of the War*, 97.

6. Lamers, *The Edge of Glory*, 34.

7. Philip Henry Sheridan, *Personal Memoirs of P. H. Sheridan* (1888; reprint, Wilmington, N.C.: Broadfoot Publishing, 1992), 1:469.

8. Jacob D. Cox, *Atlanta* (New York: Charles Scribner's Sons, 1882), 35.

9. Early, *Narrative of the War*, 378.

10. Matthew Forney Steele, *American Campaigns* (Washington, D.C.: Combat Forces Press, 1951), 224.

11. King and Derby, *Camp-Fire Sketches*, 270.

12. Bruce Catton, *Never Call Retreat* (New York: Pocket Books, 1967), 252.

13. Robertson, *Stonewall Jackson*, 652.

14. Sears, *Chancellorsville*, 237.

15. Ibid., 237.

16. George Ward Nichols, *The Story of the Great March* (1865; reprint, Williamstown, Mass.: Corner House, 1972), 126.

17. John G. Barrett, *Sherman's March through the Carolinas* (Chapel Hill: University of North Carolina Press, 1956), 47.

18. Catton, *Never Call Retreat*, 399.

19. Lewis, *Sherman*, 490.

20. Ibid., 484.

21. Ibid., 490.

10. Procuring the Maps

1. Barrett, *Sherman's March through the Carolinas*, 38.

2. Arthur Hecht, "Route Maps of the U.S. Postal Service of the Eighteenth and Nineteenth Centuries," *American Philatelist*, November 1979: 983.

3. Miller, *Mapping for Stonewall*, 37.

4. Hecht, "Route Maps," 983.

5. Ambrose Bierce Papers, Elkhart County Historical Museum, Bristol, Ind.

6. Shiman, undated, untitled paper, no pagination.

11. Preparing the Map and the Map Memoir

1. Ambrose Bierce Papers, Elkhart County Historical Museum, Bristol, Ind.

12. Map Reproduction

1. "Arms and the Map: Military Mapping by the Army Map Service," *Print Magazine* 4, no. 2 (1946): 9.

Epilogue

1. Dana, *Recollections of the Civil War*, 267.

2. Barrett, *Sherman's March through the Carolinas*, 33.

3. Joseph H. Ewing, *Sherman at War* (Dayton, Ohio: Morningside, 1992), 122.

Bibliography

Alexander, Edward Porter. *Fighting for the Confederacy: The Personal Recollections of General Edward Porter Alexander.* Edited by Gary W. Gallagher. Chapel Hill and London: University of North Carolina Press, 1989.

Alexander, Ted. "Gettysburg Cavalry Operations, June 27–July 3, 1863." *Blue and Gray Magazine* 6 (October 1988): 8.

———. "Twenty-five Miles behind Confederate Lines: Ulric Dahlgren's Greencastle Forays." *Blue and Gray Magazine* 6 (October 1988): 23.

Allan, William. *History of the Campaign of Gen. T. J. (Stonewall) Jackson in the Shenandoah Valley of Virginia from November 4, 1861, to June 17, 1862.* 1880. Reprint, Dayton, Ohio: Morningside, 1974.

Ambrose, Stephen E. *Duty, Honor, Country: A History of West Point.* Baltimore: Johns Hopkins Press, 1966.

Anderson, James S. "The March of the Sixth Corps to Gettysburg." *War Papers Being Read before the Commandery of the State of Wisconsin, Military Order of the Loyal Legion of the United States* 4 (1914): 77–84.

"Arms and the Map: Military Mapping by the Army Map Service." *Print Magazine* 4, no. 2 (1946): 3–16.

"The Atlanta Campaign: Atlanta Is Ours and Fairly Won." *Atlanta Historical Journal* 28, no. 3 (Fall 1984): special issue.

Bache, Hartman. *Report of the Secretary of War,* 37th Cong., 2d sess., Sen. Exec. Doc. 1. Washington, D.C.: U.S. Government Printing Office, 1861.

Barrett, John G. *Sherman's March through the Carolinas.* Chapel Hill: University of North Carolina Press, 1956.

Bates, David Homer. *Lincoln in the Telegraph Office: Recollections of the United States Military Telegraph Corps during the Civil War.* 1907. Reprint, Lincoln and London: University of Nebraska Press, 1995.

Bierce, Ambrose. *Ambrose Bierce's Civil War.* Edited by William McCann. 1956. Reprint, Washington, D.C.: Regnery Publishing, 1996.

Billings, John D. *Hard Tack and Coffee: Or the Unwritten Story of Army Life.* Boston: George M. Smith, 1887.

Blackford, Susan Leigh, comp. *Letters from Lee's Army.* New York: Charles Scribner's Sons, 1947.

Blackford, W. W. *War Years with Jeb Stuart.* 1945. Reprint, Baton Rouge: Louisiana State University Press, 1993.

Boatner, Mark M., III. *The Civil War Dictionary.* New York: David McKay, 1959.

Britton, Wiley. *Memoirs of the Rebellion on the Border.* 1863. Reprint, Lincoln and London: University of Nebraska Press, 1993; Chicago: Cushing, Thomas, 1882.

Brown, Andrew. *Geology and the Gettysburg Campaign.* Harrisburg: Pennsylvania Department of Internal Affairs, n.d.

Brown, J. Willard. *The Signal Corps, U.S.A. in the War of the Rebellion.* 1896. Reprint, New York: Arno Press, 1974.

Buell, Thomas B. *The Warrior Generals: Combat Leadership in the Civil War.* New York: Crown Books, 1997.

Busseret, David, ed. *From Sea Charts to Satellite Images.* Chicago and London: University of Chicago Press, 1990.

Cadle, Farris W. *Georgia Land Surveying History and Law.* Athens, Ga.: University of Georgia Press, 1991.

Caldwell, J. F. J. *The History of a Brigade of South Carolinians.* 1866. Reprint, Dayton, Ohio: Morningside, 1984.

Calkins, Christopher M. *Thirty-six Hours before Appomattox: April 6 and 7, 1865.* 1980.

Canan, H. V. "Maps for the Civil War." *Armor* 55, no. 5 (September–October 1956): 34–42.

Casler, John O. *Four Years in the Stonewall Brigade.* Dayton, Ohio: Morningside, 1971.

Castel, Albert. *Decision in the West: The Atlanta Campaign of 1864.* Lawrence, Kans.: University Press of Kansas, 1992.

Catton, Bruce. *Never Call Retreat.* New York: Pocket Books, 1967.

Church, Frank L. *Civil War Marine: A Diary of the Red River Expedition, 1864.* Edited by James P. Jones and Edward F. Keuchel. Washington, D.C.: U.S. Marine Corps, 1975.

Coddington, Edwin B. *The Gettysburg Campaign: A Study in Command.* New York: Charles Scribner's Sons, 1968.

Conrad, W. P., and Ted Alexander. *When War Passed This Way.* Greencastle, Pa.: Besore Memorial Library, 1982.

Cox, Jacob D. *Atlanta.* New York: Charles Scribner's Sons, 1882.

Cozzens, Peter. *No Better Place to Die: The Battle of Stones River.* Urbana and Chicago: University of Illinois Press, 1990.

Crist, Robert Grant. "Highwater 1863: The Confederate Approach to Harrisburg." *Pennsylvania History* 30 (April 1963): 158–83.

Crozier, Emmet. *Yankee Reporters, 1861–65.* New York: Oxford University Press, 1956.

Cullum, George W. *Biographical Register of the Officers and Graduates of the U.S. Military Academy.* Vol. 2. Boston and New York: Houghton Mifflin; Cambridge: Riverside Press, 1891.

Daily Argus (Mount Vernon, N.Y.), November 20, 1908.

Dana, Charles A. *Recollections of the Civil War.* 1898. Reprint, Lincoln and London: University of Nebraska Press, 1996.

Dawson, Francis W. *Reminiscences of Confederate Service, 1861–1865.* Baton Rouge and London: Louisiana State University Press, 1980.

Douglas, Henry Kyd. *I Rode with Stonewall.* Chapel Hill: University of North Carolina Press, 1940.

Downey, Fairfax. *Clash of Cavalry.* New York: David McKay, 1959.

Dustin, Fred. *The Custer Tragedy: Events Leading up to and Following the Little Big Horn Campaign of 1876.* El Segundo, Calif.: Upton and Sons, 1987.

Dwight, Theodore F., ed. *The Virginia Campaign of 1862 under General Pope.* 1886. Reprint, Wilmington, N.C.: Broadfoot Publishing, 1989.

Early, Jubal A. *Narrative of the War between the States.* 1912. Reprint, New York: DaCapo, 1991.

Eby, Cecil D., Jr. *"Porte Crayon": The Life of David Hunter Strother.* Chapel Hill: University of North Carolina Press, 1960.

Edmonds, David C. *The Guns of Port Hudson.* 2 vols. Lafayette, La.: Acadiana Press, 1983.

English, John Alan. "Confederate Field Communications." Master's thesis, Duke University, Durham, N.C., 1964.

Ewing, Joseph H. *Sherman at War.* Dayton: Morningside, 1992.

Fatout, Paul. "Ambrose Bierce, Civil War Topographer." *American Literature* 26, no. 3 (November 1954): 391–400.

———. *Ambrose Bierce: The Devil's Lexicographer.* Norman, Okla.: University of Oklahoma Press, 1951.

Fellman, Michael. *Citizen Sherman: A Life of William Tecumseh Sherman.* New York and Toronto: Random House, 1995.

Fishel, Edwin C. *The Secret War for the Union.* Boston and New York: Houghton Mifflin, 1996.

Fisher, Horace Cecil. *A Staff Officer's Story: The Personal Experiences of Colonel Horace Newton Fisher in the Civil War.* Boston: 1960.

Foote, Shelby. *The Civil War, a Narrative: Red River to Appomattox.* New York: Random House, 1974.

Foster, Wilbur F. "Battle Field Maps in Georgia." *Confederate Veteran* 20 (1912): 369–70.

Frassanito, William A. *Early Photography at Gettysburg.* Gettysburg, Pa.: Thomas Publications, 1995.

———. *Grant and Lee: The Virginia Campaigns, 1864–1865.* New York: Charles Scribner's Sons, 1983.

Freeman, Douglas Southall. *R. E. Lee: A Biography.* 4 vols. New York and London: Charles Scribner's Sons, 1935.

French School of War. *The Military Map.* London: Macmillan, 1916.

Frost, Dr. Lawrence A. "Balloons over the Peninsula: Fitz-John Porter and George Custer Become Reluctant Aeronauts." *Blue and Gray Magazine* 2, no. 3 (December–January 1984–85): 6–12.

Fuller, J. F. C. *The Generalship of Ulysses S. Grant.* 1929. Reprint, New York: DaCapo Press, 1991.

Furgurson, Ernest B. *Chancellorsville, 1863: The Souls of the Brave.* New York: Alfred A. Knopf, 1992.

Gallagher, Gary W., ed. *Lee: The Soldier.* Lincoln and London: University of Nebraska Press, 1996.

Glatthaar, Joseph T. *The March to the Sea and Beyond: Sherman's Troops in the Savannah and Carolinas Campaign.* New York and London: New York University Press, 1986.

Goetzmann, William H. *Army Exploration in the American West, 1803–1863.* New Haven and London: Yale University Press, 1959.

Grant, Ulysses S. *Memoirs and Selected Letters: Personal Memoirs of U. S. Grant/Selected Letters, 1839–1865.* 1885. Reprint, New York: Library of America, 1990.

Greenwood, W. Bart, comp. *The American Revolution, 1775–1783: An Atlas of Eighteenth Century Maps and Charts.* Washington, D.C.: Naval History Division, 1972.

Gregory, Herbert E., ed. *Military Geology and Topography.* New Haven: Yale University Press, 1918.

Griffith, Paddy. *Battle Tactics of the Civil War.* New Haven and London: Yale University Press, 1989.

Grimsley, Daniel A. *Battles in Culpeper County, Virginia, 1861–1865.* Culpeper, Va.: Raleigh Travers Green, 1900.

Guthorn, Peter J. *British Maps of the American Revolution.* Monmouth Beach, N.J.: Philip Freneau Press, 1972.

Hagerman, Edward. *The American Civil War and the Origins of Modern Warfare.* Bloomington and Indianapolis: Indiana University Press, 1988.

Halpine, Charles Graham [Private Miles O'Reilly]. *Baked Meats of the Funeral.* New York: Carleton, 1866.

Harding, Walter. *The Days of Henry Thoreau: A Biography.* 1962. Reprint, New York: Dover Publications, 1982.

Harley, J. B., Barbara Bartz Petchenik, and Laurence W. Towner. *Mapping the American Revolutionary War.* Chicago and London: University of Chicago Press, 1978.

Harrington, Fred Harvey. *Fighting Politician: Major General N. P. Banks.* Westport, Conn.: Greenwood Press, 1948.

Hart, B. H. Liddell. *Sherman: Soldier, Realist, American.* 1929. Reprint, New York: Frederick A. Praeger, 1958.

Hayden, Frederick Stansbury. *Aeronautics in the Union and Confederate Armies.* 2 vols. 1941. Reprint, New York: Arno Press, 1980.

Hecht, Arthur. "Route Maps of the U.S. Postal Service of the Eighteenth and Nineteenth Centuries." *American Philatelist,* November 1979.

Henderson, G. F. R. *Stonewall Jackson and the American Civil War.* 2 vols. 1898. Reprint (2 vols. in 1), New York: DaCapo, 1988.

Hennessy, John J. *Return to Bull Run: The Campaign and Battle of Second Manassas.* New York: Simon and Schuster, 1993.

Hewitt, Lawrence Lee. *Port Hudson: Confederate Bastion on the Mississippi.* Baton Rouge and London: Louisiana State University Press, 1987.

Hirshson, Stanley P. *Grenville M. Dodge: Soldier, Politician, Railroad Pioneer.* Bloomington and London: Indiana University Press, 1967.

Historical Publication Committee. *Encounter at Hanover: Prelude to Gettysburg.* Hanover, Pa.: Hanover Chamber of Commerce, 1963.

History of Franklin County, Pennsylvania. Chicago: Warner, Beers, 1887.

Hoke, Jacob. *The Great Invasion of 1863, or General Lee in Pennsylvania.* Dayton, Ohio: W. J. Shuey, 1887.

Hoogenboom, Ari. "Spy and Topog Duty Has Been . . . Neglected." *Civil War History* (1964): 368–70.

Hotchkiss, Jed. *Make Me a Map of the Valley: The Civil War Journal of Stonewall Jackson's Topographer.* Edited by Archie P. McDonald. Dallas: Southern Methodist University Press, 1973.

Howard, Oliver Otis. *Autobiography of Oliver Otis Howard.* 2 vols. New York: Baker and Taylor, 1907.

Humphreys, Henry H. *Andrew Atkinson Humphreys: A Biography.* Philadelphia: John C. Winston, 1924.

Huston, James A. *The Sinews of War: Army Logistics, 1775–1953.* Washington, D.C.: Office of the Chief of Military History, 1966.

Hyde, Thomas W. *Following the Greek Cross or, Memories of the Sixth Army Corps.* Boston and New York: Houghton Mifflin; Cambridge: Riverside Press, 1894.

Jacobs, M. *Notes of the Rebel Invasion of Maryland and Pennsylvania and the Battle of Gettysburg July 1st, 2d, and 3d, 1863.* Philadelphia: J. B. Lippincott, 1864.

Jaynes, Gregory. *The Killing Ground: Wilderness to Cold Harbor.* Editors of Time-Life Books. Alexandria, Va.: Time-Life Books, 1986.

Johnson, Robert Underwood, and Clarence Clough Buel, eds. *Battles and Leaders of the Civil War.* 4 vols. 1887–88. Reprint, Secaucus, N.J.: Castle, n.d.

Kennedy, Frances H., ed. *The Civil War Battlefield Guide.* Boston: Houghton Mifflin, 1990.

King, W. C., and W. P. Derby, comps. *Camp-Fire Sketches and Battle-Field Echoes of 61–5.* Springfield, Mass.: King, Richardson, 1888.

Klein, Frederic Shriver. *Just South of Gettysburg: Carroll County, Maryland in the Civil War.* Westminster, Md.: Historical Society of Carroll County, Maryland, 1963.

Krick, Robert K. *Conquering the Valley: Stonewall Jackson at Port Republic.* New York: William Morrow, 1996.

———. *Stonewall Jackson at Cedar Mountain.* Chapel Hill and London: University of North Carolina Press, 1990.

Ladd, David L., and Audrey J. Ladd, eds. *The Bachelder Papers: Gettysburg in Their Own Words.* 3 vols. Dayton, Ohio: Morningside, 1994–95.

Lamers, William M. *The Edge of Glory: A Biography of William S. Rosecrans, U.S.A.* New York: Harcourt, Brace and World, 1961.

Leasure, Daniel. "Personal Observations and Experiences in the Pope Campaign in Virginia." *Glimpses of the Nation's Struggle: Papers Read before the Minnesota Commandery of the Military Order of the Loyal Legion of the United States.* 1887. Reprint, Wilmington, N.C.: Broadfoot Publishing, 1992.

Lefebvre, M. *Military Landscape Sketching.* Translated by Captain W. V. Judson. Washington Barracks, D.C.: Press of the Engineer School, 1905.

Lewis, Lloyd. *Sherman: Fighting Prophet.* New York: Harcourt, Brace, 1932.

Lilley, David A. "Anticipating the Atlas to Accompany the Official Records: Post-War Mapping of Civil War Battlefields." *Lincoln Herald* 84 (Spring 1982): 37–42.

———. "The Antietam Battlefield Board and Its Atlas: Or the Genesis of the Carman-Cope Maps." *Lincoln Herald* 82 (Summer 1980): 380–87.

———. "Mapping in North America, 1775 to 1865, Emphasizing Union Military Topography in the Civil War." Master's thesis, George Mason University, Fairfax, Va., 1982.

Long, Captain James T. *Gettysburg: How the Battle Was Fought.* Harrisburg: E. K. Meyers, 1891.

Lowe, David W. "From the Rapid Ann to Coal Harbor: Post-War Topographical Survey of Civil War Battlefields." Mimeographed. 1996.

Lyman, Theodore. "On the Uselessness of the Maps Furnished to the Staff of the Army of the Potomac Previous to the Campaign of May, 1864." *Papers of the Military Historical Society of Massachusetts.* Vol. 4, *The Wilderness Campaign May–June 1864.* 1905. Reprint, Wilmington, N.C.: Broadfoot Publishing, 1989.

McCarthy, Carlton. *Detailed Minutiae of Soldier Life in the Army of Northern Virginia, 1861–1865.* 1882. Reprint, Lincoln and London: University of Nebraska Press, 1993.

M'Cauley, I. H. *Historical Sketch of Franklin County, Pennsylvania.* Chambersburg, Pa.: John M. Pomeroy, 1878.

McClellan, H. B. *The Life and Campaigns of Major-General J. E. B. Stuart, Commander of the Cavalry of the Army of Northern Virginia.* Boston and New York: Houghton Mifflin, 1885.

McClure, A. K. *Lincoln and Men of War Times.* Philadelphia: Rolley and Reynolds, 1962.

McIntosh, David Gregg. "Review of the Gettysburg Campaign." *Southern Historical Society Papers* 37, 74ff. (no place, no date).

McKinney, Francis F. *Education in Violence: The Life of George H. Thomas and the History of the Army of the Cumberland.* Detroit: Wayne State University Press, 1961.

McWilliams, Carey. *Ambrose Bierce: A Biography.* 1929. Reprint, Archon Books, 1967.

Major, Duncan K., and Roger S. Fitch. *Supply of Sherman's Army during the Atlanta Campaign.* Fort Leavenworth, Kans.: Army Service Schools Press, 1911.

Malone, Dumas. *Dictionary of American Biography.* 20 vols. New York: Charles Scribner's Sons, 1928.

Martin, David G. *Gettysburg, July 1.* Conshohocken, Pa.: Combined Books, 1995.

Marzio, Peter C. *The Democratic Art.* Fort Worth, Tex.: Amon Carter Museum of Western Art, 1979.

Matson, R. C. *Elements of Mapping.* Houghton, Mich.: Michigan College of Mining and Technology, 1940.

Meade, George Gordon. *The Life and Letters of George Gordon Meade.* 2 vols. New York: Charles Scribner's Sons, 1913.

———. *With Meade at Gettysburg.* Philadelphia: War Library and Museum of the Military Order of the Loyal Legion of the United States, 1930.

Mendell, G. H. *A Treatise on Military Surveying.* London: Trübner, 1864; New York: Van Nostrand, 1864.

Miller, William J. *Mapping for Stonewall.* Washington, D.C.: Elliot and Clark, 1993.

Mitchel, Frederick A. *Ormsby Macknight Mitchel: Astronomer and General.* Boston and New York: Houghton Mifflin, 1887.

Moore, Edward A. *The Story of a Cannoneer under Stonewall Jackson.* 1907. Reprint, Freeport, N.Y.: Books for Libraries Press, 1971.

Moore, Frank, ed. *The Rebellion Record: A Diary of American Events.* 11 vols. and supplement. New York: Putnam's, 1861–63; Van Nostrand, 1864–68.

Moore, John G. "Mobility and Strategy in the Civil War." *Military Analysis of the Civil War.* Millwood, N.Y.: KTO Press, 1977.

Morrison, James L., Jr. *The Best School in the World: West Point, the Pre–Civil War Years, 1833–1866.* Kent, Ohio: Kent State University Press, 1986.

Mosby, John S. *Stuart's Cavalry in the Gettysburg Campaign.* New York: Moffat, Yard, 1908.

Motts, Sarah E. *Personal Experiences of a House That Stood on the Road.* Carlisle, Pa.: Hamilton Library Association, 1941.

Muntz, A. Philip. "Union Mapping in the American Civil War." *Imago Mundi* 17 (1963): 90–94.

Murray, Louise Welles. "Historical Calendar of Collections of General John S. Clark." Auburn, N.Y.: Cayuga Museum, 1930.

Musham, Harry Albert. *The Techniques of the Terrain.* New York: Reinhold Publishing, 1944.

Myers, Frank M. *The Comanches: A History of White's Battalion, Virginia Cavalry.* 1871. Reprint, Marietta, Ga.: Continental Book, 1956.

National Archives and Records Administration. *A Guide to Civil War Maps in the National Archives.* Washington, D.C.: National Archives Trust Fund Board, 1986.

Neal, W. A., comp. *Illustrated History of the Missouri Engineers and the Twenty-fifth Infantry Regiments.* Elkhart, Ind.: W. A. Neal, 1888.

Neese, George M. *Three Years in the Confederate Horse Artillery.* 1911. Reprint, Dayton, Ohio: Morningside, 1983.

Nelson, Christopher. *Mapping the Civil War.* Washington, D.C.: Starwood Publishing, 1992.

Nettesheim, Daniel D. "Topographical Intelligence and the American Civil War." Master's thesis, United States Army Command and General Staff College, Fort Leavenworth, Kans., 1978.

Nevin, David. *Sherman's March: Atlanta to the Sea.* Editors of Time-Life Books. Alexandria, Va.: Time-Life Books, 1986.

Nichols, Edward J. *Toward Gettysburg: A Biography of General John F. Reynolds.* University Park: Pennsylvania State University Press, 1958.

Nichols, George Ward. *The Story of the Great March.* 1865. Reprint, Williamstown, Mass.: Corner House, 1972.

Nichols, James L. *Confederate Engineers.* Tuscaloosa: Confederate Publishing, 1957.

———. "Confederate Map Supply." *Military Engineer* 46, no. 309 (January–February 1954): 28–32.

Nolan, Alan T. *The Iron Brigade.* 1961. Reprint, Berrien Springs, Mich.: Hardscrabble Books, 1983.

Nye, Wilbur Sturtevant. *Here Come the Rebels!* Baton Rouge: Louisiana

State University Press, 1965.

Oates, Stephen B. *To Purge This Land with Blood: A Biography of John Brown.* New York: Harper and Row, 1970.

O'Conner, Richard. *Ambrose Bierce: A Biography.* Boston and Toronto: Little, Brown, 1967.

Olmsted, Frederick Law. "After Gettysburg." *Pennsylvania Magazine of History and Biography* 75 (October 1951): 436–46.

———. *The Cotton Kingdom: A Traveller's Observations on Cotton and Slavery in the American Slave States.* Edited by Arthur M. Schlesinger, Sr. New York: Modern Library, 1984.

Osborne, Charles C. *Jubal: The Life and Times of General Jubal A. Early, CSA, Defender of the Lost Cause.* Baton Rouge and London: Louisiana State University Press, 1992.

Owen, William Miller. *In Camp and Battle With the Washington Artillery of New Orleans.* Boston: Ticknor, 1885.

Pearson, Henry Greenleaf. *James S. Wadsworth of Geneseo.* New York: Charles Scribner's Sons, 1913.

Peltier, Louis C., and G. Etzel Pearcy. *Military Geography.* Princeton, N.J.: Van Nostrand, 1966.

Pennypacker, Isaac R. *General Meade.* New York: Appleton, 1901.

Perkins, J. R. *Trails, Rails, and War: The Life of General G. M. Dodge.* New York: Arno Press, 1981.

Pfanz, Harry W. *Gettysburg: The Second Day.* Chapel Hill and London: University of North Carolina Press, 1987.

Porter, Admiral David Dixon. *Incidents and Anecdotes of the Civil War.* New York: Appleton, 1885.

Porter, Horace. *Campaigning with Grant.* 1897. Reprint, New York: Blue and Gray Press, 1984.

Quint, Alonzo Hall. *The Potomac and the Rapidan: Army Notes, from the Failure at Winchester to the Reenforcement of Rosecrans, 1861–3.* Boston: Crosby and Nichols, 1864.

Raisz, Erwin. *Principles of Cartography.* New York, San Francisco, Toronto, and London: McGraw-Hill, 1962.

Rhoads, James Berton. "Civil War Maps and Mapping." *Military Engineer* 327 (January–February, 1957): 38–43.

Ristow, Walter W. *American Maps and Mapmakers.* Detroit: Wayne State University Press, 1985.

Robertson, James I., Jr. *General A. P. Hill: The Story of a Confederate Warrior.* New York: Random House, 1987.

———. *Stonewall Jackson: The Man, the Soldier, the Legend.* New York: Macmillan, 1997.

Roebling, Washington Augustus. *Wash Roebling's War.* Edited by Earl Schenck Miers. Neward, Del.: Curtis Paper, 1961.

Roper, Peter W. *Jedediah Hotchkiss: Rebel Mapmaker and Virginia Businessman.* Shippensburg, Pa.: White Mane Publishing, 1992.

Ross, Fitzgerald. *Cities and Camps of the Confederate States.* Urbana: University of Illinois Press, 1958.

Sanchez-Saavedra, E. M. *A Description of the Country: Virginia's Cartographers and Their Maps, 1607–1881.* Richmond: Virginia State Library, 1975.

Schaff, Philip. "The Gettysburg Week." *Scribner's Magazine* 16, no. 3 (July 1894): 21–30.

Schildt, John W. *Roads from Gettysburg.* Chewsville, Md.: Published by the author, 1979.

———. *Roads to Gettysburg.* Parsons, W.Va.: McClain Printing, 1978.

Schuyler, Hamilton. *The Roeblings.* Princeton, N.J.: Princeton University Press, 1931.

Scott, Colonel H. L. *Military Dictionary.* 1861. Reprint, New York: Greenwood Press, 1968.

Sears, Stephen W. *Chancellorsville.* Boston: Houghton Mifflin, 1996.

Sheridan, Philip Henry. *Personal Memoirs of P. H. Sheridan.* 2 vols.

1888. Reprint, Wilmington, N.C.: Broadfoot Publishing, 1992.

Sherman, William T. *Memoirs of General William T. Sherman.* 2 vols. 1875. Reprint (2 vols. in 1), New York: DaCapo, 1984.

Shotwell, Randolph Abbott. *The Papers of Randolph Abbot Shotwell.* 2 vols. Edited by J. G. DeRoulhac Hamilton. Raleigh, N.C.: North Carolina History Commission, 1929.

Skinker, Thomas Keith. *Samuel Skinker and His Descendants.* St. Louis: Published by the author, 1923.

Smith, Donald L. *The Twenty-fourth Michigan of the Iron Brigade.* Harrisburg: Stackpole, 1962.

Smith, R. Pearsall. "Communication Regarding the Published County Maps of the United States." *Proceedings of the American Philosophical Society* 9 (March 1864): 350–52.

Smith, Robert Ross. "Ox Hill: The Most Neglected Battle of the Civil War, September 1862." *Fairfax County and the War between the States—Ox Hill.* 1961. Reprint, Fairfax County, Va.: Office of Comprehensive Planning, 1987.

Steele, Matthew Forney. *American Campaigns.* 2 vols. Washington, D.C.: Combat Forces Press, 1951.

Steinman, D[avid] B[arnard]. *The Builders of the Bridge: The Story of John Roebling and His Son.* New York: Arno Press, 1972.

Stephenson, Richard W., comp. *Civil War Maps: An Annotated List of Maps and Atlases in the Library of Congress.* Washington, D.C.: Library of Congress, 1989.

Stevens, George T. *Three Years in the Sixth Corps.* Albany, N.Y.: S. R. Gray, 1866.

Stevens, Hazard. *The Life of Isaac Ingalls Stevens.* 2 vols. Boston and New York: Houghton Mifflin, 1900.

Strother, David Hunter. *A Virginia Yankee in the Civil War: The Diaries of David Hunter Strother.* Edited by Cecil D. Eby, Jr. Chapel Hill: University of North Carolina Press, 1961.

Sullivan, James W. *Boyhood Memories of the Civil War, 1861–65: Invasion of Carlisle.* Carlisle, Pa.: Hamilton Library Association, 1933.

Talcott, T. M. R. "General Lee's Strategy at the Battle of Chancellorsville." *Southern Historical Society Papers* 34 (January–December 1906): 1–27.

Tanner, Robert G. *Stonewall in the Valley.* Mechanicsburg, Pa.: Stackpole Books, 1996.

Taylor, Emerson Gifford. *Gouverneur Kemble Warren: The Life and Letters of an American Soldier 1830–1882.* Boston and New York: Houghton Mifflin, 1932.

Taylor, J. E. *With Sheridan up the Shenandoah Valley in 1864: Leaves from a Special Artist's Sketch Book and Diary.* Dayton, Ohio: Morningside, 1989.

Taylor, Richard. *Destruction and Reconstruction: Personal Experiences of the Late War.* 1879. Reprint, New York, London, and Toronto: Longmans, Green, 1955.

Taylor, Walter H. *General Lee: His Campaigns in Virginia, 1861–1865.* 1906. Reprint, Lincoln and London: University of Nebraska Press, 1994.

Thompson, Gilbert. *The Engineer Battalion in the Civil War.* 2 vols. Washington Barracks, D.C.: Press of the Engineer School, 1910.

Time-Life Books, ed. *Spies, Scouts, and Raiders: Irregular Operations.* Alexandria, Va.: Time-Life Books, 1985.

Todd, William. *The Seventy-ninth Highlanders: New York Volunteers in the War of the Rebellion, 1861–1865.* Albany, N.Y.: Press of Brandow, Barton, 1886.

Trass, Adrian G. *From the Golden Gate to Mexico City: The U.S. Army Topographical Engineers in the Mexican War, 1846–1848.* Washington, D.C.: Center of Military History, 1993.

Tremain, Henry Edwin. *Two Days of War: A Gettysburg Narrative and Other Excursions.* New York: Bonnell, Silver and Bowers, 1905.

Trout, Robert J. *They Followed the Plume: The Story of J. E. B. Stuart and His Staff.* Mechanicsburg, Pa.: Stackpole Books, 1993.

Trowbridge, John Townsend. *The South: A Tour of Its Battlefields and Ruined Cities.* 1866. Reprint, New York: Arno Press, 1969.

Tucker, Glenn. *High Tide at Gettysburg.* 1958. Reprint, Dayton, Ohio: Morningside, 1973.

————. *Lee and Longstreet at Gettysburg.* 1968. Reprint, Dayton, Ohio: Morningside, 1982.

United States Coast and Geodetic Survey, comp. *Military and Naval Service of the United States Coast Survey, 1861–1865.* Washington, D.C.: U.S. Government Printing Office, 1916.

United States Congress. *Report of the Secretary of War.* 37th Cong., 2d sess., Ex. Doc. No. 1. 2 vols. Washington, D.C.: U.S. Government Printing Office, 1861.

United States Military Academy. *Annual Reunion Report of the Association of the Graduates of the United States Military Academy at West Point, New York.* Published annually by the United States Military Academy.

United States War Department. *Atlas to Accompany the Official Records of the Union and Confederate Armies.* 1891–95. Reprint, New York: Fairfax, 1983.

————. *Basic Field Manual: Field Service Pocketbook, Sketching.* Washington, D.C.: U.S. Government Printing Office, 1939.

————. *Revised Regulations for the Army of the United States, 1861.* 1861. Reprint, Gettysburg, Pa.: Civil War Time Illustrated, 1974.

————. *The War of the Rebellion: A Compilation of the Official Records of the Union and Confederate Armies.* 128 vols. Washington, D.C.: U.S. Government Printing Office, 1880–1901.

Upton, Emory. *The Military Policy of the United States.* Washington, D.C.: U.S. Government Printing Office, 1904.

Urwin, Gregory J. W. *Custer Victorious: The Civil War Battles of General George Armstrong Custer.* 1983. Reprint, Lincoln and London: University of Nebraska Press, 1990.

Vandiver, Frank E. *Jubal's Raid: General Early's Famous Attack on Washington in 1864.* New York, Toronto, and London: McGraw-Hill, 1960.

————. *Mighty Stonewall.* New York, Toronto, and London: McGraw-Hill, 1957.

Van Horne, Thomas B. *History of the Army of the Cumberland.* 2 vols. 1875. Reprint, Wilmington, N.C.: Broadfoot Publishing, 1988.

Verner, Captain William Willoughby. *Rapid Field Sketching and Reconnaissance.* London: W. H. Allen, 1889.

Walker, Paula C. "Lieutenant Gouverneur Kemble Warren, the Map of the Trans-Mississippi West, and the Smithsonian Institution." Mimeographed. Fort Wayne, Ind.: 1995.

Warner, Ezra J. *Generals in Blue: Lives of the Union Commanders.* Baton Rouge and London: Louisiana State University Press, 1964.

————. *Generals in Gray: Lives of the Confederate Commanders.* Baton Rouge and London: Louisiana State University Press, 1959.

Warren, Gouverneur K. *General G. K. Warren's Report of the Operations in which He Took Part Connected with the Chancellorsville Campaign, May 12, 1863.* Joint Committee on the Conduct of the War, 38th Cong., 2d sess., Serial Set 5, 1212. Washington, D.C.: U.S. Government Printing Office, 1863.

Watkins, Sam R. *"Co. Aytch": A Sideshow of the Big Show.* 1882. Reprint, New York: Collier Books, 1962.

Webb, Alexander S. *The Peninsula: McClellan's Campaign of 1862.* 1881. Reprint, Wilmington, N.C.: Broadfoot Publishing, 1989.

Weigley, Russel F. *Quartermaster General of the Union Army.* New York: Columbia University Press, 1959.

Weller, Jac. "The Logistics of Nathan Bedford Forrest." *Military Analysis of the Civil War.* Editors of Military Affairs. Millwood, N.Y.: KTO Press, 1977.

Wert, Jeffery D. *The Controversial Life of George Armstrong Custer.* New York: Simon and Schuster, 1996.

West, John C. *A Texan in Search of a Fight.* 1901. Reprint, Waco, Tex.: Texian Press, 1969.

Wiley, Bell Irvin. *The Life of Billy Yank: The Common Soldier of the Union.* Baton Rouge and London: Louisiana State University Press, 1978.

Wilkeson, Frank. *Recollections of a Private Soldier in the Army of the Potomac.* 1886. Reprint, Freeport, N.Y.: Books for Libraries Press, 1972.

Williams, Alpheus S. *From the Cannon's Mouth: The Civil War Letters of General Alpheus S. Williams.* Edited by Milo M. Quaife. 1959. Reprint, Lincoln and London: University of Nebraska Press, 1995.

Williams, George F. *Bullet and Shell: The Civil War as the Soldier Saw It.* 1882. Reprint, Stamford, Conn.: Longmeadow Press, 1992.

Williams, Kenneth P. *Lincoln Finds a General.* 5 vols. New York: Macmillan, 1949–59.

Williams, T. Harry. *Lincoln and His Generals.* New York: Vintage Books, 1952.

Williams, Thomas J. C. *A History of Washington County, Maryland.* 1906. Reprint, Baltimore: Regional Publishing, 1968.

Wilson, Herbert M. *Topographic Surveying.* London: Chapman and Hall, 1910; New York: John Wiley, 1910.

Wilson, James Harrison. *Under the Old Flag.* 2 vols. 1912. Reprint, Westport, Conn.: Greenwood Press, 1971.

Wilt, Napier. "Ambrose Bierce and the Civil War." *American Literature* 1, no. 3 (November 1929): 260–85.

Woman's Club of Mercersburg, Pennsylvania. *Old Mercersburg.* New York: Frank Alaben Genealogical Company, 1912.

Woodbury, Augustus. *A Narrative of the Campaign of the First Rhode Island Regiment in the Spring and Summer of 1861.* Providence: Sidney S. Rider, 1862.

Worsham, John H. *One of Jackson's Foot Cavalry: His Experience and What He Saw during the War, 1861–1865.* New York: Neale Publishing, 1912.

Acknowledgments

MY WIFE, Michiko, first suggested this atlas and has been assisting with it ever since. Nancy Whitin, director of the History Book Club, thought an atlas such as this was a good idea. That, in and of itself, made it a good idea. I am grateful to both of them.

There are a goodly number of people—it's amazing how many—whose willing and unstinting help not only made the project more enjoyable but made it possible. They are curators, archivists, librarians, historians, and photographers.

Noel Harrison, Fredericksburg and Spotsylvania National Military Park, placed the resources of the park at my long-distance disposal. He also introduced me to the work of two of his National Park Service colleagues: the late David A. Lilley, whose articles and unpublished master's thesis were invaluable and whose groundwork when contemplating a similar atlas obviated many fruitless letters and telephone calls; and David W. Lowe, an expert on Federal topographical engineer Nathaniel Michler.

Les Pockell, Book-of-the-Month Club, was instrumental in arranging for the publication of this atlas, and Lisa Thornbloom performed the often complicated task of copyediting it.

Sue Marcus, Interlibrary Loan librarian at Olean Public Library, provided access to virtually every article and book requested.

Stephanie Przybylek, Cayuga Museum, Auburn, New York, allowed liberal access to the John S. Clark maps and manuscripts and was admirably patient with the efforts to reproduce them.

James Corsaro, New York State Library, likewise was liberal in providing access to the G. K. Warren Papers.

New York State Museum photographer Greg Troup took on an assignment that was faltering and efficiently and expertly brought it to completion.

Daniel Lorello, New York State Archives, provided insights and advice on New York State's archives and maps.

James Flatness and Edward Redmond, Library of Congress, Geography and Map Division, deftly made the nearly impossible and certainly imposing job of reviewing voluminous material quite easy and pleasant. Bebe Overmiller, Photoduplication at the Library of Congress, processed a complicated order quickly and efficiently.

Dr. Patrick J. Kelly, Adelphi University, ably translated the obscure nineteenth-century German military script in which Ph. I. Schopp wrote his memoir of the Battle of Freeman's Ford, Virginia.

Leslie Fields, the Gilder Lehrman Collection at the Pierpont Morgan Library, cheerfully and constantly helped to make this important collection accessible.

Paul Romaine and Sandra Trenholme, the Gilder Lehrman Collection, were enthusiastic and encouraging and facilitated the use of the collection.

The Gilder Lehrman Institute of American History granted a Gilder Lehrman Fellowship at the J. Pierpont Morgan Library.

Tina Mellott, curator and director, the Elkhart County Historical Museum, worked patiently to allow access to and permission to use the Ambrose Bierce materials on loan at the museum. Her associate Diana Zornow also was helpful.

D. W. Strauss, Elkhart, Indiana, granted permission to study and reproduce the Ambrose Bierce materials he holds.

Webster Wheelock, New York City, examined the William H. Paine Papers at the New-York Historical Society.

Dr. Philip L. Shiman generously supplied copies of his invaluable articles and works in progress on Civil War mapping and engineers.

Nan Card, Hayes Presidential Center, provided valuable suggestions and access to the Hayes collection.

Anne Sindelar, Western Reserve Historical Society, facilitated the use of their Civil War materials and arranged to have them reproduced.

Rebecca Ebert, Handley Regional Library, offered constant help and advice concerning the Jed Hotchkiss images in the library's collection.

John E. White, Southern Historical Collection, University of North Carolina at Chapel Hill, promoted the collection and provided plenty of space to examine it.

Janie C. Morris, Special Collections Library, Duke University, and her associates Lisa Stark and Christina Favretto were very generous in making their Civil War collection available, to the extent of taking items down from the walls.

Frank Boles and Evelyn Leasher, Clarke Historical Library, Central Michigan University, gave unrestricted access to the Orlando M. Poe Papers.

Michael D. Sherbon, Pennsylvania Historical and Museum Commission, patiently provided data and oversaw reproduction of a number of images from the state collections.

Georgia Barnhill, American Antiquarian Society, provided images quickly and with a minimum of bother.

Cindy Gilley, Do You Graphics, did a superb job ferreting out images and overseeing their reproduction at the National Archives.

Megaera Ausman, United States Postal Service historian, provided material on postal route maps.

R. L. Murray, Civil War historian, brought the John S. Clark maps at the Cayuga Museum to my attention.

Tammy Gobert, Rensselaer Polytechnic Institute Libraries, offered guidance on the Roebling Papers.

Albert King and Edward Skipworth, Rutgers University Libraries, furnished materials from the Roebling Papers.

D. Scott Hartwig, Gettysburg National Military Park, provided suggestions and directions.

Joseph Struble, George Eastman House, recollected a photograph of a map photographic reproduction apparatus in the field.

Stephen W. Sears, Civil War historian, gave advice, perspective, and encouragement.

Alan Aimone, Judith A. Sibley, and Alicia Mauldin, United States Military Academy, provided materials from the West Point Archives.

Mike Musick and Cindi Fox, National Archives, offered assistance locating and reproducing material.

John P. Chalmers, Chicago Public Library, graciously allowed use of the James B. McPherson map and several other items in Special Collections.

Paula C. Walker, G. K. Warren specialist, gave insights and background on Warren.

Shirley Garnett, Mount Vernon Public Library, located a detailed newspaper obituary of John C. Babcock.

Richard W. Stephenson, consultant in the history of cartography, consistently offered support and encouragement and thereby instilled confidence.

Rist Bonnefond and Cynthia McInerney, Kents Hill School, granted permission to use the Lincoln/Howard map.

Marianne M. McKee, Library of Virginia, answered questions about published Virginia maps and arranged for reproduction of two of them.

Dr. Martin K. Gordon, historian, U.S. Army Corps of Engineers, graciously provided introductions to Civil War map scholars and copies of publications of the Office of the Chief of Engineers.

Charles Hubbard and Steven Wilson, Lincoln Memorial University, facilitated the use of the Lincoln/Howard map and provided the image.

Connie Slaughter and Jeff Patrick, Wilson's Creek National Battlefield, provided information on the plat of the Jacob Shults(z) farm.

Dr. John J. McCusker, Trinity University, explained the conversion rates and modern equivalent of shillings and dollars in 1862.

Jeff Giambrone, Old Court House Museum, located photographs of Champion's Hill, Mississippi.

George R. Farr, Chemung County Historical Society, offered information on Lucius T. Stanley, Rodney E. Harris, and Robert M. McDowell.

Robert C. Hansen, National Oceanographic and Atmospheric Administration, furnished copies of materials relating to NOAA's predecessor, the United States Coast Survey.

Leah Wood, United States Civil War Center, helped locate possible resources.

Ruth Shaw, Jamestown Community College, and Robert Taylor, Olean Public Library, gave brief seminars on lithographic reproduction processes.

Michael J. Winey, United States Army Military History Institute, helped locate wartime photographs of topographical engineers.

Peter W. Roper, biographer of Jed Hotchkiss; Catherine Delano Smith, *Imago Mundi* editor; and Tony Campbell, British Library Map Collection, supplied interest and encouragement in London, England.

Fred Mende, Gettysburg scholar, offered insights into Longstreet's countermarch on July 2, 1863.

Vickie Weyss, New York State Library, represented the staff of that fine and constantly accommodating institution.

Susan Winter Trail, Antietam National Battlefield, explained the term "slackwater navigation."

John W. Carson, Civil War student, noted and forwarded relevant material.

Marion Springer, Steuben County, New York, Historian's Office, and Elizabeth Blodgett of Prattsburg, New York, supplied information about Rodney E. Harris.

Terri Hudgins, Museum of the Confederacy, helped with photographs of Confederate topographical engineers.

Joseph Hart and Wayne Wright, New York State Historical Association Research Library, provided information about Rodney E. Harris.

Luther Hanson, Quartermaster Museum, Fort Lee, Virginia, located and allowed the use of photographs.

Robert J. Trout, Confederate cavalry historian, offered information about W. W. Blackford.

Steven J. Brown, Gabor Boritt, Zachary Kent, and Robert Lescher gave professional advice and courtesies.

Eric Himmel, Dirk Luykx, and Nicole Columbus of Harry N. Abrams, Inc., were brave, patient, reassuring, helpful, and kind.

Index

Index

Credits

The author and publisher would like to thank the following collectors, historical societies, libraries, and museums for generously permitting the reproduction of maps, photographs, and portraits (numbers refer to pages):

8: Kent's Hill School, through the Cooperation of the Abraham Lincoln Library and Museum, Harrogate, Tennessee
14: Quartermaster Museum, Fort Lee, Virginia
16-20: Archives, Library of Virginia, Richmond
32: Smithsonian Institution, Washington, D.C.
40: Reproduced from the Collections of the Library of Congress, Washington, D.C.
41: U.S. Army Military History Institute, Carlisle, Pennsylvania
47: Reproduced for William Trevelyan Miller's *The Photographic History of the Civil War*, volume 9, 1911. Photographed by Blumenthal's
63: Special Collections and Preservation Division,

Chicago Public Library
70: Courtesy George Eastman House, Rochester, New York
74: Special Collections Library, Duke University, Durham, North Carolina
75-76: Geography and Map Division, Library of Congress, Washington, D.C.
77: Western Reserve Historical Society, Cleveland, Ohio
78: Geography and Map Division, Library of Congress, Washington, D.C.
79: Archives, Handley Regional Library, Winchester, Virginia
80-81: Geography and Map Division, Library of Congress, Washington, D.C.
82: Gilmer Papers, Southern Historical Collection, Library of the University of North Carolina, Chapel Hill
83: Geography and Map Division, Library of Congress, Washington, D.C.

84: Cartographic and Architectural Branch, National Archives, Washington, D.C.
85: Gilder Lehrman Collection, on deposit at the Pierpont Morgan Library, New York. Photographed by Schecter Lee
86: Cartographic and Architectural Branch, National Archives, Washington, D.C.
87: Geography and Map Division, Library of Congress, Washington, D.C.
88: Archives, Handley Regional Library, Winchester, Virginia
89-90: Western Reserve Historical Society, Cleveland
91: Gilmer Papers, Southern Historical Collection, Library of the University of North Carolina, Chapel Hill
92: Geography and Map Division, Library of Congress, Washington, D.C.
93-94: Gilmer Papers, Southern Historical Collection, Library of the University of North

Credits

Carolina, Chapel Hill

95: National Archives, Washington, D.C.

96: Special Collections Library, Duke University, Durham, North Carolina

97: Cartographic and Architectural Branch, National Archives, Washington, D.C.

98: Geography and Map Division, Library of Congress, Washington, D.C.

99–100: Reproduced from the original map in the John S. Clark Collection, Cayuga Museum, Auburn, New York. Photographed by Greg Troup

101: Cartographic and Architectural Branch, National Archives, Washington, D.C.

102: Geography and Map Division, Library of Congress, Washington, D.C.

103: Cartographic and Architectural Branch, National Archives, Washington, D.C.

104: Manuscripts and Special Collections, New York State Library, Albany. Photographed by Greg Troup

105: Geography and Map Division, Library of Congress, Washington, D.C.

106: Gilmer Papers, Southern Historical Collection, Library of the University of North Carolina, Chapel Hill

107: Cartographic and Architectural Branch, National Archives, Washington, D.C.

108: Gilmer Papers, Southern Historical Collection, Library of the University of North Carolina, Chapel Hill.

109–110: Cartographic and Architectural Branch, National Archives, Washington, D.C.

111–112: Gilmer Papers, Southern Historical Collection, Library of the University of North Carolina, Chapel Hill

113–115: Manuscripts and Special Collections, New York State Library, Albany. Photographed by Greg Troup

116: Geography and Map Division, Library of Congress, Washington, D.C.

117: Manuscripts and Special Collections, New York State Library, Albany. Photographed by Greg Troup

118: Geography and Map Division, Library of Congress, Washington, D.C.

119: McLaws Papers, Southern Historical Collection, Library of the University of North Carolina, Chapel Hill

120: Gilder Lehrman Collection, on deposit at the Pierpont Morgan Library, New York. Photographed by Schecter Lee

121–123: Cartographic and Architectural Branch, National Archives, Washington, D.C.

124: Geography and Map Division, Library of Congress, Washington, D.C.

125–127: Manuscripts and Special Collections, New York State Library, Albany. Photographed by Greg Troup

128–129: Archives, Handley Regional Library, Winchester, Virginia. Photographed by Rodney Lee Gibbons, C.P.P.

130–135: Manuscripts and Special Collections, New York State Library, Albany. Photographed by Greg Troup

136: Huntingfield Map Collection, Maryland State Archives, Annapolis

137–138: Pennsylvania State Archives, Pennsylvania Historical and Museum Commission

139: Geography and Map Division, Library of Congress, Washington, D.C.

140: Library of the American Antiquarian Society, Worcester, Massachusetts. Photographed by Les Gardner

141: Cartographic and Architectural Branch, National Archives, Washington, D.C.

142–143: Geography and Map Division, Library of Congress, Washington, D.C.

144–145: Clarke Historical Library, Central Michigan University, Mount Pleasant. Photographed by Brian Allen Roberts

146: Cartographic and Architectural Branch, National Archives, Washington, D.C.

147: Rutherford B. Hayes Presidential Center, Fremont, Ohio. Photographed by Gilbert Gonzalez

148: Manuscripts and Special Collections, New York State Library, Albany. Photographed by Greg Troup

149–150: Collection, Elkhart County Historical Museum, Bristol, Indiana. Collection on loan from D. W. Strauss. Photography courtesy Troyer Studios

151: Special Collections Library, Duke University, Durham, North Carolina

152: Clarke Historical Library, Central Michigan University, Mount Pleasant. Photographed by Brian Allen Roberts

153: Special Collections Library, Duke University, Durham, North Carolina

154: Gilmer Papers, Southern Historical Collection, Library of the University of North Carolina, Chapel Hill

155: Clarke Historical Library, Central Michigan University, Mount Pleasant. Photographed by Brian Allen Roberts.

156: Manuscripts and Special Collections, New York State Library, Albany. Photographed by Greg Troup

157: Geography and Map Division, Library of Congress, Washington, D.C.

158–159: Manuscripts and Special Collections, New York State Library, Albany. Photographed by Greg Troup

160: Geography and Map division, Library of Congress, Washington, D.C.

161: George H. Thomas Papers, National Archives, Washington, D.C.

163–170: Reproduced from the original maps in the John S. Clark Collection, Cayuga Museum, Auburn, New York. Photographed by Greg Troup

171–172: Collection, Elkhart County Historical Museum, Bristol, Indiana. Collection on loan from D. W. Strauss. Photography courtesy Troyer Studios

173: Geography and Map Division, Library of Congress, Washington, D.C.

174: Gilder Lehrman Collection, on deposit at the Pierpont Morgan Library, New York. Photographed by Schecter Lee

175–176: Rutherford B. Hayes Presidential Center, Fremont, Ohio. Photographed by Gilbert Gonzalez

177–181: Gilder Lehrman Collection, on deposit at the Pierpont Morgan Library, New York. Photographed by Schecter Lee

182–184: Geography and Map Division, Library of Congress, Washington, D.C.

185: Rutherford B. Hayes Presidential Center, Fremont, Ohio. Photographed by Gilbert Gonzalez

186: Special Collections and Preservation Division, Chicago Public Library. Photography by Garry Henderson

187–188: Gilder Lehrman Collection, on deposit at the Pierpont Morgan Library, New York. Photographed by Schecter Lee

189: Collection, Elkhart County Historical Museum, Bristol, Indiana. Collection on loan from D. W. Strauss. Photography courtesy Troyer Studios

190: Special Collections and Preservation Division, Chicago Public Library

191: Geography and Map Division, Library of Congress, Washington, D.C.

192–204: Gilder Lehrman Collection, on deposit at

the Pierpont Morgan Library, New York. Photgraphed by Schecter Lee

205: Special Collections Library, Duke University, Durham, North Carolina

206–208: Cartographic and Architectural Branch, National Archives, Washington, D.C.

209: Geography and Map Division, Library of Congress, Washington, D.C.

210: Cartographic and Architectural Branch, National Archives, Washington, D.C.

211: Gilmer Papers, Southern Historical Collection, Library of the University of North Carolina, Chapel Hill

212–218: Cartographic and Architectural Branch, National Archives, Washington, D.C.

219: Geography and Map Division, Library of Congress, Washington, D.C.

220–221: Gilmer Papers, Southern Historical Collection, Library of the University of North Carolina, Chapel Hill

222: Geography and Map Division, Library of Congress, Washington, D.C.

223: Gilder Lehrman Collection, on deposit at the Pierpont Morgan Library, New York. Photographed by Schecter Lee

224: Cartographic and Architectural Branch, National Archives, Washington, D.C.

225: Geography and Map Division, Library of Congress, Washington, D.C.

226: Lockett Papers, Southern Historical Collection, Library of the University of North Carolina, Chapel Hill

227: Special Collections Library, Duke University, Durham, North Carolina

228–231: Cartographic and Architectural Branch, National Archives, Washington, D.C.

232: Collection, Elkhart County Historical Museum, Bristol, Indiana. Collection on loan from D. W. Strauss. Photography courtesy Troyer Studios

234: Courtesy Civil War Library and Museum, Philadelphia. Photographed by Blake Magner

236: Collection, Elkhart County Historical Museum, Bristol, Indiana. Collection on loan from D. W. Strauss. Photography courtesy Troyer Studios.

237: Museum of the Confederacy, Richmond, Virginia

239: *Samuel Skinker and His Descendants,* Thomas Keith Skinker. Photographed by Blumenthal's

240: Tioga Point Museum, Athens, Pennsylvania

241: Monroe County Historical Commission, Michigan

242: Division of Military and Naval Affairs, New York State Adjutant General's Office, Albany, New York, and U.S. Army Military History Institute, Carlisle, Pennsylvania

243: Archives, Handley Regional Library, Winchester, Virginia

245: Roger D. Hunt Collection and U.S. Army Military History Institute, Carlisle, Pennsylvania

246: Massachusetts Commandery, Military Order of the Loyal Legion, and U.S. Army Military History Institute, Carlisle, Pennsylvania

247: Roger D. Hunt Collection and U.S. Army Military History Institute, Carlisle, Pennsylvania

248: Courtesy Civil War Library and Museum, Philadelphia. Photographed by Blake Magner

249: Massachusetts Commandery, Military Order of the Loyal Legion, and U.S. Army Military History Institute, Carlisle, Pennsylvania

250: Archives and Special Collections, Rensselaer Polytechnic Institute, Troy, New York

251: Western Reserve Historical Society, Cleveland, Ohio

252: Massachusetts Commandery, Military Order of the Loyal Legion, and U.S. Army Military History Institute, Carlisle, Pennsylvania